ИЗДАТЕЛЬСТВО МИР

Ю. В. Новожилов, Ю. А. Яппа

ЭЛЕКТРОДИНАМИКА

Издательство «Наука»
Главная редакция физико-математической литературы
Москва

ELECTRODYNAMICS

Yu. V. Novozhilov
and Yu. A. Yappa

Translated from the Russian by
V. I. Kisin

Mir Publishers · Moscow

First published 1981

Revised from the 1978 Russian edition

На английском языке

TO THE READER

Mir Publishers would be grateful for your comments on the content, translation and design of this book.

We would also be pleased to receive any other suggestions you may wish to make.

Our address is:
Mir Publishers
2 Pervy Rizhsky Pereulok
I-110, GSP, Moscow, 129820
USSR

Printed in the Union of Soviet Socialist Republics

CONTENTS

PREFACE TO THE RUSSIAN EDITION

This book originated from our experience in teaching electrodynamics as part of a series of lectures in theoretical physics. The lectures were read at the Leningrad State University for all students of the physics faculty, both future theoreticians and experimenters. The subject matter follows from what the students learned about electricity and magnetism in the general physics course. On the other hand, an electrodynamics course must serve as the basis for many special disciplines, such as plasma physics, propagation of radiowaves, electromagnetic methods in geophysics, the accelerator theory, and others. These factors have determined the content of this book and the manner of presentation.

With the current high level of instruction in the general physics course the students starting their study of electrodynamics have a considerable knowledge of the facts that are generalized in the Maxwell equations. This has made it possible for us to proceed from the equations and set the objective of showing how different problems in electrodynamics follow from the equations when the properties of the material media are taken into account. The approach via the Maxwell equations enables the reader to come closer to formulating in the most direct and modern way quite a number of problems currently under intensive discussion in scientific literature. Of these, this book considers some aspects of magnetohydrodynamics, the motion of charged particles in a nonuniform electromagnetic field, and the basis for the phenomenological equations of superconductivity. We have also included the concept of spatial dispersion and stated the basic ideas of nonlinear optics. Understandably, the presentation of these topics cannot claim to be complete.

In accordance with the purpose of this book, we assume that the mathematical grounding corresponds to that of a student who has completed his second year in the physics faculty of a university. This should be sufficient to understand all of the material of the book. The Appendices contain the basic facts about vector and tensor analyses, which are used throughout the book, and also some properties of the Dirac delta function.

Particular attention is paid to the application of the special theory of relativity. In Chapter 2 we briefly consider the fundamentals of special relativity, while in other sections of the book we apply this theory to specific problems. Notably, the relativistic theory of radiation by a point charge is treated in great detail. The reader more interested in nonrelativistic aspects of electrodynamics can skip these sections, which are marked with asterisks.

We have found it impossible to incorporate topics that need the quantum theory and/or statistical physics for their interpretation. Consequently, we do not discuss the electrodynamics of material media from the viewpoint of their microscopic structure. But the thermodynamical aspect of the interaction of electromagnetic fields with media is brought in wherever possible.

To keep the size of the book within limits we have found it necessary to exclude the specific problems of mathematical physics that originate in electrodynamics. For the same reason the book contains no exercises. The reader can find appropriate problems on classical electrodynamics in the well-known book by V. V. Batygin and I. N. Toptygin: *Problems in Electrodynamics* (2nd edition, Academic Press, New York, 1978). We should also like to note that the presentation is concise and hence the material requires attentive reading. The reader is advised to do all the intermediate calculations himself.

The overall number of topics discussed here is rather limited, since the book corresponds to the course taught at the Leningrad State University. It is our hope, however, that the material covered will show how diversified the field of electrodynamics is and how the different branches are connected with the common source, the Maxwell equations.

We express our gratitude to our science editor Prof. S. V. Izmailov and to the reviewers Profs. V. I. Grigor'ev and V. G. Solov'ev. Their many critical remarks and helpful suggestions greatly improved our manuscript.

In style and subject matter this book reflects the pedagogical principles of our teacher, Academician Vladimir A. Fock, to whom it is dedicated.

Yu. V. Novozhilov and Yu. A. Yappa

PREFACE TO THE ENGLISH EDITION

In preparing the original Russian edition for translation the text has been carefully checked and the necessary corrections have been made. We hope that this book will help those studying electrodynamics to go on from the general physics course of electricity and magnetism to the special literature on the subject and to more advanced textbooks, such as the acclaimed *Classical Electrodynamics* by J. D. Jackson, *Classical Electricity and Magnetism* by W.K.H. Panofsky and M. Phillips, and *The Classical Theory of Fields* and *Electrodynamics of Continuous Media* by L. D. Landau and E. M. Lifshitz. Many ideas from these books have been used in our exposition.

A note about units. In Chapter 1 the electrodynamic equations are written in a form that ensures an easy transition from Heaviside-Lorentz units to SI units and back. Next, in dealing with the electrodynamics of isotropic, homogeneous media we use Heaviside-Lorentz units (Chapters 2 to 6). In the remaining chapters we keep to SI units except for some special cases.

This book may, and should, prompt critical remarks concerning both the subject matter and the manner of presentation. Comments of this kind will be appreciated.

Yu. V. Novozhilov and Yu. A. Yappa

Indices. In Chapters 1-6 Latin-letter indices i, j, ... assume values 0, 1, 2, 3, and Greek-letter indices α, β, ... assume values 1, 2, 3. In Chapters 7-9 Latin-letter indices assume values 1, 2, 3. In § 33 Greek-letter indices enumerate linear contours. In § 34 Latin-letter indices enumerate generalized forces and generalized currents.

Latin letters

A vector potential; $A_i^{i'}$—coefficients of linear transformations.
a 3-dimensional acceleration; a_{ij}—strain tensor (§ 32), Onsager coefficients (§ 34).
B magnetic induction.
$\vec{b} \equiv d\vec{w}/d\tau$.
C capacitance of a capacitor, eikonal (§ 21).
c velocity of light in vacuo; c_{ik}—capacitance coefficients (§ 30); c_{ijkl}—elasticity moduli (§ 32).
D electric induction.
$d_{i,kl}$ piezoelectric coefficients (§ 32).
E electric field strength; $\mathscr{E}$—energy of a pointlike mass (§ 6), total radiated energy (§ 16).
e electron charge; $\mathbf{e}_i$—basis vectors in linear space; $\vec{e}_i$—basis vectors in Minkowski space.
F mechanical force; $\mathbf{F}_{\mathrm{rad}}$—force of radiative reaction; $\mathbf{F}_{\mathrm{ext}}$—external force applied to a charge; $\vec{F}$, F^i—4-dimensional Minkowski force; F^{ik}—stress tensor of electromagnetic field; F—free energy density (§§ 22, 31); $\tilde{F}$—thermodynamic potential (§ 31); $\mathscr{F}$—total free energy.
f bulk density of force; f^{ik}—induction tensor of electromagnetic field (§ 7).
G Green's function (§ 11); $\mathscr{G}$—Gibbs' potential (§ 32).
g momentum density of electromagnetic field; g_{ij}—metric tensor.
H magnetic field strength; $\mathscr{H}$—Hamiltonian (§§ 8, 28).
h Planck constant (§ 22); $\mathbf{h} \equiv d\mathbf{a}/dt$ (§ 23).
I total current (§ 12); I_1, I_2—invariants of electromagnetic field (§ 7); $I(\omega)$ — energy of radiation with frequency ω.
i surface current density.
$\mathscr{I}_i$ generalized currents (§ 34).
j current density; $\mathbf{j}^{\mathrm{ext}}$—density of current produced by external electromotive forces; j_n, j_s—ordinary and superconducting currents (§ 37).
K inertial reference frame, radiance (§ 22), kinetic energy, $\vec{\mathscr{K}}$, $\mathscr{K}^i$—4-dimensional Newtonian force.
k wave number, Boltzmann constant; $\hat{k}$—complex wave number; $\mathbf{k}, \vec{k}$—wave vector.
L self-inductance of a coil; $L_{\alpha\beta}$—induction coefficients; $\mathscr{L}$—Lagrangian.

M magnetization; $\mathbf{M}_0$—residual magnetization; M^{ik}—total 4-dimensional angular momentum.

m magnetic moment; $\tilde{\mathbf{m}}$—angular momentum density (§ 3); $\tilde{m}_{kmj}$—density of 4-dimensional angular momentum (§ 10); m_0—rest mass of a particle.

N mechanical moment of forces; N—Nernst coefficient (§ 34), number of particles (§ 39).

n unit vector of outward normal; n—refractive index (§§ 19, 39); $\hat{n}$—complex refractive index (§ 39).

P electric polarization; $\vec{P}$, P^i—energy-momentum vector; P_λ—canonical momentum of field oscillator (§ 22).

p mechanical momentum density, momentum of a pointlike mass, electric dipole moment (§ 11); p—pressure (§ 22); p_m—magnetic pressure (§ 35); p_λ—canonical momentum of field oscillator (§ 22).

Q quantity of heat; $Q_{\alpha\beta}$—quadrupole moment (§§ 11, 16); Q_α—canonical coordinate of field oscillator (§ 22).

q electric charge; **q**—heat flux density (§ 34); q_λ—canonical coordinate of field oscillator (§ 22).

R active resistance of linear contour (§ 33); **R**—radius vector, $\mathbf{R} \equiv \mathbf{r} - \mathbf{r}'$.

r radius vector, $\mathbf{r}'$—radius vector of a source; r_L—Larmor radius; r_{ik}—tensor of resistance of anisotropic medium (§ 34).

S Poynting vector; S—entropy density; $\mathscr{S}$—action integral (§ 9), total entropy (§ 22).

s propagation vector (§ 39); s_{ik}—potential coefficients (§ 30); s_{ijkl}—elasticity coefficients (§ 32).

T temperature, time; T^{ik}—4-dimensional energy-momentum tensor of electromagnetic field; $T_{\alpha\gamma}$—total stress tensor.

t time, t'—source time.

U potential energy (§ 28); internal energy density (§§ 22, 31).

u 3-dimensional velocity; $\vec{u}$, u^i—4-dimensional velocity.

V 3-dimensional volume; electromotive force in a circuit (§ 33).

v 3-dimensional velocity; $\mathbf{v}_\infty$—velocity of electric drift in a medium with infinite electric conductivity (§ 35); $\mathbf{v}_{su}$—velocity of superconducting electrons (§ 37); $\mathbf{v}_{gr}$—group velocity (§ 39).

$\mathscr{W}$ radiative energy.

w total energy density; w_ν—spectral energy density; $\vec{w}$, w^i—4-dimensional velocity.

X_i generalized forces (§ 34).

$\vec{x}$, x^i 4-dimensional radius vector in Minkowski space.

Remark. § 15 and Appendix D operate with auxiliary symols $\vec{p}$, W, B, V, defined differently.

Greek letters

α dimensional coefficient in the Maxwell equations.

$\beta \equiv v/c$, $\boldsymbol{\beta} \equiv \mathbf{v}/c$.

$\gamma \equiv (1-\beta^2)^{-1/2}$.

Γ natural width of spectral line; $\tilde{\Gamma}$—total width of spectral line.

δ_i^k Kronecker delta; $\delta(x-\xi)$—delta function.

ε dielectric permittivity of a medium, infinitesimal parameter (§ 27); ε_0—electric constant; ε'—relative dielectric permittivity (§ 29); $\boldsymbol{\varepsilon}$—polarization vector of a plane wave; $\varepsilon_{\alpha\beta\gamma}$—unit pseudoscalar (Levi-Civita symbol).

ζ chemical potential (§ 31); azimuthal angle.

η differential thermo-e.m.f. (§ 34).

ϑ polar angle in spherical system of coordinates.

$\varkappa$ mass density (§§ 35 and 40); $\varkappa \equiv dT/dt'$ (§ 14); $\varkappa_0$—invariant density of rest mass (§ 8).

λ surface charge density, wavelength; λ_L—London penetration depth (§ 37).

μ magnetic permeability of a medium; μ'—relative magnetic permeability (§ 29); μ_0—magnetic constant; $\boldsymbol{\mu}$—magnetic moment due to orbital motion (§ 26).

ν frequency; $\boldsymbol{\nu}$—normal to interface between two media (§ 18).

$\boldsymbol{\Pi}$ Hertz vector (§§ 2, 38); Π—Peltier coefficient (§ 34).

$\vec{\pi}$, π^k 4-momentum of charge.

ρ bulk density of charge, 4-dimensional distance between observer and source (§ 15); ρ_0—invariant charge density.

σ 2-dimensional surface, electric conductivity, Stefan-Boltzmann constant (22.13), scattering cross section (§ 25); σ_{abs}—absorption cross section (§ 25); σ_{kmj}—spin tensor of electromagnetic field (§ 10).

Σ 3-dimensional hypersurface; $\Delta\Sigma$—cross-sectional area of a contour (§ 33).

τ proper time, relaxation time (§ 35); τ—Thompson coefficient (§ 34); τ_{ij}—stress tensor of anisotropic medium (§ 32); $\boldsymbol{\tau}$—double layer density (§ 36).

Φ magnetic flux; $\vec{\Phi}$, Φ^k—4-dimensional potential.

φ scalar potential; $\boldsymbol{\varphi}$—surface force density.

χ_{el} electric susceptibility; χ_m—magnetic susceptibility.

ψ magnetic scalar potential, function defining gauge transformation (§ 2), any of the Cartesian components of a vector (§ 20).

Ω 4-dimensional volume in Minkowski space; $d\Omega$—element of solid angle (identical to $d\omega$ in § 15 and Appendix D).

ω cyclic frequency; ω_L—Larmor frequency (§ 26).

ω_{ij} coefficients of the infinitesimal Lorentz transformation.

CHAPTER 1

THE BASICS OF MAXWELL'S ELECTRODYNAMICS

§ 1. The Maxwell equations. Electromagnetic units

1.1. The experimental and theoretical study of physical phenomena led to the idea of the electromagnetic field as a physical reality (object), something with definite properties. It is created by sources—electric charges, currents, permanent magnets—and is the cause of the interaction of sources. The field created by a source can be measured by the effect it produces on other sources. To define a field quantitatively, it is necessary to measure the force with which the field acts on specific sources, called test sources. A *test source* is a source whose dimensions are negligibly small and whose field is so weak that it does not affect the results of the measurement. Hence a field can be measured at any point in space. Moreover, a field can, generally speaking, be time dependent. The force measured at time t with the help of a test source that is placed at a point with radius vector $\mathbf{r}$ will be denoted $\mathbf{F}(\mathbf{r}, t)$.

The properties of an electromagnetic field manifest themselves in "pure form" when the action of its sources is studied in vacuo. A *material medium* consists of the simplest (but not in the sense of their internal structure!) entities—atoms, electrons, molecules. These entities always possess definite electromagnetic properties. The properties of these entities together with their arrangement in space in relation to each other, and the state of their motion with respect to each other produce a specific reaction of the medium to the "external" electromagnetic field.

From the macroscopic viewpoint, the discreteness of matter can usually be ignored and matter can be described as a continuous distribution of field sources. The distribution may change if there is an electromagnetic field created by outside sources and/or if the thermodynamic properties of the medium change. The electromagnetic properties of material media vary greatly, but all are described, as the reader will soon see, by only two macroscopic quantities: electric polarization and magnetization.

The fundamental laws of the electromagnetic field with due regard for the macroscopic properties of material media are formulated mathematically in the Maxwell equations, which will be studied in this section. This very general formalism must, if possible, be independent of any specific assumptions about the microscopic structure of the media. It stands to reason that one of the main tasks of a physical theory is to explain observed facts from the microscopic viewpoint. More than that, it is the microscopic theory that often makes it possible to predict new physical phenomena and ways to observe them. However, the phenomenological method of description, our choice for this book, has its advantages. It uses none but those characteristics of phenomena that can be measured, at least in principle, by macroscopic instruments. Any microscopic theory, for its part, must inevitably lead to definite conclusions about the phenomenological characteristics and foretell and explain their behavior. This is true not only of the classical electron theory but of the present-day quantum theory of matter. To quote Niels Bohr,

> ...the unambiguous interpretation of any measurement must be essentially framed in terms of the classical physics theories, and we may say that in this sense the language of Newton and Maxwell will remain the language of physicists for all time.[1]

Any mathematical reflection of the laws of nature, including the Maxwell equations, must be invariant with respect to certain groups of transformations of the physical quantities interconnected by this reflection. Above all, the principle of relativity must hold, that is, with fixed initial and boundary conditions the Maxwell equations must bring the same results in any inertial reference frame. The very concept of the electromagnetic field as a physical object, which does not depend on the choice of the inertial reference frame, may be defined only when the relativity principle is explicitly taken into account. In each given inertial frame the electromagnetic field "splits" into two fields, quite distinct in their properties, an electric and a magnetic. It is these fields that are measured. We will leave the study of the relativity principle and its corollaries for Chapter 2 and confine ourselves here to examining the Maxwell equations in an arbitrary inertial frame. In an inertial reference frame the equations must be independent of the orientation of the spatial axes. We will see that the Maxwell equations are indeed relations between three-dimensional vectors (3-vectors), so that the invariance with respect to the rotations of the spatial axes in three-dimensional space does hold.[2]

Thus, in what follows we will establish the main laws governing electric and magnetic fields in an inertial reference frame, which we

---

[1] N. Bohr: "Maxwell and modern theoretical physics", *Nature*, **128** (1931), p. 692.

[2] A brief outline of the elements of vector analysis is given in Appendices A and B.

choose arbitrarily but which, after the choice is made, remains fixed. (An inertial reference frame is one of a class of frames which move with respect to each other with constant velocities.) But first we must agree on the units of measurement and on the dimensions of the quantities used in electrodynamics. This problem will also be considered.

1.2. We assume first that the field under study is *static*. By definition, in such a field the force created by a given distribution of sources and acting on the test source is time independent, that is, it can be represented by a function **F** (**r**) of the three-dimensional radius vector **r**. (The vector gives the position of the pointlike test source in space.) Static electric and magnetic fields are generated by sources with different physical properties and have different structures. To determine them it is important to consider the case when these sources are themselves pointlike. Here also the fields may have different symmetry, which means that the functions **F** (**r**) may differ, but at least the electric field on the one hand and the magnetic on the other are created by sources that ensure the simplest possible configurations of the fields. For the electric field this source is the point charge. In vacuo the field is spherically symmetric.

Experiment shows that if various pointlike charges are placed at some point in space near an arbitrary pointlike source (of an electric field), then to each of these charges there can be assigned a number q such that for any two such charges the following relationship holds: $F_\alpha^{(1)} : F_\alpha^{(2)} = q_1 : q_2$ ($\alpha = 1, 2, 3$). It so happens that the number q (called *quantity of charge*, or simply *charge*) characterizes the physical properties of the point charge under consideration: it does not depend either on the properties of the source that produces the field acting on the charge or on the point in space where the charge is situated. For different point charges this "label" can take on positive or negative values.

Such a definition of the quantity of charge says nothing about the dimensions it must have. It is obvious, however, that any charge can be chosen as the unit charge. The force that a given source of an electric field exerts on a unit point charge is called the *electric field strength* **E** (**r**):

$$\mathbf{F}(\mathbf{r}) = q\mathbf{E}(\mathbf{r}) \tag{1.1}$$

The results of experimental studies of the properties of static electric fields generated by a variaty of sources can be expressed by equations involving vector **E** (**r**). We will now turn to these equations.

First, let s be an arbitrary closed circuit. Then

$$\oint_s \mathbf{E} \cdot d\mathbf{s} = 0 \tag{1.2}$$

In other words, when a point charge is moved along an open circuit in a static electric field, the work spent on, or produced by, transferring the charge does not depend on the form of the circuit but only on its initial and final points. Equation (B.25) makes it possible to find a differential equation for the vector function **E** (**r**):

$$\operatorname{curl} \mathbf{E} = 0 \tag{1.3}$$

(**E** (**r**) must satisfy this equation at any point in space). Equations (1.2) and (1.3) hold for a static field both when the charges are in vacuo and when material media are present.

If the source of the electric field is not pointlike, the field can, in general, be represented as produced by a continuous distribution of charges with a density ρ (**r**).[3] The total electric charge q of the source inside volume V is then given by the equation

$$q = \int_V \rho(\mathbf{r})\, dV \tag{1.4}$$

The field **E** created by a charge q in vacuo satisfies

$$\varepsilon_0 \oint_\sigma \mathbf{E} \cdot \mathbf{n}\, d\sigma = q \tag{1.5}$$

irrespective of how this charge is distributed in a volume V. (Here σ is any closed two-dimensional surface bounding V, and **n** is the unit outward normal at the area element $d\sigma$.) The factor ε_0 is introduced so as to account for the different dimensions and units of measurements of electric quantities.

The definition (1.1) for the electric field strength shows that the product of the dimensions of q and **E** must be equal to the dimensions of force **F**. If we multiply both sides of (1.5) by a quantity q' with the dimensions of charge, we get a condition which the dimensions of charge must satisfy (the brackets denote the dimensions of the quantity inside them):

$$[q]^2 = [F]\,[L]^2\,[\varepsilon_0] \tag{1.6}$$

Later in our exposition the reader will see that the choice of dimensions for ε_0, called the *electric constant* (also the *permittivity of empty space*), may be different, and so may be the dimensions of charge, $[q]$. In certain experiments (for instance, involving electrolysis), the basic unit (standard) of electric charge may be established, at least in principle, independently of the measurement of the field strength by mechanical means. The dimensions of ε_0 can then be determined via (1.6), and its numerical value will depend on the choice of the basic units of mechanical quantities and electric charge.

[3] In what follows we will also consider charge distributions over two-dimensional surfaces with a surface density λ; we will not dwell on this possibility here, since it does not alter our reasoning to any extent.

In the particular case where the charge, q, is pointlike, we choose a surface σ in the form of a sphere with radius r and with its centre at the point where the charge is. The symmetry of the problem shows that at all points on the sphere $\mathbf{E}$ must be directed along the radius, that is $\mathbf{E} = E\mathbf{n}$, and that everywhere on the sphere E is the same. From (1.5) we immediately get

$$\mathbf{E} = \frac{q\mathbf{n}}{4\pi\varepsilon_0 r^2} \tag{1.7}$$

which is the Coulomb law.

Now, in the general case we can contract σ to any point inside V and, using (B.22), arrive at $\varepsilon_0 \operatorname{div} \mathbf{E} = \lim_{V\to 0} V^{-1} \int_V \rho(\mathbf{r})\, dV$. That is, for a wide class of functions $\rho(\mathbf{r})$,

$$\varepsilon_0 \operatorname{div} \mathbf{E} = \rho(\mathbf{r}) \tag{1.8}$$

Hence, instead of the integral relations (1.2) and (1.5) we may consider the system of differential equations (1.3) and (1.8).

The reader must bear in mind that, if there is a distribution of electric charges with density $\rho(\mathbf{r})$, then, since charges are of two types (positive and negative), the condition $q=0$ does not inevitably lead to $\mathbf{E} = 0$. This property of the electric field makes it quite different from Newton's gravitational field. In the latter both laws (1.2) and (1.5) hold, but $q = 0$ always yields $\mathbf{E} = 0$.

The electric field equations are linear, hence the *superposition principle* can be applied. In relation to our problem this principle states that any linear combination (with constant coefficients) of solutions of (1.3) and (1.8) is also a solution. Later the reader will see that the Maxwell equations in the general form likewise possess this property for a broad class of material media.

1.3. Let us now turn to the case where there is a material medium surrounding the electric charge (see (1.5)). The medium may fill a definite volume or all of space. We assume that the charge inside a closed surface σ is the same as in vacuo, so that the right-hand side of (1.5) remains the same. Experience shows that the field $\mathbf{E}(\mathbf{r})$ created by a charge in a medium differs from the field $\mathbf{E}_0(\mathbf{r})$ of the same charge in vacuo. Hence, for $\mathbf{E}(\mathbf{r})$, (1.5) does not hold.[4] If we want the field $\mathbf{E}_0(\mathbf{r})$ to remain unchanged in the presence of the medium, we must take a charge of another magnitude. To describe these experimental facts mathematically, we introduce a vector field $\mathbf{D}(\mathbf{r})$ called the *electric induction* (also *electric displacement*). This field is chosen in such a way that its flux through a closed sur-

[4] In our case the charge density can change inside the volume V, provided that $\int_V \rho(\mathbf{r})\, dV = \int_V \rho_0(\mathbf{r})\, dV$, where ρ_0 is the charge density in vacuo, and ρ in the medium.

face is equal to the magnitude of the electric charge inside the surface:

$$\oint_\sigma \mathbf{D}\cdot\mathbf{n}\,d\sigma = q \tag{1.9}$$

(the notations are the same as in (1.5)). The differential equation for $\mathbf{D}(\mathbf{r})$ is similar to (1.8):

$$\operatorname{div}\mathbf{D} = \rho \tag{1.10}$$

To characterize the properties of the medium, we introduce a new vector, the *electric polarization* $\mathbf{P}$:

$$\mathbf{D} = \varepsilon_0\mathbf{E} + \mathbf{P} \tag{1.11}$$

In vacuo, $\mathbf{P} = 0$. As we have mentioned before, Eqs. (1.2) and (1.3) remain valid in the presence of a medium.

Thus, the static electric field is determined in the general case by the system of equations (1.2) and (1.9) (or in differential form by (1.3) and (1.10)).

The electric charge has a fundamental property firmly established by numerous experiments: the change in the quantity of charge inside a region of volume V bounded by the surface σ is always equal to the flux of charge through this surface. This is known as the *law of charge conservation.* If we denote by $\mathbf{j}(\mathbf{r}, t)$ the electric current density, that is, the quantity of the charge flowing at point $\mathbf{r}$ in the direction of $\mathbf{j}$ across a unit surface perpendicular to $\mathbf{j}$ per unit time, then charge conservation is written

$$\frac{dq}{dt} = \frac{d}{dt}\int_V \rho(\mathbf{r}, t)\,dV = -\oint_\sigma \mathbf{j}(\mathbf{r}, t)\cdot\mathbf{n}\,d\sigma \tag{1.12}$$

In particular, when the surface σ is fixed, we can write $(d/dt)\int_V dV = \int_V (\partial\rho/\partial t)\,dV$. Just as we did in deriving (1.8) and (1.10), we can find the law of charge conservation in differential form:

$$\frac{\partial\rho}{\partial t} + \operatorname{div}\mathbf{j} = 0 \tag{1.13}$$

An important case from the physical viewpoint is when an electric current $\mathbf{j}$ passes through an electrically neutral medium. Let us consider an enclosure of volume V in the form of a closed tube (a "doughnut"). Suppose the tube is filled with an electric charge of one sign whose distribution is fixed in the inertial reference frame of the tube. Through this distribution another distribution of charge of opposite sign flows parallel to the tube's axis in such a way that the total charge inside any section of the tube does not change with time (often this total charge is taken to be zero). Then (1.12) yields

$\int_{\sigma_1} \mathbf{j}\cdot\mathbf{n}_1\, d\sigma_1 = \int_{\sigma_2} \mathbf{j}\cdot\mathbf{n}_2\, d\sigma_2$ for any two cross sections σ_1 and σ_2 of the tube (here the positive direction of one of the unit normals, $\mathbf{n}_1$ or $\mathbf{n}_2$, is changed to the opposite). The current density $\mathbf{j}$ can, in principle, depend also on time if the time rate of flow of charge changes in all the cross sections according to one law. But if $\mathbf{j}$ does not depend on time, the current is called constant. The generalization for the case of an alternating current can also be carried out. In this case the quantity of charge along any section of the tube varies, and we must use (1.12) for each section (and not for the tube as a whole). The integral $I = \int_{\sigma} \mathbf{j}\cdot\mathbf{n}\, d\sigma$ is called the *intensity of* (electric) *current* through surface σ.

1.4. Electric currents, that is, moving electric charges, interact via the *magnetic field* that each of them creates. If, as in the case we have just examined, the current density is independent of time, such a current is a source of a *static* magnetic field. Apart from electric currents, ferromagnetic media (permanent magnets) can generate magnetic fields. The fact that fields created by currents and those created by magnets have identical properties was proved by studying the interaction of currents and magnets.[5]

What makes the magnetic field different from the electric field is that there are no pointlike sources for which the corresponding magnetic fields are spherically symmetric. The simplest source of the magnetic field is called a magnetic moment. To describe it we must imagine a vector $\mathbf{m}$ at the point in space where the source is.[6] For instance, the behavior of an infinitely small current-carrying loop can be described by a vector $\mathbf{m}$ whose length depends on the current I in the loop and whose direction is along the normal to the loop's plane and connected with the direction of I via the right-hand screw rule. Vector $\mathbf{m}$ is called the *magnetic moment* of a closed current.

How does a magnetic field act on such a test source? The loop experiences a mechanical torque $\mathbf{N}$ whose magnitude depends on the magnetic moment of the test source and the magnetic field. The intensity of the magnetic field, $\mathbf{B}$, is defined by the equation

$$\mathbf{N} = \mathbf{m} \times \mathbf{B} \tag{1.14}$$

We still do not know how to determine $\mathbf{m}$ and $\mathbf{B}$; we do not even know their dimensions. What we do know is that the product of dimensions $[m]$ and $[B]$ must have the dimensions of torque (or, which is the same, the dimensions of work). For historical reasons the magnetic

[5] We have in mind, first of all, the classical experiments of Oersted and Ampère carried out in 1819-20.

[6] To be more exact, $\mathbf{m}$ and the vector $\mathbf{B}$ that we will soon introduce are pseudovectors. For the discussion of this aspect see Section 2.

field $\mathbf{B}$ is usually called the *magnetic induction* rather than the magnetic field strength.

The properties of $\mathbf{B}$ can be described by equations that express the main properties of the magnetic field. We will write the first of these equations for the case where the magnetic field sources are in vacuo. If s is any closed circuit and σ is an arbitrary two-dimensional surface bounded by this circuit, the first main equation of magnetostatics states that

$$\frac{\alpha}{\mu_0}\oint_s \mathbf{B}\cdot d\mathbf{s}=\int_\sigma j_n\, d\sigma \tag{1.15}$$

that is, the circulation of $\mathbf{B}$ along s is numerically equal to the total current flowing through surface σ (or, in other words, the total current inside circuit s). The factor α/μ_0 is a dimensional constant and appears because the dimensions of the right-hand side of (1.15) are definite if we specify the dimensions of charge, but the dimensions of $\mathbf{B}$ still remain unknown. Later we will see why it is convenient to write the dimensional constant as a ratio of two quantities (μ_0 is called the *magnetic constant*, or *permeability of empty space*).

The second main equation of magnetostatics expresses in mathematical form our previous statement that there are no pointlike sources of magnetic field in nature. In other words, there are no pointlike magnetic charges and the magnetic field cannot be spherically symmetric. For the same reason there are no magnetic fields that can be created by a distribution of magnetic charges whose total charge is nonzero. This is why, instead of (1.5), for the magnetic field we have

$$\oint_\sigma \mathbf{B}\cdot\mathbf{n}\, d\sigma=0 \tag{1.16}$$

where σ is an arbitrary closed surface.

If a given distribution of currents is not in vacuo but in a material medium, the created field $\mathbf{B}$ ceases to satisfy (1.15). However, we can define a new vector, the *magnetic field strength* (or simply the *magnetic field*)

$$\mathbf{H}=\frac{1}{\mu_0}\mathbf{B}-\mathbf{M} \tag{1.17}$$

where $\mathbf{M}$ is the *magnetization* of the medium (equal to zero in vacuo). We introduce this new vector quantity so that the equation

$$\alpha\oint_s \mathbf{H}\cdot d\mathbf{s}=\int_\sigma j_n\, d\sigma \tag{1.18}$$

holds in a medium as well as in vacuo. As for (1.16), it remains valid for a medium. From the above we see, for one, that the magnetic constant μ_0 relates the different (in principle) dimensions of **B** and **H** just as ε_0 relates the dimensions of **D** and **E**.

Applying (B.22) to (1.16) and Stokes' theorem (B.26) to (1.18), we easily find the two main equations of electrostatics in differential form:

$$\operatorname{div} \mathbf{B} = 0 \tag{1.19}$$

$$\alpha \operatorname{curl} \mathbf{H} = \mathbf{j} \tag{1.20}$$

1.5. We now turn to the general case of fields that act in media and are time dependent.

First, let us see under what condition the relationship (1.10) between the electric induction **D** and the charge density ρ, derived initially for the static case, will also be true when **D** and ρ depend on time. If we assume that this is so and take the time derivative of (1.10), we arrive at a relation between the different derivatives: $\operatorname{div}(\partial \mathbf{D}/\partial t) = \partial\rho/\partial t$. If we also take into account charge conservation (1.13), the last equation can be written $\operatorname{div}(\mathbf{j} + \partial\mathbf{D}/\partial t) = 0$. As (B.13) shows, this is true if there is a vector field $\mathbf{H}(\mathbf{r}, t)$ such that

$$\alpha \operatorname{curl} \mathbf{H} = \mathbf{j} + \frac{\partial \mathbf{D}}{\partial t} \tag{1.21}$$

Now, if we assume that (1.21) is true and repeat the above reasoning in reverse order, we get $(\partial/\partial t)(\operatorname{div} \mathbf{D} - \rho) = 0$. Hence, if (1.10) is true for some moment of time, according to (1.21) it is valid for all other moments. Equation (1.21) is a generalization of (1.20) to the case where the fields are time dependent. The term $\partial\mathbf{D}/\partial t$, first introduced by Maxwell, is called the *density of the displacement current* (Maxwell called it the displacement current).

It remains to us to formulate the last main equation of the electromagnetic field theory. Let us consider a circuit s bounded by an arbitrary open surface σ. The line integral $\oint \mathbf{E}_s \cdot d\mathbf{s}$ is called the *electromotive force* around circuit s. The magnetic flux through a surface σ is defined in the following manner: $\Phi = \int_\sigma \mathbf{B} \cdot \mathbf{n}\, d\sigma$ (this explains why **B** is sometimes called the *magnetic flux density*). Faraday observed that the induced electromotive force around the circuit s is proportional to the time rate of change of the magnetic flux through the surface σ. Hence the *law of electromagnetic induction* is summed up in the mathematical expression

$$\beta \oint_s \mathbf{E} \cdot d\mathbf{s} = -\frac{d}{dt} \int_\sigma B_n \, d\sigma \tag{1.22}$$

The time derivative in the right-hand side is determined both by the time dependence of $\mathbf{B}(\mathbf{r}, t)$ and by the possible time dependence of the limits of integration. The latter may occur if the shape of the circuit s is a function of time or if s moves in relation to the sources of the magnetic field. The factor β depends, as above, on the choice of units for the electric and magnetic fields (in the left- and right-hand sides) and ensures the equality of the dimensions on both sides; it is the last arbitrary constant we introduce here. From (1.5) we see that the dimensions of the integral in the left-hand side of (1.22) are $[\varepsilon_0]^{-1}[q][L]^{-1}[\beta]$. On the other hand, from (1.12) and (1.15) it follows that the dimensions of the right-hand side are $[\mu_0/\alpha][q][L][T]^{-2}$. These two dimensions will be the same if $[\mu_0\varepsilon_0/\alpha\beta] = [L/T]^{-2}$.[7]

If the time dependence in the right-hand side of (1.22) is completely due to the change in the integrand, then $(d/dt)\int \mathbf{B}\cdot\mathbf{n}\,d\sigma = \int (\partial\mathbf{B}/\partial t)\cdot\mathbf{n}\,d\sigma$. In perfect analogy to the way we derived (1.20), we arrive at the law of electromagnetic induction in differential form:

$$\beta \operatorname{curl} \mathbf{E} + \partial\mathbf{B}/\partial t = 0 \tag{1.23}$$

By taking the divergence of both sides of (1.23), we get $(\partial/\partial t)\operatorname{div}\mathbf{B} = 0$. It follows that if (1.19) is true for a specified moment of time, it is true for all other moments.

Now we consider (1.8), (1.19), (1.21) and (1.23) in vacuo. (In this case we must set $\mathbf{B} = \mu_0\mathbf{H}$ and $\mathbf{D} = \varepsilon_0\mathbf{E}$.) We will also assume that the sources of the field are outside the region where the field is being studied (so we can set $\rho = 0$ and $j = 0$). By taking the curl of both sides of (1.21) and using (B.21), (1.19) and (1.23), we get a partial differential equation of the second order for $\mathbf{H}$, which in Cartesian coordinates is

$$\left(\nabla^2 - \frac{\mu_0\varepsilon_0}{\alpha\beta}\frac{\partial^2}{\partial t^2}\right)\mathbf{H} = 0 \tag{1.24}$$

In the same way, with the help of (1.8) and (1.21) we can conclude from (1.23) that $\mathbf{E}$ too satisfies an equation of type (1.24) with the very same factor of the time derivative.

From the physical viewpoint this conclusion is of major importance. It shows that both fields, $\mathbf{H}$ and $\mathbf{E}$, can propagate in vacuo with a velocity $c = (\alpha\beta/\varepsilon_0\mu_0)^{1/2}$, since this is the meaning of the factor of the time derivative in the wave equation (1.24). We have already established that the dimensions of the right-hand side in the expression for c must be those of velocity. This result connects all the di-

---

[7] We note that the dimensions of electric and magnetic field quantities are considered as yet independent. For one, the dimensions of magnetic moment do not depend on those of electric charge. Equations (1.1), (1.14) and (1.22) yield the following result: $[\beta] = ([q]/[m])\,L^2/T$.

mensional constants introduced in the Maxwell equations with a quantity that can be actually measured. This fact can serve as a basis for classifying the different systems of units used in measuring electromagnetic quantities.

1.6. All customary systems of units may be obtained if we put $\alpha = \beta$, so that

$$c = \alpha/(\varepsilon_0\mu_0)^{1/2} \tag{1.25}$$

In turn, α is given this or that value. We must bear in mind the following two main cases:

(1) $\alpha = c$. Then the electric and magnetic constants are related thus: $(\varepsilon_0\mu_0)^{1/2} = 1$. The system of units will be fully defined if we assume that both ε_0 and μ_0 are dimensionless and that $\varepsilon_0 = \mu_0 = 1$. Such a system is called the *Gaussian system*. It follows from definitions (1.11) and (1.17) that in this system the dimensions of all four field quantities, **E**, **D**, **B** and **H**, are viewed as identical. The dimensions of charge prove to be derived from the dimensions of the different mechanical quantities (see (1.6)).

(2) α is dimensionless and of unit size. In this case $\varepsilon_0\mu_0 = 1/c^2$. Here three ways of introducing systems of units are widely used.

First, we can assume ε_0 to be dimensionless and of unit size. Then $\mu_0 = 1/c^2$. The corresponding system is called the *electrostatic system of units* (esu).

Second, we can assume, in contrast, μ_0 to be dimensionless and of unit size. Then $\varepsilon_0 = 1/c^2$. The system is called the *electromagnetic system of units* (emu).

Finally, as mentioned before (p. 16), at the very start we can refrain from relating the dimensions of electric charge to those of mechanical quantities and assume that the unit of charge is established via independent measurements. For example, to electrolyze one gram-equivalent mass of a substance it is necessary to pass a certain amount of electricity through the electrolyte. This amount can be used to introduce a unit of charge and is taken to be 96 485 C (*coulomb*). However, since it is inconvenient to reproduce such a standard, the unit of current intensity, the *ampere* (A), is taken as a basic unit, and 1C = 1 A × 1 s. In addition to the second (s), there are two other basic mechanical units in this system: the meter (m) and the kilogram (kg), the last being the unit of mass. These four basic units are part of the *International System of Units* (abbreviated SI from the French “Le Système International d'Unités”) approved at the Eleventh General Conference on Weights and Measures (1960). In Section 29 we will see that ε_0 and μ_0 can have assigned to them definite numerical values.

From the practical viewpoint, the SI system is undoubtedly the most convenient. But in many problems of physics, particularly those that deal with the interaction of sources in vacuo, the Maxwell equations appear most natural when they are written in the Gaussian system. Note also that everywhere so-called rationalized units have been used.[8] The unrationalized units differ only in the normalization of the field quantities, for which charge density ρ in (1.10), current density $\mathbf{j}$ in (1.20), electric polarization $\mathbf{P}$ in (1.11), and magnetization $\mathbf{M}$ in (1.17) must each have a factor 4π.

1.7. Thus the Maxwell equations can be presented in the following form:

$$\operatorname{div} \mathbf{D} = \rho \tag{M.1}$$

$$\operatorname{div} \mathbf{B} = 0 \tag{M.2}$$

$$\operatorname{curl} \mathbf{E} = -\frac{1}{\alpha}\frac{\partial \mathbf{B}}{\partial t} \tag{M.3}$$

$$\operatorname{curl} \mathbf{H} = \frac{1}{\alpha}\left(\mathbf{j} + \frac{\partial \mathbf{D}}{\partial t}\right) \tag{M.4}$$

Whether $\alpha = 1$ or $\alpha = c$ and $\mu_0 = \varepsilon_0 = 1$ depends on whether we use the SI or the Gaussian system.

The integral counterparts of equations (M.1)-(M.4) are

$$\oint \mathbf{D}\cdot\mathbf{n}\, d\sigma = q \tag{M'.1}$$

$$\oint \mathbf{B}\cdot\mathbf{n}\, d\sigma = 0 \tag{M'.2}$$

$$\oint \mathbf{E}\cdot d\mathbf{s} = -\frac{1}{\alpha}\frac{d}{dt}\int \mathbf{B}\cdot\mathbf{n}\, d\sigma \tag{M'.3}$$

$$\oint \mathbf{H}\cdot d\mathbf{s} = \frac{1}{\alpha}\int\left(\mathbf{j} + \frac{\partial \mathbf{D}}{\partial t}\right)\cdot\mathbf{n}\, d\sigma \tag{M'.4}$$

We can arrive at (M′.4) by integrating both sides of (M.4) over an arbitrary open surface σ bounded by a circuit s and then using Stokes's theorem (B.26).

It is the system of the Maxwell equations (M) or (M′) that must now be considered the complete starting formulation of the laws of the electromagnetic field. For example, charge conservation in the form of the continuity equation (1.13) can be found from the system (M) if we find the time derivative of (M.1) and use (M.4). From the physical viewpoint it is more natural, however, to consider charge conservation independent of system (M) and then, as we have already done, use this law to prove that (M.1) does not change with the passage of time. In this case the basic equations are (M.3), (M.4), the continuity equation (1.13), and the equations (M.1) and (M.2) taken at a definite moment of time (the initial moment).

We can also write (M.1) and (M.4) in a somewhat different form if we use the definitions of electric polarization (1.11) and magneti-

---

[8] The rationalized Gaussian system of units is also known as the Heaviside-Lorentz system.

zation (1.17):

$$\operatorname{div}\mathbf{E}=\frac{1}{\varepsilon_0}(\rho-\operatorname{div}\mathbf{P}) \tag{M.1'}$$

$$\operatorname{curl}\mathbf{B}-\frac{\mu_0\varepsilon_0}{\alpha}\frac{\partial\mathbf{E}}{\partial t}=\frac{\mu_0}{\alpha}\left(\mathbf{j}+\frac{\partial\mathbf{P}}{\partial t}\right)+\mu_0\operatorname{curl}\mathbf{M} \tag{M.4'}$$

For given sources ρ and $\mathbf{j}$ the system comprising (M.1′), (M.2), (M.3), (M.4′) determines the unknown functions $\mathbf{E}(\mathbf{r}, t)$ and $\mathbf{B}(\mathbf{r}, t)$ only when vectors $\mathbf{P}$ and $\mathbf{M}$, which characterize the medium, are known. These characteristics, in turn, are determined by the fields created by the sources in the medium and also by the physical properties of the medium. For this reason it is only in vacuo, where $\mathbf{P}=0$ and $\mathbf{M}=0$, that the abovementioned system of equations defines the electromagnetic field in terms of the properties of the sources. In the general case, the system is an incomplete set of relationships between the different physical quantities.

In principle, if we want to solve the Maxwell equations, we must know the functional relationships of the type $\mathbf{P}=\mathbf{P}(\mathbf{E})$ and $\mathbf{M}=\mathbf{M}(\mathbf{B})$ or, which is the same, $\mathbf{D}=\mathbf{D}(\mathbf{E})$ and $\mathbf{H}=\mathbf{H}(\mathbf{B})$. For each medium these relationships can be found only by studying the specific electromagnetic properties of the medium. Such studies make it possible to divide the media into broad classes. Let us consider some of the simplest cases; we will deal with generalizations in Section 4.

The first is the class of *isotropic* media, for which

$$\mathbf{P}=\varepsilon_0\chi_e\mathbf{E}, \qquad \mathbf{M}=\chi_m\mathbf{H} \tag{1.26}$$

where χ_e is the *dielectric susceptibility* and χ_m is the *magnetic susceptibility*. We can rewrite (1.26) as

$$\mathbf{D}=\varepsilon\mathbf{E}, \qquad \mathbf{B}=\mu\mathbf{H} \tag{1.27}$$

where

$$\varepsilon=\varepsilon_0(1+\chi_e), \qquad \mu=\mu_0(1+\chi_m) \tag{1.28}$$

Factor ε is called the *permittivity* and factor μ the *permeability* of a medium. They can be functions of spatial coordinates and time.

Let us substitute (1.27) into (M) and assume that ε and μ are constant and ρ and $\mathbf{j}$ vanish. Then, in a manner similar to the way in which we derived (1.24), we get an equation which states that, if we apply the operator $\nabla^2-(\varepsilon\mu/\alpha^2)(\partial^2/\partial t^2)$ to $\mathbf{E}$ or $\mathbf{H}$, we get zero. The factor of the time derivative must be assumed to be the reciprocal of the square of the electromagnetic field velocity of propagation. A medium in which ε and μ are constant is called *homogeneous*. Equation (1.25) yields

$$\left(\frac{\varepsilon\mu}{\varepsilon_0\mu_0}\right)^{1/2}=\frac{c}{v} \tag{1.28'}$$

Often to characterize a medium (not necessarily homogeneous) the *relative permittivity* $\varepsilon' = \varepsilon/\varepsilon_0$ and the *relative permeability* $\mu' = \mu/\mu_0$ are used.

For *anisotropic* media we have the more general linear relationships

$$D_\alpha = \varepsilon_{\alpha\beta} E_\beta, \quad B_\alpha = \mu_{\alpha\beta} H_\beta \tag{1.29}$$

where $\varepsilon_{\alpha\beta} = \varepsilon_0 (\delta_{\alpha\beta} + \chi_{(e)\alpha\beta})$ and $\mu_{\alpha\beta} = \mu_0 (\delta_{\alpha\beta} + \chi_{(m)\alpha\beta})$ are tensors of rank 2 with respect to three-dimensional spatial rotations.[9] The reader must bear in mind that relationships of type (1.27) and (1.29) are valid only in a definite range of values for **E** and **B**. Outside this range, that is, for strong or weak fields, they may break down. Whether these relationships can be applied in a particular case depends also on the thermodynamic state of the medium (for one, its temperature). It is one of the goals of the microscopic theory of matter and, in the final analysis, of quantum theory to derive such "material relationships" and to determine their domain of application.

Experiments have shown that for any media χ_e is always greater than zero but that χ_m can be either positive or negative. When it is positive, the medium is called *paramagnetic*; when negative, *diamagnetic*. In all the above equations, electric polarization **P** and magnetization **M** vanish when **E** and **H** vanish. This condition does not hold for magnetization of *ferromagnetic* substances and polarization of *ferroelectric* substances. The explanation is that the material relationships for such media are not of the simple linear type (1.29).

A characteristic feature of ferroelectrics and ferromagnetics is the phenomenon of hysteresis (see Section 36) due to which **P** or **M** can be nonzero at $\mathbf{E} = 0$ or, respectively, $\mathbf{B} = 0$. If we denote such residual magnetization by $\mathbf{M}_0$, we see that (M.4′) implies that curl $\mathbf{M}_0$ can be the source of field **B**, which constitutes the phenomenological description for the action of permanent magnets.[10] We note, in passing, that above a certain temperature (the *Curie point*) ferromagnetics change their properties to paramagnetic.

Besides the above classification of media, the division of media into conductors and insulators is of primary importance in electrodynamics. For a wide range of values of **E** there is a simple law that connects the electric field and the current density, *Ohm's law*:

$$j_\alpha = \sigma_{\alpha\beta} E_\beta + j_\alpha^{\text{ext}} \tag{1.30}$$

where $\sigma_{\alpha\beta}$ is the *conductivity tensor*, and $\mathbf{j}^{\text{ext}}$ is the part of the current density that may be considered fixed and independent of field **E**. This part of the current density may be due to sources of a nonelec-

[9] It follows from thermodynamics that $\varepsilon_{\alpha\beta}$ and $\mu_{\alpha\beta}$ must be symmetric tensors (see Section 32).

[10] For a detailed description see Section 36.

tromagnetic nature, which act because other types of energies are transformed into the work of displacing the electric charges (for instance, in a chemical cell and an electric generator). Such sources may be considered external with respect to those that produce field **E**. Suppose that $\mathbf{j}^{\text{ext}} = 0$, $\sigma_{\alpha\beta} = \sigma\delta_{\alpha\beta}$, and σ and ε are constants (we assume that (1.27) is also valid). If we apply the abovementioned conditions and (M.1) to the continuity equation (1.13), we get

$$\operatorname{div}\left(\varepsilon\frac{\partial \mathbf{E}}{\partial t}+\sigma\mathbf{E}\right)=0 \tag{1.31}$$

For this equation to be true it is sufficient to assume that $\mathbf{E} = \mathbf{E}_0 \exp\{-\sigma(t-t_0)/\varepsilon\}$. From (M.1) it follows that the charge density depends on time in a similar way: $\rho = \rho_0 \exp\{-\sigma(t-t_0)/\varepsilon\}$. The factor ε/σ plays the role of the relaxation time T characteristic of the medium under observation. By T we denote a typical time for the duration of processes studied in experiments of a definite type. At $\varepsilon/\sigma \gg T$ the medium is called a *dielectric* (or insulator) and at $\varepsilon/\sigma \ll T$ a *conductor*. Only in the first case will the medium keep its charge for a sufficiently long time; in the second the charges introduced into the medium will travel to other regions very soon.[11] Charge conservation implies that, if a conductor fills the entire space, the charges travel to infinity, but if it is immersed in a dielectric, the charges can gather at the conductor-insulator boundary. This produces a surface charge. At any rate, the relaxation condition is applicable for $\mathbf{H} \approx 0$; this can easily be seen from the reasoning preceding (1.21).

§ 2. The potentials of the electromagnetic field. Gauge invariance. Hertz vectors

2.1. Being invariant with respect to rotations of the spatial axes in three-dimensional space, the basic electric and magnetic quantities act differently under spatial reflection (or inversion). The experimental data on the properties of electric charge does not contradict the assumption that q is a scalar. If we now bring in the charge conservation law (1.12), it follows that the current density **j** has the properties of a polar vector. The electric field **E** is also a polar vector. (To see this, look at (M.1) in vacuo.) Equation (M.3) implies that magnetic induction **B** and, by virtue of (1.17), magnetic field **H** and magnetization **M** must be pseudovectors, since the curl operator turns any polar vector into a pseudovector.[12] In other words, we can define

[11] We see that this classification of media is to a certain extent a matter of convention. In practice, substances with a sufficiently large number of charge carriers (metals, for instance) may be considered conductors.

[12] See Appendix B, the text following (B.20).

the positive directions for these quantities only after we agree as to which of the two possible orientations of the coordinate systems (left- or right-handed) will be considered positive. Under spatial reflection this direction changes to the opposite.[13]

What plays an important role in studying the Maxwell equations is the possibility of expressing the electric and magnetic fields in terms of auxiliary fields, the potentials. To determine the transformation properties of these auxiliary fields under spatial rotation and reflection, we must use the abovementioned properties of the electric and magnetic fields.

Here is how the electromagnetic potentials are introduced. First, if we look at (B.13), we see that (M.2) can be solved if we find a vector function $\mathbf{A}(\mathbf{r}, t)$, called the *vector potential*, such that

$$\mathbf{B} = \operatorname{curl} \mathbf{A} \tag{2.1}$$

If we then substitute (2.1) into (M.3), we find

$$\operatorname{curl}\left(\mathbf{E} + \frac{1}{\alpha}\frac{\partial}{\partial t}\mathbf{A}\right) = 0$$

If we now turn to (B.12), we can express the electric field as

$$\mathbf{E} = -\operatorname{grad}\varphi - \frac{1}{\alpha}\frac{\partial \mathbf{A}}{\partial t} \tag{2.2}$$

where φ is known as the *scalar potential*. With (2.1) and (2.2) we find that $\mathbf{A}$ must be a polar vector and φ a true scalar (not a pseudoscalar).

Equations (2.1) and (2.2), which express the vector functions $\mathbf{B}$ and $\mathbf{E}$ in terms of $\mathbf{A}$ and φ, are independent of any specific assumptions about the properties of the medium in which $\mathbf{B}$ and $\mathbf{E}$ act. They also show that for given electric and magnetic fields there is an infinite number of ways in which we may choose the potentials.

Indeed, (B.12) implies that, if the function $\mathbf{A}(\mathbf{r}, t)$ satisfies (2.1) with a given left-hand side, any function

$$\mathbf{A}' = \mathbf{A} + \operatorname{grad}\psi \tag{2.3}$$

where $\psi(\mathbf{r}, t)$ is an arbitrary differentiable function, is also a solution of (2.1). We can see from (2.2) that $\mathbf{E}(\mathbf{r}, t)$ likewise remains unchanged if we simultaneously transform from φ to

$$\varphi' = \varphi - \frac{1}{\alpha}\frac{\partial \psi}{\partial t} \tag{2.4}$$

Such transformations of potentials, (2.3) and (2.4), that do not change the solutions $\mathbf{B}$ and $\mathbf{E}$ of the Maxwell equations (and, hence, do

[13] Note that in (1.14) the cross product, the torque $\mathbf{N}$, is a pseudovector. In mechanics the pseudovector nature of $\mathbf{N}$ is due to the fact that torque is the cross product of two polar vectors. Since in our case $\mathbf{B}$ is a pseudovector, $\mathbf{m}$ must also be considered a pseudovector.

not change the equations themselves) are called *gauge transformations.*

To derive equations for the potentials, we use the Maxwell equations (M.1) and (M.4), which connect the electromagnetic field with its sources. Let us consider the case of an isotropic, homogeneous medium, for which (1.27) is valid and ε and μ are constant. If we bear in mind (1.27), substitute (2.1) and (2.2) into (M.1) and (M.4), and use (B.21), we get, in Cartesian coordinates,

$$\nabla^2\mathbf{A}-\frac{1}{v^2}\frac{\partial^2\mathbf{A}}{\partial t^2}=-\frac{\mu}{\alpha}\mathbf{j}+\operatorname{grad}\left(\operatorname{div}\mathbf{A}+\frac{\varepsilon\mu}{\alpha}\frac{\partial\varphi}{\partial t}\right) \tag{2.4$_1$}$$

$$\nabla^2\varphi=-\frac{1}{\varepsilon}\rho-\frac{1}{\alpha}\frac{\partial}{\partial t}\operatorname{div}\mathbf{A} \tag{2.4$_2$}$$

(Here, as in (1.28′), $v^2=\alpha^2/\varepsilon\mu$.) We can considerably simplify the above equations for the potentials if we assume that

$$\operatorname{div}\mathbf{A}+\frac{\varepsilon\mu}{\alpha}\frac{\partial\varphi}{\partial t}=0 \tag{2.5}$$

which is known as the *Lorentz condition*. Indeed, in this case

$$\left(\nabla^2-\frac{1}{v^2}\frac{\partial^2}{\partial t^2}\right)\mathbf{A}=-\frac{\mu}{\alpha}\mathbf{j} \tag{2.6$_1$}$$

$$\left(\nabla^2-\frac{1}{v^2}\frac{\partial^2}{\partial t^2}\right)\varphi=-\frac{1}{\varepsilon}\rho \tag{2.6$_2$}$$

Let us now show that the potentials can always be found to satisfy the Lorentz condition. This possibility is due to the freedom of choice of potentials provided by gauge invariance. Indeed, suppose that in some way we have found such functions $\mathbf{A}$ and φ that for given $\mathbf{B}$ and $\mathbf{E}$ satisfy (2.1) and (2.2) but do not satisfy (2.5). Then let us make a gauge transformation according to (2.3) and (2.4) and demand that the new potentials $\mathbf{A}'$ and φ' satisfy the Lorentz condition (2.5). We substitute $\mathbf{A}'$ and φ' into (2.5) and express them in terms of $\mathbf{A}$ and φ. We get an equation for the gauge function ψ:

$$\left(\nabla^2-\frac{1}{v^2}\frac{\partial^2}{\partial t^2}\right)\psi=-\left(\operatorname{div}\mathbf{A}+\frac{\varepsilon\mu}{\alpha}\frac{\partial\varphi}{\partial t}\right) \tag{2.7}$$

Since we have assumed that $\mathbf{A}$ and φ are known, the right-hand side of this inhomogeneous wave equation may be considered given. Hence, if we take any solution of (2.7) as the gauge function ψ, we will not change the physics of the problem and yet shift to potentials $\mathbf{A}'$ and φ' that satisfy (2.5). But, with very general assumptions, (2.7) has an infinite number of solutions.[14] Moreover, if we assume that the starting potentials satisfy the Lorentz condition, then (2.7)

[14] Here we do not consider the role played by the initial and boundary conditions.

becomes a homogeneous wave equation for ψ, which also has an infinite number of solutions satisfying the superposition principle. We see from this that the potentials which satisfy the Lorentz condition can be determined only to within a gauge transformation of a special type, with the gauge function ψ being an arbitrary solution of the homogeneous wave equation. Such gauge transformations constitute a subgroup of the group of all gauge transformations, and the potentials in this restricted class are said to belong to the Lorentz gauge.

The above reasoning shows that the Lorentz condition is not the only possible restriction on the choice of potentials. For instance, another gauge commonly used is the so-called *Coulomb gauge*. In this gauge

$$\operatorname{div} \mathbf{A} = 0 \tag{2.8}$$

Equations (2.4_1) and (2.4_2) then transform to

$$\left(\nabla^2 - \frac{1}{v^2}\frac{\partial^2}{\partial t^2}\right)\mathbf{A} = -\frac{\mu}{\alpha}\mathbf{j} + \frac{\varepsilon\mu}{\alpha}\operatorname{grad}\frac{\partial\varphi}{\partial t} \tag{2.9$_1$}$$

$$\nabla^2\varphi = -\frac{1}{\varepsilon}\rho \tag{2.9$_2$}$$

We can calculate the term with grad $(\partial\varphi/\partial t)$ in the right-hand side of (2.9_1) if we know the solution to (2.9_2). This term constitutes an "addition" to the current density $\mathbf{j}$ (for more details see Section 13). The gauge function ψ that provides for transforming from arbitrary potentials to those satisfying the Coulomb gauge must satisfy a condition similar to (2.7), namely

$$\nabla^2\psi = -\operatorname{div}\mathbf{A} \tag{2.9$_3$}$$

If we know how the sources are distributed, that is, if the functions $\rho(\mathbf{r}, t)$ and $\mathbf{j}(\mathbf{r}, t)$ are given, then, by solving either (2.6) or (2.9) for the potentials, we can calculate the fields $\mathbf{B}$ and $\mathbf{E}$ created by these sources by applying (2.1) and (2.2). The resulting formulas, however, are not the most general. If there are some sources outside the region for which we obtained our solution, then these sources can also generate fields in the region. These new fields must satisfy, in the given region, the Maxwell equations (M) with $\mathbf{j} = 0$ and $\rho = 0$. Then the general solution for the Maxwell equations is expressed as a sum of the particular solution for the given values of ρ and $\mathbf{j}$ and the general solution for the homogeneous Maxwell equations, which we must expect for a system of linear partial differential equations.

Reasoning in the same way as we did at the beginning of this section, we see that for $\mathbf{j} = 0$ and $\rho = 0$ the solution of the Maxwell equations (M) can be sought for in the form

$$\mathbf{D} = -\operatorname{curl}\mathbf{A}^*, \quad \mathbf{H} = -\operatorname{grad}\varphi^* - \frac{1}{\alpha}\frac{\partial\mathbf{A}^*}{\partial t} \tag{2.10}$$

Here $\mathbf{A}^*$ should be considered a pseudovector and φ^* a pseudoscalar. Equations (M) then yield

$$\left(\nabla^2-\frac{1}{v^2}\frac{\partial^2}{\partial t^2}\right)\mathbf{A}^*=0, \quad \left(\nabla^2-\frac{1}{v^2}\frac{\partial^2}{\partial t^2}\right)\varphi^*=0$$

if, as we have done before, the Lorentz condition

$$\operatorname{div}\mathbf{A}^*+\frac{\varepsilon\mu}{\alpha}\frac{\partial\varphi^*}{\partial t}=0 \tag{2.11}$$

is introduced. In the case where (1.27) holds true with constant ε and μ, we get the general solution of the Maxwell equations in the following form:

$$\mathbf{B}=\operatorname{curl}\mathbf{A}-\frac{\mu}{\alpha}\frac{\partial\mathbf{A}^*}{\partial t}-\mu\operatorname{grad}\varphi^*$$

$$\mathbf{E}=-\operatorname{grad}\varphi-\frac{1}{\alpha}\frac{\partial\mathbf{A}}{\partial t}-\frac{1}{\varepsilon}\operatorname{curl}\mathbf{A}^* \tag{2.12}$$

In a region without sources, the potential φ, as can be seen from (2.6_2), satisfies a homogeneous wave equation. If the Lorentz condition holds, the gauge function ψ also satisfies a homogeneous wave equation (i.e. (2.7) with a zero right-hand side). We can thus choose the gauge function ψ so that $\varphi = 0$. The electromagnetic field is then determined only in terms of the vector potential:

$$\mathbf{B}=\operatorname{curl}\mathbf{A}, \quad \mathbf{E}=-\frac{1}{\alpha}\frac{\partial\mathbf{A}}{\partial t}, \quad \operatorname{div}\mathbf{A}=0$$

$$\left(\nabla^2-\frac{1}{v^2}\frac{\partial^2}{\partial t^2}\right)-\mathbf{A}=0 \tag{2.12'}$$

We note, in passing, that the dimensions of the scalar potential are those of energy per unit charge (see (2.2)), while the dimensions of the vector potential in the SI system are those of momentum per unit charge.

2.2. Let us now consider one more method of solving the Maxwell equations, which is particularly useful when they are written as (M.1′), (M.2), (M.3), (M.4′) and when $\mathbf{j}$ and ρ are zero. We will assume that electric polarization $\mathbf{P}$ and magnetization $\mathbf{M}$ can be written as

$$\mathbf{P}=\varepsilon_0\chi_e\mathbf{E}+\mathbf{P}_0, \quad \mathbf{M}=\chi_m\mathbf{H}+\mathbf{M}_0 \tag{2.13}$$

with $\mathbf{P}_0$ and $\mathbf{M}_0$ reflecting some special properties of the medium and not the presence of the external field. For instance, $\mathbf{P}_0\neq 0$ for ferroelectric and pyroelectric substances, and $\mathbf{M}_0\neq 0$ for ferromagnetics. It is also convenient to use the vectors $\mathbf{P}_0$ and $\mathbf{M}_0$ in describing the sources of an electromagnetic field to be found in the theory of antennas and waveguides. With (1.28), that is, $\mathbf{B}=\mu\mathbf{H}+\mu_0\mathbf{M}_0$ and $\mathbf{D}=\varepsilon\mathbf{E}+\mathbf{P}_0$, we can rewrite the system of the Maxwell equations (ε and μ constant) as

$$\operatorname{curl}\mathbf{H}-\frac{\varepsilon}{\alpha}\frac{\partial\mathbf{E}}{\partial t}=\frac{1}{\alpha}\frac{\partial\mathbf{P}_0}{\partial t},\qquad \operatorname{div}\mathbf{E}=-\frac{1}{\varepsilon}\operatorname{div}\mathbf{P}_0 \tag{2.14$_1$}$$

$$\operatorname{curl}\mathbf{E}+\frac{\mu}{\alpha}\frac{\partial\mathbf{H}}{\partial t}=-\frac{\mu_0}{\alpha}\frac{\partial\mathbf{M}_0}{\partial t},\qquad \operatorname{div}\mathbf{H}=-\frac{\mu_0}{\mu}\operatorname{div}\mathbf{M}_0 \tag{2.14$_2$}$$

Let us first focus on (2.14_1). We assume that there is an auxiliary vector field $\mathbf{\Pi}$ (called a *Hertz vector*) such that

$$\mathbf{A}=\frac{\varepsilon\mu}{\alpha}\frac{\partial\mathbf{\Pi}}{\partial t},\qquad \varphi=-\operatorname{div}\mathbf{\Pi} \tag{2.15}$$

The Lorentz condition (2.5) is satisfied automatically. If we also assume that $\mathbf{M}_0=0$ (or $\mathbf{B}=\mu\mathbf{H}$), then magnetic field $\mathbf{H}$ is determined via (2.1). We use (2.1), (2.2) and (2.15) to reduce the first equation in (2.14_1) to

$$\frac{\partial}{\partial t}\left(\operatorname{curl}\operatorname{curl}\mathbf{\Pi}-\operatorname{grad}\operatorname{div}\mathbf{\Pi}+\frac{1}{v^2}\frac{\partial^2\mathbf{\Pi}}{\partial t^2}\right)=\frac{1}{\varepsilon}\frac{\partial\mathbf{P}_0}{\partial t}$$

which shows that the expression in the parentheses can differ from $(1/\varepsilon)\,\mathbf{P}_0$ only in a term that is an arbitrary function of spatial coordinates, $\mathbf{f}\,(\mathbf{r})$. Let this arbitrary function be zero. Then a similar substitution shows that the second equation in (2.14_1) is satisfied automatically. The last step is to use a formula from vector analysis, (B.21). We finally get a wave equation for $\mathbf{\Pi}$:

$$\left(\nabla^2-\frac{1}{v^2}\frac{\partial^2}{\partial t^2}\right)\mathbf{\Pi}=-\frac{1}{\varepsilon}\mathbf{P}_0 \tag{2.16}$$

The electromagnetic field that satisfies the above conditions is known as the *electric-type field.*

We turn now to (2.14_2) and assume that $\mathbf{P}_0=0$. Then $\mathbf{D}=\varepsilon\mathbf{E}$, and to determine $\mathbf{D}$ and $\mathbf{H}$ we use (2.10). We introduce the pseudovector $\mathbf{\Pi}^*$ (also known as the Hertz vector) in the following way:

$$\mathbf{A}^*=\frac{\varepsilon\mu}{\alpha}\frac{\partial\mathbf{\Pi}^*}{\partial t},\qquad \varphi^*=-\operatorname{div}\mathbf{\Pi}^* \tag{2.17}$$

(obviously, $\mathbf{A}^*$ and φ^* satisfy the Lorentz condition). After substituting (2.10) and (2.17) into (2.14_2) and reasoning as we did in the previous paragraph, we get a wave equation for $\mathbf{\Pi}^*$:

$$\left(\nabla^2-\frac{1}{v^2}\frac{\partial^2}{\partial t^2}\right)\mathbf{\Pi}^*=-\frac{\mu_0}{\mu}\mathbf{M}_0 \tag{2.18}$$

The electromagnetic field obtained in these conditions is known as the *magnetic-type field.*

§ 3. Laws governing the variation of energy, momentum, and angular momentum

3.1. Let us start with *energy*. We have formulated the Maxwell equations in a very general way. They can now be used to formulate the laws for the variation of the quantities that characterize the mechanical properties of the electromagnetic field. These properties manifest themselves when the field acts on the sources in material media. Due to the fact that these quantities are considered simultaneously, we can try to find them on dimensional grounds.

If the dimensions of electric charge, $[q]$, have been established, then (1.1) together with the Maxwell equations provides us with the dimensions of the electric and magnetic fields:

$$[D] = [q]\cdot[L^{-2}], \quad [H] = [D]\cdot[L]\cdot[T]^{-1}\,[\alpha]^{-1} \tag{3.1}$$

If we take into account the dimensions of **E** and **B**, we see that

$$[ED] = [HB] = [F]\cdot[L]^{-2} \tag{3.2}$$

that is, such bilinear quantities have the dimensions of energy density regardless of the system of electromagnetic units. We can therefore expect that the field energy, $\mathscr{E}$, will be connected with these quantities. As to the law of the variation of field energy, it can be found from the Maxwell equations as automatically as, for instance, the law of the variation of a particle's energy from Newton's equations of motion.

Let us multiply both sides of (M.3) scalarly by **H** and both sides of (M.4) scalarly by **E** and subtract one from the other. With the help of (B.19) this difference is transformed to

$$\alpha\,\mathrm{div}\,(\mathbf{E}\times\mathbf{H})+\mathbf{E}\frac{\partial\mathbf{D}}{\partial t}+\mathbf{H}\frac{\partial\mathbf{B}}{\partial t}+\mathbf{j}\cdot\mathbf{E}=0 \tag{3.3}$$

Note that (3.2) implies that $\alpha\mathbf{E}\times\mathbf{H}$ has the dimensions of $[\mathscr{E}]\cdot[L]^{-2}\,[T]^{-1}$. Consider a region V confined by a closed surface σ. If we integrate (3.3) over V and apply Gauss's theorem to the first term, we get the integral counterpart of (3.3):

$$\alpha\oint_\sigma(\mathbf{E}\times\mathbf{H})\cdot\mathbf{n}\,d\sigma+\int_V\left(\mathbf{E}\frac{\partial\mathbf{D}}{\partial t}+\mathbf{H}\frac{\partial\mathbf{B}}{\partial t}\right)dV+\int_V\mathbf{j}\cdot\mathbf{E}\,dV=0 \tag{3.4}$$

We analyze this equation starting with the second term. This term can be interpreted, with the help of (3.2), as the quantity that expresses the time rate of change of the field energy within the region. The term $\mathbf{E}\cdot\partial\mathbf{D}/\partial t$ is determined only by the properties of the electric field. It is natural, then, to expect that it is equal to the time rate of change in the volume density of "electric energy", $\partial w_e/\partial t$. In the same way $\mathbf{H}\cdot\partial\mathbf{B}/\partial t$ expresses the time rate of change in the volume density of the "magnetic energy", $\partial w_\mathrm{m}/\partial t$. The sum of the two is the time rate of change in the volume density of the total field energy $\partial w/\partial t$. (In deriving (3.4) we did not use any reference to the specific forms of $\mathbf{D}(\mathbf{E})$ and $\mathbf{B}(\mathbf{H})$.) However, $đw_e = \mathbf{E}\,d\mathbf{D}$, $đw_m = \mathbf{H}\,d\mathbf{B}$, and $đw = đw_e + đw_m$ are not, generally speaking, complete differentials and hence no explicit expressions for w_e, w_m, and w can be found. But if (1.29) holds and neither $\varepsilon_{\alpha\beta}$ nor $\mu_{\alpha\beta}$ depends on time, then, by using the fact that permittivity and permeability are symmetric in their indices, it is easy to show that

$$\mathbf{E}\cdot\frac{\partial\mathbf{D}}{\partial t}=\frac{1}{2}\frac{\partial}{\partial t}(\mathbf{E}\cdot\mathbf{D}),\quad \mathbf{H}\cdot\frac{\partial\mathbf{B}}{\partial t}=\frac{1}{2}\frac{\partial}{\partial t}(\mathbf{H}\cdot\mathbf{B})$$

that is

$$w_e=\frac{1}{2}\,\mathbf{E}\cdot\mathbf{D},\quad w_m=\frac{1}{2}\,\mathbf{H}\cdot\mathbf{B},\quad w=w_e+w_m \tag{3.5}$$

Then the second term in the left-hand side of (3.4) can be interpreted as the time derivative of the total (electromagnetic) field energy within the region V. Such an interpretation provides the key to an understanding of the physical meaning of the other two terms in (3.4) as well. Indeed, the surface integral expresses the electromagnetic energy flux through σ, and the density and direction of this flux at each point of σ are determined by the *Poynting vector*

$$\mathbf{S} = \alpha\mathbf{E}\times\mathbf{H} \tag{3.6}$$

Finally, the last term in (3.4) is equal to the work per unit time spent by the electromagnetic field in producing conduction currents in the region. To some extent this term is responsible for the heat effects that usually appear (see Section 33). For this reason this term is called the *Joule heat*, though rather imprecisely.

Thus (3.4) expresses the law of conservation of energy for the electromagnetic field, and (3.3) the same law for the energy's density; namely, the energy can change only owing to the production of the Joule heat and the energy flux through σ. The reader must bear in mind, however, that the field energy can change into other forms—mechanical energy, chemical energy, and heat—and can be produced when these forms of energy change into each other. For this reason, if such processes do take place in the region under consideration, this should be accounted for by the inclusion of an additional term in the left-hand side, $d\mathscr{E}^{\text{ext}}/dt$, which expresses the participation of the energy of nonelectromagnetic sources in the overall energy balance. However, some of these factors, namely those that are related to the work produced by nonelectromagnetic sources in moving the charges, are already accounted for by the Joule heat, $\mathbf{j}\cdot\mathbf{E}$. Indeed, we showed in Section 1 that current density $\mathbf{j}$ can usually be represented as $\mathbf{j} = \sigma\mathbf{E} + \mathbf{j}^{\text{ext}}$, where $\mathbf{j}^{\text{ext}}$ is due to the action on the charges of nonelectromagnetic (external) forces. When Ohm's law is valid, then the following notation is often used: $\mathbf{j}^{\text{ext}} =$

$= \sigma\mathbf{E}^{\text{ext}}$. Actually this is simply a definition of the vector $\mathbf{E}^{\text{ext}}$. In the rest of this section we will assume that the current density has the abovementioned structure.

3.2. We now turn to *momentum* and *angular momentum*. First let us see how to obtain the law of the variation of (electromagnetic) field momentum using the Maxwell equations.

We start with the vector (curl $\mathbf{E}$) $\times$ $\mathbf{D}$. The definitions of the curl of a vector (see (B.10) and (B.11)) and the vector product, (B.3), yield

$$(\operatorname{curl}\mathbf{E})\times\mathbf{D}\,|_\alpha=\varepsilon_{\alpha\beta\gamma}\varepsilon_{\beta\delta\zeta}\frac{\partial E_\zeta}{\partial x^\delta}D_\gamma$$

The summing over β can be achieved with the first of the formulas (B.1). This yields

$$(\operatorname{curl}\mathbf{E})\times\mathbf{D}\,|_\alpha=\frac{\partial E_\alpha}{\partial x^\gamma}D_\gamma-\frac{\partial E_\gamma}{\partial x^\alpha}D_\gamma$$

$$=\frac{\partial}{\partial x^\gamma}(E_\alpha D_\gamma-\delta_{\alpha\gamma}\mathbf{E}\cdot\mathbf{D})+\delta_{\alpha\gamma}\mathbf{E}\cdot\frac{\partial\mathbf{D}}{\partial x^\gamma}-E_\alpha\operatorname{div}\mathbf{D}\quad(3.7)$$

Next we assume that (1.29) holds and that $\varepsilon_{\alpha\beta}$ and $\mu_{\alpha\beta}$ do not depend on spatial coordinates.[15] In this case, as we have just seen, $\mathbf{E}\cdot(\partial\mathbf{D}/\partial x^\gamma)=(1/2)\,(\partial/\partial x^\gamma)\,(\mathbf{E}\cdot\mathbf{D})$. If in addition we apply (M.1), we get

$$(\operatorname{curl}\mathbf{E})\times\mathbf{D}\,|_\alpha=\frac{\partial}{\partial x^\gamma}\left(E_\alpha D_\gamma-\frac{1}{2}\delta_{\alpha\gamma}\mathbf{E}\cdot\mathbf{D}\right)-\rho E_\alpha\qquad(3.8)$$

Clearly

$$T^{(\mathrm{e})}_{\alpha\gamma}=E_\alpha D_\gamma-\frac{1}{2}\delta_{\alpha\gamma}\mathbf{E}\cdot\mathbf{D}\qquad(3.9)$$

is a tensor of rank 2 with respect to three-dimensional rotations and is not necessarily symmetric under our assumptions $(E_\alpha D_\gamma)=$ $=\varepsilon_{\gamma\delta}E_\alpha E_\delta\neq\varepsilon_{\alpha\delta}E_\gamma E_\delta)$.

In a similar manner, by using (1.29) we come to the result

$$(\operatorname{curl}\mathbf{H})\times\mathbf{B}\,|_\alpha=\frac{\partial T^{(\mathrm{m})}_{\alpha\gamma}}{\partial x^\gamma}\qquad(3.10)$$

(we have kept in mind that div $\mathbf{B}=0$). Here

$$T^{(\mathrm{m})}_{\alpha\gamma}=H_\alpha B_\gamma-\frac{1}{2}\delta_{\alpha\gamma}\mathbf{H}\cdot\mathbf{B}\qquad(3.11)$$

The tensor $T_{\alpha\gamma}=T^{(e)}_{\alpha\gamma}+T^{(m)}_{\alpha\gamma}$ is known as the *Maxwell stress tensor*, $T^{(e)}_{\alpha\gamma}$ being its electric part and $T^{(m)}_{\alpha\gamma}$ its magnetic part.

[15] The reader must not confuse the permittivity $\varepsilon_{\alpha\beta}$ with the pseudoscalar $\varepsilon_{\alpha\beta\gamma}$.

Let us multiply vectorially the Maxwell equation (M.4) by **B** and (M.3) by **D**. If we add the two vector products, then with the help of (3.8), (3.9) and (3.10) we get

$$\frac{\partial T_{\beta\gamma}}{\partial x^{\gamma}}=\alpha^{-1}\mathbf{j}\times\mathbf{B}\,|_{\beta}+\rho E_{\beta}+\frac{1}{\alpha}\frac{\partial}{\partial t}(\mathbf{D}\times\mathbf{B})_{\beta} \tag{3.12}$$

To clarify the physical meaning of the above equation we can use the fact that $\rho\mathbf{E}$ is the bulk density of the force that the electric field exerts on the charges. By analogy, since the vector $\alpha^{-1}\mathbf{j}\times\mathbf{B}$ has the same dimensions and is fully determined by the distribution of currents, $\mathbf{j}$, and the magnetic induction, $\mathbf{B}$, it should be interpreted as the force with which the magnetic field acts on the currents.[16] The total bulk density of the forces is called the *Lorentz force density*:

$$\mathbf{f}=\alpha^{-1}\mathbf{j}\times\mathbf{B}+\rho\mathbf{E} \tag{3.13}$$

According to the definition of mechanical momentum $\mathbf{p}$ we must assume that

$$\mathbf{f}=\partial\mathbf{p}/\partial t \tag{3.14}$$

But the third term in the right-hand side of (3.12) has a similar structure. Since this term is fully determined by the vectors **B** and **D**, we can draw the conclusion that the vector product

$$\mathbf{g}=\alpha^{-1}\mathbf{D}\times\mathbf{B} \tag{3.15}$$

is the *electromagnetic momentum density*. Equation (3.12) then assumes the form

$$\frac{\partial T_{\beta\gamma}}{\partial x^{\gamma}}=\frac{\partial}{\partial t}(\mathbf{p}+\mathbf{g})\,|_{\beta} \tag{3.16}$$

When conditions (1.27) are satisfied, vector **g** is proportional to the Poynting vector **S**:

$$\mathbf{g}=\mathbf{s}/v^{2} \tag{3.17}$$

(Here we have used (1.25) and (1.28′); obviously, for a field in vacuo $v=c$.)[17]

---

[16] It stands to reason that such an interpretation must be verified with specific examples (see, for example, Section 12).

[17] The above definitions of Maxwell's stress tensor and field momentum in the case of a medium were first introduced by H. Minkowski. Often they are replaced by the definitions of M. Abraham, according to which the stress tensor is symmetric also in anisotropic media and the field momentum is the vector $\mathbf{S}/c^2$ both in vacuo and in material media. But then, in addition to the Lorentz force we must introduce the so-called *Abraham force*. We will not dwell on these questions here since they play no important role in our further exposition. An interested reader can refer to V. L. Ginzburg: *Theoretical Physics and Astrophysics*, Pergamon Press, Oxford, 1979, Chapter 12.

Now let us study the left-hand side of (3.12). We integrate it over volume V, which is confined by surface σ. We transform the result into a surface integral by means of Gauss's integral theorem. The integrand of the surface integral is

$$\varphi_\beta = T_{\beta\gamma} n^\gamma \tag{3.18}$$

The dimensions of $\boldsymbol{\varphi}$ are those of force per unit area. This fact clarifies the meaning of the stress tensor $T_{\beta\gamma}$; namely, this tensor establishes a linear relationship between the force vector at a point on surface σ and the unit normal to that surface at the same point. If the shape of σ does not depend on time, we can equate the abovementioned surface integral of the left-hand side of (3.12) to the volume integral of the right-hand side:

$$\oint_\sigma \boldsymbol{\varphi}\, d\sigma = \frac{d}{dt} \int_V (\mathbf{p} + \mathbf{g})\, dV \tag{3.19}$$

This is the law of the variation of momentum: the time rate of change of the total momentum (field plus charges plus currents) inside a certain volume is equal to the total force acting on the boundary surface or, which is the same, the momentum flux. Following the same reasoning as we did in connection with the law of the variation of energy (see the end of Subsection 3.1), we see that in V there can be variations in other types of momentum, since the charges and currents may be influenced by forces of a nonelectromagnetic nature. In this case we must add a new term, $d\mathbf{P}^{\text{ext}}/dt$, to the right-hand side of (3.19) and a surface force density $\boldsymbol{\varphi}^{\text{ext}}$ to the left-hand side integrand. Both terms account for this "external" influence. The reader must bear in mind that (3.19) does not express the equilibrium between surface and volume forces if both types of forces are due only to the action of the electromagnetic field. Mechanical equilibrium may be achieved only as a result of the interaction of electromagnetic and external (that is, nonelectromagnetic) forces.

After defining field momentum we see that a field must also possess angular momentum. Indeed, we know from the general principles of mechanics that the angular momentum density due to the field, the charges, and the currents is determined via the formula $\widetilde{\mathbf{m}} = \mathbf{r} \times (\mathbf{p} + \mathbf{g})$. (Here $\mathbf{r}$ does not depend on time, since it is the radius vector of a fixed volume element.) If we multiply both sides of the differential law of the variation of momentum by $\mathbf{r}$ vectorially, we come to the following equation:

$$\varepsilon_{\beta\gamma\delta} x^\gamma \frac{\partial T^{\delta\zeta}}{\partial x^\zeta} = \frac{\partial}{\partial t} \varepsilon_{\beta\gamma\delta} x^\gamma (p+g)^\delta = \frac{\partial \widetilde{m}^\beta}{\partial t} \tag{3.20}$$

But

$$\varepsilon_{\beta\gamma\delta} x^\gamma \frac{\partial T^{\delta\zeta}}{\partial x^\zeta} = \varepsilon_{\beta\gamma\delta} \frac{\partial}{\partial x^\zeta} (x^\gamma T^{\delta\zeta}) - \delta_{\gamma\zeta} \varepsilon_{\beta\gamma\delta} T^{\delta\zeta} = \frac{\partial}{\partial x^\zeta} (\varepsilon_{\beta\gamma\delta} x^\gamma T^{\delta\zeta}) - \varepsilon_{\beta\gamma\delta} T^{\delta\gamma}$$

If the Maxwell stress tensor is symmetric ($T^{\delta\gamma} = T^{\gamma\delta}$), the last term is identically zero. Then

$$\frac{\partial \pi_{\beta\zeta}}{\partial x^{\zeta}} = \frac{\partial \widetilde{m}_{\beta}}{\partial t} \tag{3.21}$$

with $\pi_{\beta\zeta} = \varepsilon_{\beta\gamma\delta} x^{\gamma} T^{\delta\zeta}$. The physical meaning of the tensor $\pi_{\beta\zeta}$ is clarified if we integrate the left-hand side of (3.21) over V and use Gauss's integral theorem as presented in (B.23′):

$$\int \frac{\partial \pi^{\beta\zeta}}{\partial x^{\zeta}} dV = \oint \varepsilon_{\beta\gamma\delta} x^{\gamma} T^{\delta\zeta} n_{\zeta}\, d\sigma = \oint \varepsilon_{\beta\gamma\delta} x^{\gamma} \varphi^{\delta}\, d\sigma = \oint \mathbf{r} \times \boldsymbol{\varphi}\,|_{\beta}\, d\sigma$$

(Here we have used (3.18).) The result is simply the total angular momentum of surface forces acting on the surface σ that encloses volume V (or, which is the same, the angular momentum flux). Our final result is the law of the variation of angular momentum in integral form:

$$\frac{d}{dt} \int \widetilde{\mathbf{m}}\, dV = \oint \mathbf{r} \times \boldsymbol{\varphi}\, d\sigma \tag{3.22}$$

Here also we may have to account for "external" forces.

In Section 9 we will study the conservation laws from a more general viewpoint. We will use the variational principle and study the connection between these laws and the invariance of field quantities under different transformations. The reader must note that in homogeneous, isotropic media all the conservation laws have the form (3.3), (3.12) or (3.21), where the energy density is defined as in (3.5) and the stress tensor as in (3.9) and (3.11).

§ 4. Properties of the Maxwell equations. Uniqueness of the solution in bounded regions. Boundary conditions at the interface between two media

4.1. The mathematical properties of the Maxwell equations, needless to say, depend to a great extent on the assumptions that are made regarding the functions $\mathbf{P}(\mathbf{E})$ and $\mathbf{M}(\mathbf{B})$ (or $\mathbf{M}(\mathbf{H})$). In the simplest case of a homogeneous, isotropic media with sources ρ and $\mathbf{j}$ being known functions of position and time, the Maxwell equations (M), as we have seen, reduce to wave equations (2.6) for the potentials. (We will study the solution of these wave equations for given boundary and initial conditions in Section 13.) Let us now apply the curl operator to both sides of (M.3) and (M.4). If we use (B.21) and the remaining pair of the Maxwell equations, we can easily show that in Cartesian coordinates the fields satisfy two wave equations:

$$\left(\nabla^2 - \frac{1}{v^2}\frac{\partial^2}{\partial t^2}\right)\mathbf{E} = \frac{\mu}{\alpha^2}\frac{\partial \mathbf{j}}{\partial t} + \frac{1}{\varepsilon}\,\text{grad}\,\rho$$

$$\left(\nabla^2 - \frac{1}{v^2}\frac{\partial^2}{\partial t^2}\right)\mathbf{B} = -\frac{\mu}{\alpha}\,\text{curl}\,\mathbf{j} \tag{4.1}$$

The reader must bear in mind that solving (2.6) is equivalent to solving the first-order Maxwell equations, whereas (4.1) is simply a necessary corollary of the latter. In Section 1 we used a particular case of (4.1) in a region free of sources (see (1.24)).

From the physical point of view much can be gained from studying media for which the relationships (1.29) hold, with $\varepsilon_{\alpha\beta}$ and $\mu_{\alpha\beta}$ being functions of position and time. In some cases one can assume that the induction vectors, **D** and **B**, at point **r** and time t depend on the electric and magnetic field vectors, **E** and **H**, given at the same point and the same instant. Then for the field vectors one can construct second-order differential equations with variable coefficients. But usually one has to take into account dispersion phenomena, expressed in the fact that the inductions vectors at point **r** and time t depend also on the field vectors at earlier moments of time (*frequency dispersion*) and at other points in the medium (*spatial dispersion*). Then in relationships (1.29) one must assume that $\varepsilon_{\alpha\beta}$ and $\mu_{\alpha\beta}$ are operators. For instance, if $\varepsilon_{\alpha\beta}$ acts on **E**, it turns it into **D**. In many cases these operators are linear and integral, so that for **D**, (1.29) takes the form

$$D_\alpha(\mathbf{r}, t) = \int_{-\infty}^{t} dt' \int d\mathbf{r}'\,\varepsilon_{\alpha\beta}(t, t'; \mathbf{r}, \mathbf{r}')\,E_\beta(\mathbf{r}', t')$$

The integral with respect to time accounts for the causality principle, that is, the fact that the value of **D** at time t depends on the values of **E** at times $t' \leqslant t$. If no selected moment of time exists and the medium is spatially homogeneous, the function $\varepsilon_{\alpha\beta}$ can depend only on $t - t'$ and $\mathbf{r} - \mathbf{r}'$. Hence

$$D_\alpha(\mathbf{r}, t) = \int_{-\infty}^{t} dt' \int dV'\,\varepsilon_{\alpha\beta}(t - t', \mathbf{r} - \mathbf{r}')\,E_\beta(\mathbf{r}', t') \tag{4.2}$$

The Maxwell equations then turn into a set of integrodifferential equations. If we substitute into (4.2) the representation for $\mathbf{E}(\mathbf{r}', t')$ in the Fourier integral

$$\mathbf{E}(\mathbf{r}', t') = \int d\omega \int d\mathbf{k}\, e^{i(\mathbf{k}\cdot\mathbf{r}' - \omega t')}\mathbf{E}(\mathbf{k}, \omega) \tag{4.3}$$

and a similar representation for $\mathbf{D}$, then for (4.2) we obtain

$$D_\alpha(\mathbf{k},\ \omega) = \varepsilon_{\alpha\beta}(\mathbf{k},\ \omega)\, E_\beta(\mathbf{k},\ \omega) \tag{4.4}$$

where

$$\varepsilon_{\alpha\beta}(\mathbf{k},\ \omega) = \int_0^\tau d\tau \int d\mathbf{R}\, e^{-i(\mathbf{k}\cdot\mathbf{R}-\omega\tau)}\varepsilon_{\alpha\beta}(\tau,\ \mathbf{R})$$

and $\tau = t - t'$, $\mathbf{R} = \mathbf{r} - \mathbf{r}'$. We note that a relationship of type (4.4) stems from the above assumption concerning the temporal and spatial homogeneity. Quantities $\mathbf{k}$ and ω in the above formulas are considered real.

Later we will return to the problem of dispersion (Section 39), but we note in passing that it is advisable to take into account the damping of the electromagnetic field in a medium. We will see that this requires that $\mathbf{k}$ be complex-valued. For this reason the study of dispersion involves the study of the tensor $\varepsilon_{\alpha\beta}$ $(\mathbf{k},\ \omega)$ by the methods of the theory of functions of a complex variable.

Finally, if the dependence of $\mathbf{P}$ on $\mathbf{E}$ and of $\mathbf{M}$ on $\mathbf{B}$ is nonlinear, the Maxwell equations become a set of nonlinear differential equations (without dispersion) or a set of nonlinear integrodifferential equations (with dispersion). We noted in Section 1 that the nonlinearity of $\mathbf{M}$ $(\mathbf{B})$ is characteristic of ferromagnetic media and the nonlinearity of $\mathbf{P}$ $(\mathbf{E})$ is characteristic of ferroelectrics. Aside from these two types of media, the two functions become nonlinear when the fields $\mathbf{B}$ and $\mathbf{E}$ are sufficiently strong, even if for weak fields the properties of the medium are linear. Since it is now technically feasible to produce such strong fields (for instance, with laser sources), there is increasing interest in such cases. (This is the subject of *nonlinear optics*; see Section 41.) The superposition principle is not valid for nonlinear phenomena.

Thus, from the mathematical viewpoint the properties of the Maxwell equations can be very diversified depending on the media whose electromagnetic properties are to be studied. The mathematical methods used in solving the Maxwell equations are also very diversified.

4.2. We now turn to two theorems that will be needed later. One is restricted to homogeneous media, the other describes the properties of the Maxwell equations in the general case.

We wish to show that in homogeneous media the solutions of the Maxwell equations are uniquely determined by specifying the initial and boundary conditions. Let us consider a certain bounded region in space. The law of conservation of energy for homogeneous media, as shown in Section 3, is

$$\alpha \oint (\mathbf{E}\times\mathbf{H})\cdot\mathbf{n}\, d\sigma + \frac{1}{2}\frac{d}{dt}\int (\mathbf{E}\cdot\mathbf{D}+\mathbf{B}\cdot\mathbf{H})\, dV + \int \mathbf{j}\cdot\mathbf{E}\, dV = 0 \tag{4.5}$$

What is more, let us assume, to be definite, that Ohm's law in the form $\mathbf{j} = \sigma\mathbf{E}$ holds. Since the "material constants" σ, $\varepsilon_{\alpha\beta}$, and $\mu_{\alpha\beta}$ are always positive, the integrands in (4.5), $\mathbf{j}\cdot\mathbf{E}$, $\mathbf{E}\cdot\mathbf{D}$, and $\mathbf{H}\cdot\mathbf{B}$, are never less than zero. The first integral and the time derivative of the second can assume both positive and negative values. Now suppose we have found two solutions for the Maxwell equations, $\mathbf{E}_1$, $\mathbf{H}_1$ and $\mathbf{E}_2$, $\mathbf{H}_2$, which on σ take on the same values, that is, satisfy the same boundary conditions:

$$\mathbf{E}_1 \,|_\sigma = \mathbf{E}_2 \,|_\sigma, \quad \mathbf{H}_1 \,|_\sigma = \mathbf{H}_2 \,|_\sigma \tag{4.6}$$

In addition, we assume that they satisfy the same initial conditions:

$$\mathbf{E}_1 \,|_{t=t_0} = \mathbf{E}_2 \,|_{t=t_0}, \quad \mathbf{H}_1 \,|_{t=t_0} = \mathbf{H}_2 \,|_{t=t_0} \tag{4.7}$$

Since in our assumptions the superposition principle holds, the vector fields $\mathbf{E} = \mathbf{E}_1 - \mathbf{E}_2$ and $\mathbf{H} = \mathbf{H}_1 - \mathbf{H}_2$ will also be solutions for the Maxwell equations. Due to (4.6) and (4.7) this last solution satisfies zero boundary and initial conditions:

$$\mathbf{E} \,|_\sigma = 0, \ \mathbf{H} \,|_\sigma = 0, \quad \mathbf{E} \,|_{t=t_0} = 0, \quad \mathbf{H} \,|_{t=t_0} = 0 \tag{4.8}$$

But the fields $\mathbf{E}$ and $\mathbf{H}$ must satisfy the law of conservation of energy (4.5). The first term in (4.5) vanishes due to the first and second conditions in (4.8). If we integrate the remaining two terms in (4.5) with respect to time from a moment t_0 (zero time) to an arbitrary moment t, we get

$$\frac{1}{2}\int (\mathbf{E}\cdot\mathbf{D}+\mathbf{H}\cdot\mathbf{B})\, dV \Big|_{t=t_0}^{t} = \frac{1}{2}\int (\mathbf{E}\cdot\mathbf{D}+\mathbf{H}\cdot\mathbf{B})\, dV \Big|_t = -\int_{t_0}^{t} dt' \int dV\, \mathbf{j}\cdot\mathbf{E}$$

The first of these relationships is a corollary of the initial conditions in (4.8). The right-hand side of the above equation is nonpositive, the left nonnegative. Consequently the two are equal only when they are both zero, which is possible only if $\mathbf{E} = 0$ and $\mathbf{H} = 0$, that is, if $\mathbf{E}_1 = \mathbf{E}_2$ and $\mathbf{H}_1 = \mathbf{H}_2$.

What happens when a boundless region is involved? Here the properties of the electromagnetic field follow from (4.5) in the limiting case of the boundary surface σ expanding to infinity. The uniqueness of solution depends on the behavior of the integrals in this limit, and their behavior in turn depends on how $\mathbf{E}$ and $\mathbf{H}$ behave as $|\mathbf{r}| \to \infty$. If the boundary conditions correspond, for instance, to a so-called *radiation condition* (see Section 20), the solutions for the Maxwell equations are unique in a boundless region as well.

Let us now examine the case where the electromagnetic properties of a medium on one side of a surface σ, which macroscopically can be considered infinitely thin, differ from those on the other side. We will assume, for simplicity, that on one side of σ there is a medium I and on the other a medium II. The vectors **E**, **D**, **B**, **H** will then be labelled with subscripts I or II depending on the medium where their properties are being studied. It is also customary to assume that the vectors are finite on both sides of the boundary and their time derivatives change continuously. If we use the Maxwell equations in integral form (M′), we can come to certain conclusions as to how the different values of the field vectors on the two sides of the boundary are connected.

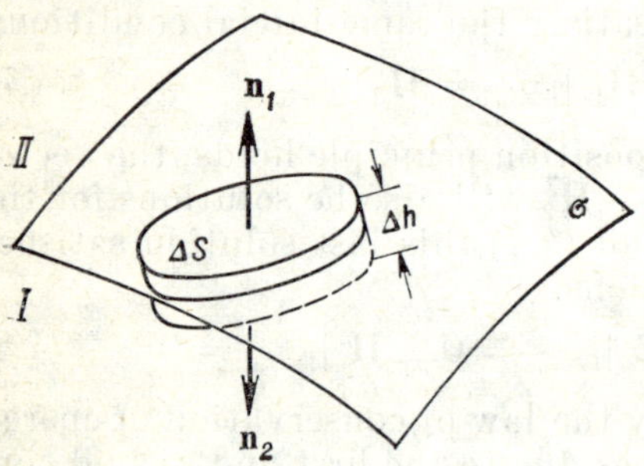

Fig. 1

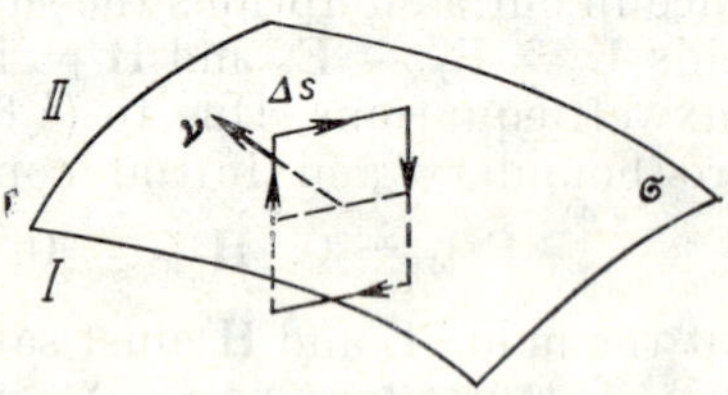

Fig. 2

First, we apply (M′.1) to an infinitely small volume enclosed by a cylindrical surface whose upper base is in medium I and whose lower base is in medium II. (Both bases are so small that they can be considered as being parallel to the part of the boundary surface σ cut out by the cylinder (Fig. 1).) Let ΔA be the area of a base of the cylinder, and Δh its altitude. The total electric charge inside the cylinder is then $q = \rho\, \Delta h\, \Delta A$ (to within higher-order terms), with ρ the charge density at some point inside the cylinder. Suppose ΔA is so small that **D** can be considered constant on each of the bases. Then, to within higher-order terms, (M′.1) yields the following relationship:

$$\Delta A\,(\mathbf{D}^{II}\cdot\mathbf{n}_1+\mathbf{D}^{I}\cdot\mathbf{n}_2)+\text{integral over lateral surface}=\rho\,\Delta h\,\Delta A$$

The lateral surface is proportional to Δh. We allow $\Delta h \to 0$ (which virtually means that the lateral surface area is considered an infinitesimal of higher order than ΔA). We will assume that $\rho \to \infty$ in the process but that $\lim_{\rho\to\infty,\ \Delta h\to 0}(\rho\,\Delta h) \equiv \lambda$ remains finite. Obviously, λ may be thought of as the *surface charge density* at the point on σ to which the cylindrical surface contracts. If at this point ρ is continuous, $\lambda = 0$. Bearing all this in mind and introducing the notation $\mathbf{n}_1 = \mathbf{n}$ and $\mathbf{n}_2 = -\mathbf{n}$, we arrive at the following result:

$$(\mathbf{D}^{II} - \mathbf{D}^{I})\cdot\mathbf{n} = \lambda \tag{4.9}$$

Here $\mathbf{n}_1$ and $\mathbf{n}_2$ are outer normals to σ used in calculating the vector flux in Gauss's integral theorem, and $\mathbf{n}$ is the direction of transition from medium I to medium II, that is, the normal unit vector to σ directed from I to II.

Now we consider (M'.4) and choose the path of integration on the left-hand side of this equation in the form of an infinitely small rectangle (Fig. 2) whose upper side is in medium II and lower side in medium I; these two sides are parallel to a tangent to σ. We denote the length of each of these sides as Δs and the length of each of the other two sides as Δl. Let $\mathbf{s}_1 = \mathbf{s}$ and $\mathbf{s}_2 = -\mathbf{s}$, where $\mathbf{s}$ is a unit vector tangent to σ. If we then think of Δl as tending to zero faster than Δs, we can write (M'.4) to within higher-order terms in the form[18]

$$\alpha\,(\mathbf{H}^{II}-\mathbf{H}^{I})\cdot\mathbf{s} = \lim_{\Delta l\to 0}\left(\frac{\partial D_\nu}{\partial t}+j_\nu\right)\Delta l$$

We define the surface current density as

$$\mathbf{i} = \lim_{\mathbf{j}\to\infty,\ \Delta l\to 0} \mathbf{j}\,\Delta l$$

The first term on the right-hand side of the modified equation (M'.4) tends to zero because of the assumption that $\partial\mathbf{D}/\partial t$ is continuous. Hence

$$\alpha\,(\mathbf{H}^{II} - \mathbf{H}^{I})\cdot\mathbf{s} = i_\nu \tag{4.10}$$

Applying the same reasoning to (M'.2) and (M'.3), we obtain

$$(\mathbf{E}^{II} - \mathbf{E}^{I})\cdot\mathbf{s} = 0 \tag{4.11}$$

$$(\mathbf{B}^{II} - \mathbf{B}^{I})\cdot\mathbf{n} = 0 \tag{4.12}$$

Equations (4.9)-(4.12) are known as the *boundary conditions*, or the *discontinuity equations*, at the interface between the different media. In this form, however, they do not define all the relationships between the field vectors in the two media. For instance, at this stage nothing can be said about the behavior of the tangential components of $\mathbf{B}$ or the normal components of $\mathbf{E}$. The lacking information may be obtained only if more is known about the properties of media I and II, for instance, in the form (1.11) and (1.17) with given electric polarizations and magnetizations. The general boundary conditions (4.9)-(4.12) then yield

$$\varepsilon_0\,(\mathbf{E}^{II}-\mathbf{E}^{I})\cdot\mathbf{n} = \lambda + (\mathbf{P}^{I}-\mathbf{P}^{II})\cdot\mathbf{n}$$

$$(\mathbf{B}^{II}-\mathbf{B}^{I})\cdot\mathbf{s} = \frac{\mu_0}{\alpha}\,i_\nu + \mu_0\,(\mathbf{M}^{II}-\mathbf{M}^{I})\cdot\mathbf{s}$$

$$(\mathbf{D}^{II}-\mathbf{D}^{I})\cdot\mathbf{s} = (\mathbf{P}^{II}-\mathbf{P}^{I})\cdot\mathbf{s}$$

$$(\mathbf{H}^{II}-\mathbf{H}^{I})\cdot\mathbf{n} = (\mathbf{M}^{I}-\mathbf{M}^{II})\cdot\mathbf{n} \tag{4.13}$$

[18] The direction of vector $\boldsymbol{\nu}$ is connected with the direction of circling the integration path according to the right-hand screw rule, as shown in Fig. 2.

For isotropic media, that is, when (1.27) holds, these relationships are simpler:

$$(\varepsilon^{II}\mathbf{E}^{II} - \varepsilon^{I}\mathbf{E}^{I})\cdot\mathbf{n} = \lambda$$
$$\alpha\left(\frac{1}{\mu^{II}}\mathbf{B}^{II} - \frac{1}{\mu^{I}}\mathbf{B}^{I}\right)\cdot\mathbf{s} = i_{\nu}$$
$$\frac{1}{\varepsilon^{II}}\mathbf{D}^{II}\cdot\mathbf{s} = \frac{1}{\varepsilon^{I}}\mathbf{D}^{I}\cdot\mathbf{s}$$
$$\mu^{II}\mathbf{H}^{II}\cdot\mathbf{n} = \mu^{I}\mathbf{H}^{I}\cdot\mathbf{n} \tag{4.14}$$

These relationships are used in the study of reflection and refraction of electromagnetic waves (cf. Section 18).

CHAPTER 2

RELATIVISTIC ELECTRODYNAMICS

§ 5. The principle of relativity. Lorentz transformations, and relativistic kinematics

5.1. Observers investigating electromagnetic field may be located in different frames of reference moving with respect to one another. The formulation of the laws of electromagnetism presented in Chapter 1 is based on the assumption that here are such frames of reference in which these laws are verified by physical measurements. Any experiment is based on the ability to measuse distances and time intervals. These measurements make it possible to determine kinematic parameters of motion of any material object: its velocity and acceleration in a given reference frame.

Since any reference frame is related to a material object and to instruments used for measurements, it is natural to define a class of *inertial* frames of reference. Any two reference frames in this class move at a constant velocity with respect to each other. The following properties characterize any inertial reference frame from the standpoint of observation of physical events.

1. An object in an inertial reference frame moves along a straight line and at a constant velocity when no forces are applied to it (in fact, this is a basis for defining the concept of the mechanical force).

2. The velocity c of propagation of the electromagnetic field in vacuo measured in an inertial reference frame is independent of the velocity of motion of the source of the field with respect to the observer (the principle of independence of the velocity of light in vacuo on the source's motion[1]).

The class of inertial reference frames occupies a special position among all possible frames of reference. Namely, laws of physics can be formulated in such a manner that they are independent of the choice of the inertial frame of reference in which they are considered. In other words, the *relativity principle* must hold: a uniform rectilinear motion of a system as a whole does not affect the processes taking place inside the system, that is the laws governing physical processes are identical in all inertial reference frames.

---

[1] This second requirement, not included into classical mechanics which did not consider electromagnetic phenomena, was first introduced by A. Einstein in 1905.

Let us consider two inertial reference frames K and K', moving at a constant velocity $\mathbf{v}$ with respect to K. Let Cartesian coordinates x^1, x^2, x^3, and time t be used in reference frame K, and Cartesian coordinates $x^{1'}$, $x^{2'}$, $x^{3'}$, and time t' in reference frame K'. In order to determine the relationship between the equations expressing a physical law in reference frames K and K', one has first of all to find the relation between the spatial and temporal measurements x^i $(i = 0, 1, 2, 3)$ and $x^{i'}$ $(i' = 0', 1', 2', 3')$; it will be convenient to use notation $x^0 = ct$ and $x^{0'} = ct'$. Mathematically, the problem consists in determining the possible form of functions

$$x^{i'} = f^{i'}(x^0, x^1, x^2, x^3) \tag{5.1}$$

The derivation must use the formulated above properties of the inertial frames and the relativity principle[2]. Functions $f^{i'}$ must be such that (5.1) could be solved for variables x^i (i.e. the inverse transformation must exist). Besides, the class of functions in which the solution is to be sought must be somehow restricted (it is natural to assume, for example, that these functions are at least twice continuously differentiable). One of the important restrictions of the possible relation between variables $x^{i'}$ and x^i is a corollary of the principle of independence of the light velocity in vacuo on the motion of the source. Indeed, if the measurements conducted in the reference frame K give distance $d\mathbf{r}$ between the point of emission and that of subsequent absorption of light, and the duration of this process dt, then the corresponding measurement in reference frame K' must give results $d\mathbf{r}'$ and dt'. But according to the abovementioned principle, the light propagation velocity c is identical in both frames. In other words, from $c^2\,dt^2 = d\mathbf{r}^2$ must follow $c^2\,dt'^2 = d\mathbf{r}'^2$, and vice versa. A mathematical solution of the problem for the type of functions $f^{i'}$ on the basis of the abovementioned physical principles[3] leads to a theorem that these functions must be linear, that is must have the form

$$x^{i'} = A^{i'}_i x^i + a^{i'}, \quad x^i = A^i_{i'} x^{i'} + a^i \tag{5.2}$$

where coefficients $A^{i'}_i$, $A^i_{i'}$, and quantities $a^{i'}$, a^i are constants. If relations (5.2) hold, the principle of independence of the light velocity on the source motion results in the equality

$$c^2\,dt'^2 - d\mathbf{r}'^2 = c^2\,dt^2 - d\mathbf{r}^2 \tag{5.3}$$

---

[2] Our presentation of the theory of relativity serves mostly as a summary for reference. It cannot substitute a detailed study of the subject; for study, special treatises must be consulted. The most rigorous presentation can be found, for example, in *The Theory of Space, Time, and Gravitation* by V. A. Fock (2nd edition), Pergamon Press, Oxford, 1963, and *Special Theory of Relativity*, by V. A. Ugarov, Mir Publishers, Moscow, 1979.

[3] See V. A. Fock, *The Theory of Space, Time, and Gravitation* (2nd edition). Pergamon Press, Oxford, 1963.

(regardless of whether this quadratic form is zero or not). The quantity $ds^2 \equiv c^2\, dt^2 - d\mathbf{r}^2$ is therefore invariant under transformation (5.2). It is called the *space-time interval* between two events, and (5.2) are called the *Lorentz transformations*. Quantities $a^{i'}$ describe the possible displacement of the reference point of space coordinates and time in one reference frame with respect to another reference frame. Assume now $a^{i'} = 0$, thus restricting the analysis to such inertial reference frames in which rectilinear paths are determined by a condition that the origins (reference points) of space coordinates coincide at a certain moment of time (common for all these reference frames).

5.2. In terms of geometry (see Appendix A) we see that the relativity principle determines the geometry of the four-dimensional space-time manifold as a pseudo-Euclidean geometry in which the invariant scalar product is given by formula (A.14), where $N = 4$ and $k = 1$ (obviously, geometric relations will be the same if we set $N - k = 1$ and $k = 3$). Interval ds^2 is thus interpreted as the square of the length of the four-dimensional "radius vector", which characterizes the "space-time distance" between the corresponding physical events, regardless of the choice of the inertial frame of reference. The coefficients of Lorentz transformations are then related by (A.11), so that in the general case the homogeneous Lorentz transformation may be a function of six linearly independent parameters (and the most general nonuniform transformation may depend on ten parameters, if $a^{i'} \neq 0$). The geometrical meaning of these parameters is obvious: as follows from the above arguments, the Lorentz transformation is a rotation in the four-dimensional pseudo-Euclidean space, keeping interval ds^2 invariant; hence, it can be decomposed into six independent rotations, according to the number of six mutually orthogonal planes of the four-dimensional space. The set of Lorentz transformations forms a group. Obviously, ordinary rotations in the three-dimensional Euclidean space (subject to conditions $d\mathbf{r}^2 = d\mathbf{r}'^2$ and $dt = dt'$) also can be included into the Lorentz transformations; they form a subgroup of the Lorentz group.

We thus find, in terms of geometry, that each inertial frame of reference must be put in correspondence with a set of basis vectors $\vec{e}_i$ in the pseudo-Euclidean space R_4^1 [4] of the above-described structure, vectors $\vec{e}_i$ are normalized as in (A.13), that is

$$\vec{e}_0^{\,2} = 1, \quad \vec{e}_\alpha^{\,2} = -1, \quad (\vec{e}_i, \vec{e}_j) = 0 \quad (\alpha = 1, 2, 3;\ i \neq j) \qquad (5.4)$$

A transition to a new inertial reference frame corresponds to a rotation of vectors $\vec{e}_i$ in space R_4^1, which transforms them into vectors $\vec{e}_i^{\,\prime}$

---

[4] Arrows over symbols denote vectors in the four-dimensional space.

in this new reference frame, that is to a transformation

$$\vec{e}_{i'} = A^i_{i'} \vec{e}_i, \quad \vec{e}_i = A^{i'}_i \vec{e}_{i'} \tag{5.5}$$

(cf. (A.2)). The coefficients of the Lorentz transformation must satisfy (A.11):

$$g_{ij} A^i_{i'} A^j_{j'} = g_{i'j'} \quad \text{or} \quad g^{ij} A^{i'}_i A^{j'}_j = g^{i'j'} \tag{5.6}$$

Here $g_{ij} = (\vec{e}_i, \vec{e}_j)$ and $g_{i'j'} = (\vec{e}_{i'}, \vec{e}_{j'})$, and therefore $g_{00} = 1$, $g_{\alpha\alpha} = -1$, $g_{ij} = 0$ $(i \neq j)$.

Any Lorentz transformation can be decomposed into three-dimensional rotations and a transformation which transforms basis vectors $\vec{e}_0, \vec{e}_1$ into vectors $\vec{e}_{0'}, \vec{e}_{1'}$, respectively, but does not affect other basis vectors (it is called the *partial* Lorentz transformation). Clearly, we are interested in the latter, since the properties of three-dimensional rotations can be considered familiar.

We shall not give the proof of this statement here[5], but we offer the following arguments. Obviously, a transformation in which $x^2 = x^{2'}$ and $x^3 = x^{3'}$, and also $(dx^0)^2 - (dx^1)^2 = (dx^{0'})^2 - (dx^{2'})^2$, conserves the space-time interval and constitutes a partial Lorentz transformation. One can readily find the coefficients of this transformation. Having accomplished this, we shall show how to find the general Lorentz transformation in terms of these coefficients.

In the two-dimensional case (5.6) take the form

$$(A^{0'}_0)^2 = (A^{1'}_0)^2 = 1, \quad (A^{1'}_0)^2 - (A^{1'}_1)^2 = -1, \quad A^{0'}_0 A^{1'}_0 = A^{0'}_1 A^{1'}_1 \tag{5.7}$$

The solution of these equations can be expressed through a parameter $\beta = A^{0'}_1 / A^{0'}_0 = A^{1'}_1 / A^{1'}_0$. As a result, equations (A.5) for the coordinate transformation take the following form:

$$x^{0'} = \frac{x^0 + \beta x^1}{\pm\sqrt{1-\beta^2}}, \quad x^{1'} = \frac{x^1 + \beta x^0}{\pm\sqrt{1-\beta^2}}, \quad x^{2'} = x^2, \quad x^{3'} = x^3 \tag{5.8}$$

where the plus or minus signs in the first two formulas can be chosen independently. Coefficients of transformation (5.8) can be written as a matrix

$$(A^{i'}_i) = \begin{bmatrix} \gamma & \gamma\beta & 0 & 0 \\ \gamma\beta & \gamma & 0 & 0 \\ 0 & 0 & 1 & 0 \\ 0 & 0 & 0 & 1 \end{bmatrix} \tag{5.8'}$$

where $\gamma \equiv (1-\beta^2)^{-1/2}$. This notation will be frequently used throughout the book.

---

[5] See, for example, P. K. Rashevsky, *Riman Geometry and Tensor Analysis*, Gostekhizdat, 1953, § 48 (in Russian).

First, let us discuss the situation in which the positive sign was selected in both formulas (5.8). We can immediately determine the physical meaning of parameter β. Consider in reference frame K the motion of a point whose coordinates in reference frame K', $x^{1'}$, $x^{2'}$, $x^{3'}$, are fixed. The equations of motion of this point in reference frame K will be obtained by writing the relations between the differentials of coordinates, following from (5.8):

$$\frac{dx^1}{dx^0}=-\beta, \quad \frac{dx^2}{dx^0}=0, \quad \frac{dx^3}{dx^0}=0$$

We see that $\beta = \mp v/c$, where v is the velocity of the point fixed in K' (i.e. of the reference frame K' itself) with respect to K. The minus sign must be chosen if this velocity is in the direction of axis x^1, and the plus sign in the opposite case. Transformations inverse to (5.8) are quite similar, with x^0 replaced by $x^{0'}$, x^1 by $x^{1'}$, and vice versa, and β by $-\beta$.

The first equation of (5.8) for the motion of a point fixed in reference frame K' takes the form

$$dx^{0'}=\sqrt{1-\beta^2}\,dx^0 \tag{5.9}$$

This expression means that when a clock placed in reference frame K' at this fixed point measures an interval of time $dx^{0'}$, a clock in reference frame K shows that the point of frame K' moves from its position in frame K, corresponding to the beginning of time interval $dx^{0'}$, to a position corresponding to the end of this interval, over time dx^0. Therefore, this formula expresses the relativistic *time dilatation* effect (slowing of clocks).

Consider now two simultaneous physical events in reference frame K, that is events for which $dx^0=0$. Denote by dx^1 the spatial distance between these events in this reference frame (this means that the beginning and the end of the spatial distance are measured at the same moment of time). Then, according to (5.8),

$$dx^1=\sqrt{1-\beta^2}\,dx^{1'} \tag{5.10}$$

where $dx^{1'}$ is the distance between the same events in reference frame K'. We have obtained an expression for the relativistic effect of scale contraction. Using this formula, as well as relations $dx^{2'} = dx^2$ and $dx^{3'} = dx^3$, we immediately obtain the expression for the transformation of the three-dimensional volume:

$$dV=\sqrt{1-\beta^2}\,dV' \tag{5.10$'$}$$

It should be emphasized that this relation has the following meaning. If a continuous set of point physical events, which do not move with respect to reference frame K', fills in this frame of reference a three-dimensional volume dV', then the measurements of three-dimensional positions of these events in reference frame K at the same moment of time by the clock in this frame will fill a three-dimensional volume dV.

Clearly, the two analyzed effects are corollaries of the invariance of the space-time interval.

Returning to the initial geometry-based arguments, we easily notice that vector $\vec{e}_1$ in the discussed particular (two-dimensional) case is directed along the relative velocity vector $\mathbf{v}$ in reference frame K, while vector $\vec{e}_{1'}$ is directed along vector $\mathbf{v}$ in reference frame K'. Assume now that axes $\vec{e}_1$, $\vec{e}_2$, $\vec{e}_3$, and axes $\vec{e}_{1'}$, $\vec{e}_{2'}$, $\vec{e}_{3'}$ undergo identical three-dimensional rotation with respect to the initial configuration. The axes of reference frame K' will thereby remain parallel to those of frame K, and vector $\mathbf{v}$ will be at the same angles with the axes of frame K as with the axes of reference frame K'. The three-dimensional radius vector $\mathbf{r}$ in frame K can be represented as a sum of two orthogonal projections: $\mathbf{r} = \mathbf{r}_{\parallel} + \mathbf{r}_{\perp}$, where $\mathbf{r}_{\parallel}$ is a component of vector $\mathbf{r}$ in the direction of vector $\mathbf{v}$, and $\mathbf{r}_{\perp}$ is a component of $\mathbf{r}$ in the two-dimensional plane orthogonal to vector $\mathbf{v}$. Similarly, $\mathbf{r}' = \mathbf{r}'_{\parallel} + \mathbf{r}'_{\perp}$. Let us denote by $\mathbf{v}_1$ a unit vector in the direction of $\mathbf{v}$. Then the partial Lorentz transformation (5.8) can be rewritten in the form

$$\mathbf{r}_{\perp} = \mathbf{r}'_{\perp}, \quad \mathbf{r}'_{\parallel} \cdot \mathbf{v}_1 = \gamma \left(\mathbf{r}_{\parallel} \cdot \mathbf{v}_1 - \frac{v}{c} x^0 \right)$$

$$x^{0\prime} = \gamma \left(x^0 - \frac{1}{c} \mathbf{r} \cdot \mathbf{v} \right) \tag{5.11}$$

(the last relation takes into account that $\mathbf{r}_{\parallel} \cdot \mathbf{v} = \mathbf{r} \cdot \mathbf{v}$). Therefore,

$$\mathbf{r}' = (\mathbf{r}'_{\parallel} \cdot \mathbf{v}_1)\, \mathbf{v}_1 + \mathbf{r}'_{\perp} = \gamma \left[(\mathbf{r} \cdot \mathbf{v}_1)\, \mathbf{v}_1 - \frac{1}{c} x^0 \mathbf{v} \right] + \mathbf{r}_{\perp}$$

that is

$$\mathbf{r}' = \mathbf{r} + (\gamma - 1) \frac{1}{v^2} (\mathbf{r} \cdot \mathbf{v})\, \mathbf{v} - \frac{\gamma}{c} \mathbf{v} x^0 = \mathbf{r} + (\gamma - 1)\, \mathbf{r}_{\parallel} - \frac{\gamma}{c} \mathbf{v} x^0$$

$$x^{0\prime} = \gamma \left(x^0 - \frac{1}{c} \mathbf{r} \cdot \mathbf{v} \right) \tag{5.12}$$

We have derived the Lorentz transformation which relates the inertial frames of reference whose spatial axes are parallel but whose relative velocity is not directed along any axis.

And finally, the most general Lorentz transformation corresponds to the case in which the spatial axes of the two reference frames are not parallel. However, this case is derived from the preceding one if vector $\mathbf{r}'$ undergoes a three-dimensional rotation S (i.e. $\mathbf{r}'$ in the left-hand side of (5.12) is substituted by $S\mathbf{r}'$).

5.3. Let us derive now the relativistic *formula for velocity summation.* We analyze a motion of a pointlike mass (in the general case, motion with acceleration) in the frames of reference K and K'. The velocity of a pointlike mass in reference frame K at a moment of time fixed by a clock in this frame is $\mathbf{u} = d\mathbf{r}/dt$. In reference frame K' the velocity of the same pointlike mass at the corresponding (according to the Lorentz transformation) point of time will be $\mathbf{u}' = d\mathbf{r}'/dt'$. By using (5.12) it is easy to find, on the one hand, relations between $d\mathbf{r}'$ and $d\mathbf{r}$, and on the other hand, between dt' and dt. The necessary formula is obtained from the ratio of these differentials:

$$\mathbf{u}' = \frac{\mathbf{u}+\mathbf{v}\left[(\gamma-1)\frac{1}{v^2}\,\mathbf{u}\cdot\mathbf{v}-\gamma\right]}{\gamma\left(1-\frac{1}{c^2}\,\mathbf{u}\cdot\mathbf{v}\right)} \tag{5.13}$$

Vector $\mathbf{u}'$ is called the relativistic sum of velocities $\mathbf{u}$ and $\mathbf{v}$. It must be remembered that the result of summation depends on the order in which velocities $\mathbf{u}$ and $\mathbf{v}$ are summed up (in the general case, the right-hand side of (5.13) changes when $\mathbf{u}$ is replaced by $\mathbf{v}$ and $\mathbf{v}$ by $\mathbf{u}$). In particular, if $\mathbf{u}$ and $\mathbf{v}$ are orthogonal, then $\mathbf{u}' = \frac{1}{\gamma}(\mathbf{u} \pm \gamma\mathbf{v})$. In the case corresponding to the partial Lorentz transformation, (5.13) takes the form

$$u'_x = \frac{u_x \pm v}{1 \pm u_x v/c^2}, \quad u'_y = \frac{u_y\sqrt{1-\beta^2}}{1 \pm u_x v/c^2}, \quad u'_z = \frac{u_z\sqrt{1-\beta^2}}{1 \pm u_x v/c^2} \tag{5.13'}$$

If v is very small compared with the velocity of light in vacuo, c, we can set $\beta \approx 0$, that is $\gamma \to 1$. In this limit formulas (5.12) and (5.13) reduce to the Galilean transformation and the velocity summation law of classical mechanics:

$$\mathbf{r}' \simeq \mathbf{r} - \mathbf{v}t, \quad t' \simeq t, \quad \mathbf{u}' \simeq \mathbf{u} - \mathbf{v} \tag{5.14}$$

If $v > c$, that is when $|\beta| > 1$, coefficients in the transformation formulas (5.14) become imaginary. It must be assumed that this fact points to an impossibility of the physical realization of inertial reference frames satisfying the condition $v > c$. The velocity of light c in vacuo is therefore found to be the limiting propagation velocity of physical processes. Indeed, we can always assume that a material carrier of a process is associated with a specific reference frame. Then the "propagation" is understood as a sequence of causality-related states. Consequently, all possible intervals ds^2 relating a given physical event to any other event are divided into three classes. If $ds^2 = 0$ then the corresponding two events can be considered as coincident with the start and end of propagation of a light ray (therefore this interval is called the *light interval*). If $ds^2 > 0$, the events can be related by a process propagating at a velocity $v < c$ (*timelike interval*). Finally, if $ds^2 < 0$, the events cannot be linked causally (*spacelike interval*). Similarly, the finite radius vectors $\vec{r}$ in space-time are classified according to their space-time length $\vec{r}$[6]. Light intervals radiating from a given point in space-time form a *light cone*. Events related to a given event by spacelike intervals are called quasi-simultaneous with this event. Obviously, quasi-simultaneous events do not affect each other. If, however, $ds^2 > 0$, then for $dt > 0$ events belong to the absolute future, and for $dt < 0$, to the absolute past with respect to the event which is at the beginning of the timelike interval. Owing to the invariance of space-time intervals, the described classification of events with respect to any one of them is invariant. It is easy to show that in terms of geometry the events timelike to a given event are inside the light cone having its apex at this event, while quasi-simultaneous events are outside of this cone.

Two concepts, those of *timelike curves* and *spacelike hypersurfaces*, are very important from the physics standpoint. The former are defined by the condition $ds^2 > 0$ for any two points on such a curve. The latter are defined by the condition $ds^2 < 0$ for any two points belonging to such a hypersurface. The timelike curves and spacelike (3-dimensional) hypersurfaces are invariant geometric entities. As follows from the above arguments, any timelike curve is a geometric representation of a possible motion of a pointlike mass (such a motion in which velocity may be variable but can never reach c). Spacelike hypersurfaces are important because on such surfaces we can set the states of physical objects arbitrarily, without worrying about the causality principle. One important example of spacelike hypersurfaces is a three-dimensional hyperplane defined by equation $t = \text{const}$ in an arbitrary frame of reference. Later on (see § 9) we shall have to integrate over spacelike hypersurfaces. This operation is essential in the study of the variational principle.

5.4. A timelike curve, which is often referred to as the *world line*, can be defined by means of equations of the type $x^i = x^i(\tau)$, where τ is, in the general case, an arbitrary parameter. This parameter is conveniently chosen proportional to the length of the segment of the curve, namely in the form

$$d\tau = \frac{1}{c}\, ds = \frac{1}{c}\sqrt{c^2\, dt^2 - d\mathbf{r}^2} = dt\sqrt{1-\beta^2} \tag{5.15}$$

In this case it is known as the *proper time* of a moving particle (pointlike mass). It is then logical to introduce the concepts of the *four-dimensional velocity* $\vec{u}$ and *four-dimensional acceleration* $\vec{w}$ defined as

[6] As before, arrows indicate four-dimensional vectors (4-vectors).

follows:

$$u^i = \frac{dx^i}{d\tau}, \quad w^i = \frac{du^i}{d\tau} \tag{5.16}$$

Once an inertial reference frame is chosen, the components of these quantities are readily expressed in terms of the three-dimensional velocity **v** and acceleration **a**:

$$u^0 = c\gamma, \qquad u^\alpha = c\beta^\alpha\gamma \tag{5.17$_1$}$$

$$w^0 = \gamma^4 (\beta\mathbf{a}), \quad w^\alpha = a^\alpha\gamma^2 + \beta^\alpha w^0 = a^\alpha\gamma^2 + \beta^\alpha\gamma^4 (\boldsymbol{\beta}\cdot\mathbf{a}) \tag{5.17$_2$}$$

The second of these equalities is easily transformed to

$$\mathbf{w} = \gamma^4 [\mathbf{a} + \boldsymbol{\beta} \times (\boldsymbol{\beta} \times \mathbf{a})] \tag{5.17$_3$}$$

(here we use an obvious notation $\boldsymbol{\beta} \equiv \mathbf{v}/c$).

As follows from (5.17$_1$),

$$\vec{u}^2 \equiv (u^0)^2 - \mathbf{u}^2 = c^2 \tag{5.18}$$

whence

$$\vec{u}\,\vec{w} = 0 \tag{5.19}$$

Besides, it is not difficult to find by using (5.17$_2$) that

$$\vec{w}^2 = (w^0)^2 - \mathbf{w}^2 = -\gamma^6 [(\boldsymbol{\beta}\cdot\mathbf{a})^2 + \gamma^{-2}\mathbf{a}^2]$$

or

$$\vec{w}^2 = -\gamma^6 [\mathbf{a}^2 - (\boldsymbol{\beta} \times \mathbf{a})^2] \tag{5.20}$$

We thus see that $\vec{w}^2 < 0$, that is $\vec{w}$ is a spacelike vector. This property of $\vec{w}$ can also be derived from (5.19) if we take into account that $\vec{u}$ is a timelike vector.

Note that in manipulations with vectors in the pseudo-Euclidean space-time, one has to distinguish between covariant and contravariant components of vectors; this is indicated in Appendix A (see, in particular, (A.17)).

5.5. Let us discuss now the choice of signs in the denominators of our fundamental transformation (5.8). Evidently, our selection of signs (plus in both cases) corresponds to a continuous set of inertial reference frames obtained by varying the numerical parameter β, with the determinant of the Lorentz transformation always remaining equal to $+1$. All inertial reference frames of this type are called reference frames with identical orientation, that is identical to orientation of the initial frame obtained for $\beta = 0$.

When $\beta = 0$, the selection of signs realizes the following situations:

$$\text{(a)} \quad x^{0'} = x^0, \ x^{1'} = -x^1$$

$$\text{(b)} \quad x^{0'} = -x^0, \ x^{1'} = x^1$$

$$\text{(c)} \quad x^{0'} = -x^0, \ x^{1'} = -x^1$$

Case (a) is the *spatial reflection* transformation, case (b) is the *time reflection*, and case (c) is the *space-time reflection*. In the first two cases the transformation determinant is -1, that is it changes discontinuously compared with the case of no reflection. In case (c) it equals $+1$, but only as a result of two discontinuous transformations. The set of necessary inertial reference frames can be obtained in each of the indicated three situations by means of the subsequent continuous variation of parameter β. Note that the reflection transformation in the three-dimensional space can also be defined by a condition $x^{\alpha'} = -x^\alpha$ ($\alpha = 1, 2, 3$), when the signs of all three spatial coordinates are changed (the inversion transformation). However, the study of reflection in the three-dimensional space shows that no new results are obtained thereby. A reflection with respect to two directions only leaves the sign of the determinant unaltered, and hence, can be reduced to the continuous rotation transformation. These arguments are easily extended to the general Lorentz transformations (5.12). The initial selection of transformations which do not change the orientation defines a group of transformations usually referred to as the *proper Lorentz group*. The set of all possible Lorentz transformations (including the abovementioned reflection transformations (a), (b), (c)) is the *complete Lorentz group*. The complete Lorentz group thus decomposes into four connected domains, that is into four components. The inertial reference frames contained within each of these components can be "enumerated" by a continuous variation of parameters of the Lorentz transformation. A transition from one of these components to any other component is realized by a discontinuous reflection transformation. It is also possible to find from formula (5.12) and from the consequent remark the physical meaning of the parameters it includes. The three of them are components of the relative velocity $\mathbf{v}$, and the remaining three are the angles of three-dimensional rotation, provided the rotation is necessary.

5.6. It is often helpful to use the *infinitesimal Lorentz transformations*. The coefficients of these transformations are written in the form

$$A_i^{i'} = \delta_i^{i'} + \omega^{i'}_{\cdot i} = \delta_i^{i'} + g_{ij}\omega^{i'j} \tag{5.21}$$

where $\omega^{i'j} \to 0$. A substitution of (5.21) into the fundamental formula (5.6) yields the condition

$$\omega_{j'i}\delta_j^{j'} + \omega_{i'j}\delta_i^{i'} = 0 \tag{5.22}$$

(we have taken into account only the terms linear with respect to infinitesimals $\omega^{i'j}$). The transformation of coordinates in space-time is written, by using (5.21), in the form

$$x^{i'} = A^{i'}_i x^i \approx x^i + \omega^{i'}_{.j} x^j$$

As coefficients $\omega^{i'}_{.j}$ are infinitesimals, and as in the zero approximation $x^{i'} = x^i$, the primed and nonprimed indices of $\omega^{i'}_{.j}$ need not be distinguished. Therefore, the final form of the transformation is

$$(x^i)' = x^i + \omega^i_{.j} x^j = x^i + g^{ii} \omega_{ij} x^j \tag{5.23}$$

and, as follows from (5.22),

$$\omega_{ij} + \omega_{ji} = 0 \tag{5.23'}$$

We find, therefore, that tensor ω_{ij} has six linearly independent components describing infinitesimal rotations in six mutually orthogonal planes of four-dimensional space-time. In the most general case, when transformations (5.2) have to be used, we need to take into account four infinitesimal translations corresponding to the vector included in (5.2). Then equation (5.23) takes the form

$$(x^i)' = x^i + g^{ii} \omega_{ij} x^j + \omega^i \tag{5.24}$$

It is not difficult to verify that coefficients $\omega_{\alpha\beta}$ correspond to three-dimensional rotations.

§ 6. Relativistic particle dynamics

It was shown in the preceding section (see p. 51) that the Galilean transformation which constitutes the foundation of classical mechanics is the limiting particular case of the Lorentz transformation for $v \ll c$. It is clear then that the concepts of classical mechanics need to be elaborated in order to make them invariant under the Lorentz transformations. Only after this will mechanics satisfy the relativity principle, that is its laws will have physical meaning independent of the choice of reference frame for which they are written. And since Newton's laws hold with very high accuracy at velocities of motion small compared with that of light, we must demand that the exact laws of mechanics transform to Newton's laws in this approximation[7].

Let us consider world lines of pointlike particles (see p. 52). Uniform motions of a pointlike mass along a straight line can be represented only by straight world lines. Hence, if a pointlike mass moves with an acceleration with respect to an inertial reference frame, its world lines will be curved; this property of the world lines must be

[7] Usually this requirement is called the Einstein correspondence principle.

observed in any inertial frame of reference. As a measure of this curvature, let us choose a value of the four-dimensional acceleration at a given point of the world line. Then, in an analogy with Newton's second law of classical mechanics, we can define the *four-dimensional force* $\vec{F}$ by an equation

$$\vec{F} = m_0 \vec{w} \tag{6.1}$$

We assume here that dynamic behavior of a particle can be characterized by a parameter m_0 having dimensions of mass and invariant under Lorentz transformations. We shall also assume, at the moment, that m_0 is independent of the proper time τ measured along the world line. Equation (6.1) is the basic postulate of relativistic mechanics; ultimate confirmation of this postulate can only be obtained in an experiment. So far it can only be said that the definition of force (6.1) is invariant with respect to transition from one inertial reference frame to another, and appears to be the simplest generalization of Newton's second law possessing this property.

Let us consider (6.1) from the standpoint of an arbitrary inertial reference frame. In this frame the four-dimensional force $\vec{F}$ determines four quantities $K^i = \gamma^{-1} F^i$. The spatial components K^α then have the form

$$K^\alpha = m_0 \frac{du^\alpha}{dt} = \frac{d}{dt} \frac{m_0 v^\alpha}{\sqrt{1-\beta^2}} = \frac{dp^\alpha}{dt} \tag{6.2}$$

We have used the independence of m_0 on τ and therefore on t, and also (5.17_1). Here

$$p^\alpha = mv^\alpha, \quad m = m_0 / \sqrt{1-\beta^2} \tag{6.3}$$

But (6.2) can be interpreted as a form of Newton's second law for a particle with mass m_0 and momentum $\mathbf{p}$, differing from classical mechanics in that the mass of a moving particle is a function of the velocity of motion with respect to a given reference frame. The value of m_0, that is mass measured in the inertial reference frame whose velocity at a given moment is equal to the velocity of the particle (this reference frame is called the instantly co-moving frame) is known as the *rest mass*. Relation (6.3) is confirmed experimentally.

Component F^0 of the four-dimensional force vector also has a physical meaning. Relation (5.19) considered simultaneously with (5.17_1) and definition (6.1) yields

$$cK^0 = \mathbf{K} \cdot \mathbf{v} \tag{6.4}$$

As vector $\mathbf{K}$ was interpreted as Newtonian force (including now the dependence of mass on velocity), we find from (6.4) that the quantity cK_0 has a meaning of work done by this force per unit time. Denoting the energy of a particle by $\mathscr{E}$ we derive from (6.1) and formula (5.17_1) for u^0 that

$$\frac{d\mathscr{E}}{dt} = cK^0 = \frac{d}{dt}\frac{m_0c^2}{\sqrt{1-\beta^2}}$$

which gives

$$\mathscr{E} = m_0c^2/\sqrt{1-\beta^2} \tag{6.5}$$

to within a constant term which we assume equal to zero.

On the other hand, if we use definition (5.16) of four-dimensional velocity u^i, then $p^\alpha = m_0u^\alpha$. Hence, 4-vector $\vec{p}$ with components

$$p^i = m_0u^i \tag{6.6}$$

is called the *four-dimensional momentum* (4-momentum). Then

$$p_0 = \mathscr{E}/c \tag{6.7}$$

and basic equation (6.1) can be rewritten in the form

$$F^i = dp^i/d\tau \tag{6.8}$$

As follows from (5.18),

$$(p^0)^2 - \mathbf{p}^2 = m_0^2c^2$$

that is,

$$\mathscr{E} = \sqrt{m_0^2c^4 + \mathbf{p}^2c^2} \tag{6.9}$$

The formula for energy, (6.5), has a corollary which is of fundamental physical importance: a particle in a reference frame in which its velocity is zero possesses the rest energy

$$\mathscr{E}_0 = m_0c^2 \tag{6.10}$$

This equation derived by Einstein is an expression of the *relation between mass and energy*. Note that this follows directly from the fact that in deriving equation (6.5) we set the constant term equal to zero. However, nonzero addition to energy (6.5) would lead to a physically meaningless result. Indeed, let us assume that $\mathscr{E}/c = m_0c/\sqrt{1-\beta^2} + a^0$, where a^0 is a constant. The expression for the three-dimensional momentum is given by (6.2) also to within of certain constants a^α. Therefore, we write $p^i = m_0u^i + a^i$. Quantities a^i must be components of some space-time vector in order for the 4-momentum $\vec{p}$ to be a physical quantity from the standpoint of the relativity principle. But the physical meaning of three-dimensional momentum $\mathbf{p}$ demands that it be essentially equal to zero when the velocity of a particle is zero, and therefore all a^α must be zero in any reference frame. But then $a^0 = 0$ in all reference frames; otherwise there would also be reference frames in which $a^\alpha \neq 0$.

The mass-energy interrelation in fact states that any change in the inertial rest mass of a particle is accompanied by a change in its energy, and vice versa. This statement makes the foundation for interpretation of a very large number of phenomena related, for example, to transitions between elementary particles. In particular, the energy of electromagnetic radiation can be transformed into the rest mass of such particles as the electron and the positron (or proton and antiproton). Likewise, the rest mass of some particles may be transformed into kinetic energy of motion of other particles or into the energy of electromagnetic radiation.

It thus follows from the mass-energy relation that in the most general case the rest mass m_0 must be a function of time. In order to generalize equation (6.1), we must take an equation of the type (6.8), where momentum is defined, as before, by relation (6.6) but in which $m_0 = m_0\,(\tau)$.

Rest mass m_0 undergoes changes most often in collisions of particles. In this case world lines of particles before a collision and particles after it are straight and form a bundle of lines passing through one point in space-time. The interactions are restricted to a very small region of space-time around this point. A physical event consisting in a collision of particles and in a transformation of particles in the course of the collision must be characterized by momentum and energy conservation. To be precise, this means that the total momentum $\mathbf{p}$ and energy $\mathscr{E}$ of particles prior to the collision (at the moment when particles are not yet interacting) must be equal to the total momentum $\tilde{\mathbf{p}}$ and energy $\tilde{\mathscr{E}}$ of particles appearing as the result of interactions during the collision. It will suffice here to assume that particles are completely defined by the values of their rest masses (although in practice it is necessary to take into account their spins, charges, etc.). By using (6.6) and (6.7), the abovementioned laws of energy and momentum conservation are joined into the unified relativistically invariant *law of energy-momentum conservation*, written in the form

$$p^i = \tilde{\mathbf{p}}^i \tag{6.11}$$

Here $p^0 = \frac{\mathscr{E}}{c} = \frac{1}{c} \sum_{l=1}^{N} \mathscr{E}_l$, where N is the number of colliding particles, $\mathscr{E}_l$ is the energy of the lth particle prior to the collision; $\tilde{p}_0 = \frac{\tilde{\mathscr{E}}}{c} = \frac{1}{c} \sum_{\tilde{l}=1}^{\tilde{N}} \tilde{\mathscr{E}}_{\tilde{l}}$, where $\tilde{N}$ is the number of particles after the collision, and $\tilde{\mathscr{E}}_{\tilde{l}}$ is the energy of the $\tilde{l}$th particle. As follows from (6.10), $\mathscr{E}_l = \sqrt{m_{0l}^2 c^4 + \mathbf{p}_l^2 c^2}$ and similarly $\mathscr{E}_{\tilde{l}} = \sqrt{m_{0\tilde{l}}^2 c^4 + \mathbf{p}_{\tilde{l}}^2 c^2}$. A collision may be referred to as elastic if $N = \tilde{N}$ and the rest masses m_{0l}

remain the same. If this is not the case, the collision is inelastic and is accompanied by an exchange of energy between particles; energy and rest mass may convert into each other. Spatial components of relation (6.11) can be recast by means of the introduced notations to the form

$$\sum_{l=1}^{N} \mathbf{p}_l = \sum_{\tilde{l}=1}^{\tilde{N}} \mathbf{p}_{\tilde{l}} \tag{6.12}$$

This equation expresses the *law of momentum conservation.*

§ 7. The relativistic Maxwell equations. The field strength transformations

7.1. In applying the relativity principle to electrodynamics, one has to demand that the physical law of charge conservation be invariant, that is, independent of the choice of inertial reference frame in which it is analyzed. It will be sufficient to assume that charge density ρ and current density $\mathbf{j}$ are related as components of *four-dimensional current vector* s^i, namely,

$$c\rho = s^0, \quad \mathbf{j}^\alpha = s^\alpha \tag{7.1}$$

Indeed, the continuity equation for electric charge (1.13) can then be written in the form

$$\frac{\partial s^i}{\partial x^i} = 0 \tag{7.2}$$

where the left-hand side is a scalar under the Lorentz transformations[8].

As follows from (7.1), the Maxwell equations (M.1) and (M.4) must somehow be brought to a unified form if we analyze transitions between different inertial reference framos. From this point of view we shall first analyze the Maxwell equations in vacuo. They will be written in the simplest form if we resort to the Gaussian system of units, that is if we assume $\alpha = c$, $\varepsilon_0 = \mu_0 = 1$, whence $\mathbf{E} = \mathbf{D}$ and $\mathbf{B} = \mathbf{H}$. Let us show that this unification of equations can be achieved if we assume that the scalar potential φ and vector potential $\mathbf{A}$ are components Φ_i of a 4-vector, namely,

$$\varphi = -\Phi_0, \quad A_\alpha = \Phi_\alpha \tag{7.3}$$

Formulas (2.1) and (2.2) determining field strengths $\mathbf{B}$ and $\mathbf{E}$ in terms of potentials now can be written in the form

$$B_\alpha = \frac{1}{2}\varepsilon_{\alpha\beta\gamma}\left(\frac{\partial \Phi_\gamma}{\partial x^\beta} - \frac{\partial \Phi_\beta}{\partial x^\gamma}\right), \quad E_\alpha = \frac{\partial \Phi_0}{\partial x^\alpha} - \frac{\partial \Phi_\alpha}{\partial x^0} \tag{7.4}$$

---

[8] We remind the reader that here and throughout the text one has to be very careful about distinguishing between covarient and contravarient components of vectors and tensors in space-time.

We can conclude now that the quantity

$$F_{ik}=\frac{\partial\Phi_k}{\partial x^i}-\frac{\partial\Phi_i}{\partial x^k} \tag{7.5}$$

transformed as an antisymmetric tensor of rank 2 under the Lorentz transformations, makes possible a complete description of the electric and magnetic fields in any inertial reference frame:

$$E_\alpha = F_{\alpha 0},\ B_1 = F_{23},\ B_2 = F_{31},\ B_3 = F_{12} \tag{7.6}$$

The last three definitions take into account the pseudovectorial nature of magnetic field **B** under reflections in the three-dimensional space and are merely a different form of the first of equations (7.4). Tensor F_{ik} is known as the *field tensor* or the *field strength tensor*[9].

When notations (7.6) and (7.1) are introduced into the Maxwell equations (M.1) and (M.4), it is easy to show that both can be written in the form

$$\frac{\partial F^{ik}}{\partial x^k}=\frac{1}{c}\ s^i \tag{7.7}$$

whose invariance under the Lorentz transformations is obvious. Note that formulas (A.19) yield the following relations:

$$F^{\alpha 0} = -F_{\alpha 0},\quad F^{\alpha\gamma}=F_{\alpha\gamma} \tag{7.8}$$

The remaining Maxwell equations (M.2) and (M.3) also take on an invariant form

$$\frac{\partial F_{ik}}{\partial x^l}+\frac{\partial F_{kl}}{\partial x^i}\ \frac{\partial F_{li}}{\partial x^k}=0 \tag{7.9}$$

This is readily verified by rewriting (7.9) in terms of "three-dimensional" notations (7.6). The left-hand side of (7.9) is a tensor of rank 3, completely antisymmetric with respect to permutations of its indices.

The system of the Maxwell equations in vacuo is thus transformed by postulates (7.1) and (7.4) into an invariant system (7.7) and (7.9)

[9] The four-dimensional potential is often defined by relations $\Phi'^0 = \varphi$, $\Phi'^\alpha = A^\alpha$. In this case $\vec{\Phi}' = -\vec{\Phi}$ in the same Minkowski metric as we have chosen in this book. Components of field tensor F'_{ik} can now be introduced by equations (7.5), using components Φ'_i instead of Φ_i. In this case $F'_{ik} = -F_{ik}$. This definition was chosen in "The Classical Theory of Fields" by L. D. Landau and E. M. Lifshitz (*Course of Theoretical Physics*, vol. 2 (4th edition), Pergamon Press, Oxford, 1975). Other authors, using potential $\vec{\Phi}'$, introduce field tensor $F''_{ik} = -F'_{ik} = F_{ik}$. Our definitions correspond to Fock's book referred to on p. 46. Comparing our formulas with formulas in other books, the reader must keep in mind that the space-time metric is often introduced via interval $ds'^2 = dr^2 - c^2\,dt^2 = -ds^2$ (see Appendix D, Item 4). This leads to a different relation between covariant and contravariant components. Obviously, the physical meaning of the relativistic form of the Maxwell equations remains the same in all these cases.

from which field tensor F_{ik} can be found. It is the introduction of this tensor which enables us to describe electromagnetic field invariantly in the sense of the relativity principle. The components of this tensor can be interpreted separately only if one chooses a system of coordinates in four-dimensional space-time, that is if one chooses an inertial frame of reference. Only then can the concepts of electric and magnetic fields be separated; later we shall discuss in detail how to realize such a separation.

By using field tensor F_{ik} we can first of all construct quantities which are quadratic functions of the components of electric and magnetic fields and are invariant under the Lorentz transformations. If we define a pseudotensor $F^*_{lm} = (1/2)\,\varepsilon_{lmik}F^{ik}$, then we find from the elementary properties of tensors (see Appendix A) that these inva iant quantities are

$$I_1 = (1/2)\,F_{ik}F^{ik} = \mathbf{B}^2 - \mathbf{E}^2 \qquad (7.10_1)$$

and

$$I_2 = (1/4)\,F_{ik}\overset{*}{F}{}^{ik} = \mathbf{E}\cdot\mathbf{B} \qquad (7.10_2)$$

The first of these quantities is invariant under any Lorentz transformation while the second is invariant only under the Lorentz transformations without reflection; under reflection its sign is reversed.

7.2. Transformations of field strengths **B** and **E** can be found by means of the general law of the field tensor transformation:

$$F^{i'k'} = A^{i'}_{i}A^{k'}_{k}F^{ik} \qquad (7.11)$$

Of course, the main factor is the form taken by this law when partial Lorentz transformation (5.8) is carried out. By using the table of coefficients (5.8′) of this transformation, as well as relations (7.6) and (7.8), we easily rewrite (7.11) in the form

$$E_{1'} = E_1, \quad E_{2'} = \gamma\,(E_2 + \lambda B_3), \quad E_{3'} = \gamma\,(E_3 - \beta B_2)$$
$$B_{1'} = B_1, \quad B_{2'} = \gamma\,(B_2 - \beta E_3), \quad B_{3'} = \gamma\,(B_3 + \beta E_2) \qquad (7.12)$$

Recalling now what was mentioned on p. 48 about the form of the general Lorentz transformation, we can understand that a correct result for an arbitrary direction of the relative velocity of reference frames will be obtained if the transformation law (7.12) is rewritten, by using three-dimensional vector notations, in a form invariant under three-dimensional rotations. The role played by directions of axes x_1 and $x_{1'}$ must be taken in this case by directions of the relative velocity **v** in the reference frames under consideration. Since in formulas (7.2) $\beta = -v_1/c$ [10], and since $v_2 = v_3 = 0$ under a partial

[10] If we assume that reference frame K' moves in the positive direction of axis x_1.

Lorentz transformation, we find

$$\beta B_3 = \frac{1}{c}\,\mathbf{v}\times\mathbf{B}|_2, \quad -\beta B_2 = \frac{1}{c}\,\mathbf{v}\times\mathbf{B}|_3$$

Similar arguments applied to the second line of (7.12) yield the required result:

$$\mathbf{E}'_{\parallel} = \mathbf{E}_{\parallel},\ \mathbf{E}'_{\perp} = \gamma\left(\mathbf{E}_{\perp} + \frac{1}{c}\,\mathbf{v}\times\mathbf{B}\right)$$

$$\mathbf{B}'_{\parallel} = \mathbf{B}_{\parallel},\ \mathbf{B}'_{\perp} = \gamma\left(\mathbf{B}_{\perp} - \frac{1}{c}\,\mathbf{v}\times\mathbf{E}\right) \qquad (7.13)$$

Here the symbol $\parallel$ marks the components of vectors $\mathbf{E}$, $\mathbf{E}'$, $\mathbf{B}$ and $\mathbf{B}'$ parallel to the direction of the relative velocity $\mathbf{v}$, and the symbol $\perp$ marks components in a plane orthogonal to $\mathbf{v}$. Formulas of inverse transformations are similar in form, with $\mathbf{E}$ replaced by $\mathbf{E}'$, $\mathbf{B}$ by $\mathbf{B}'$, and vice versa, and $\mathbf{v}$ by $-\mathbf{v}$.

If $v \ll c$, we can assume in formulas (7.13) that $\gamma \approx 1$, that is

$$\mathbf{E}'_{\parallel} = \mathbf{E}_{\parallel},\quad \mathbf{E}'_{\perp} \simeq \mathbf{E}_{\perp} + \frac{1}{c}\,\mathbf{v}\times\mathbf{B}$$

$$\mathbf{B}'_{\parallel} = \mathbf{B}_{\parallel},\quad \mathbf{B}'_{\perp} \simeq \mathbf{B}_{\perp} - \frac{1}{c}\,\mathbf{v}\times\mathbf{E} \qquad (7.13')$$

By expressing the components of vectors $\mathbf{B}$ and $\mathbf{E}$ parallel and orthogonal to velocity in terms of unit vector $\mathbf{v}/v$, we obtain also

$$\mathbf{E}' = \gamma\mathbf{E} + \frac{1-\gamma}{v^2}\,\mathbf{v}\,(\mathbf{E}\cdot\mathbf{v}) + \frac{\gamma}{c}\,\mathbf{v}\times\mathbf{B}$$

$$\mathbf{B}' = \gamma\mathbf{B} + \frac{1-\gamma}{v^2}\,\mathbf{v}\,(\mathbf{B}\cdot\mathbf{v}) - \frac{\gamma}{c}\,\mathbf{v}\times\mathbf{E} \qquad (7.14)$$

Invariants I_1 and I_2 can be used to characterize different types of electromagnetic field. Both invariants, as the field itself, are, of course, functions of spatial coordinates and time. Consequently, classification of fields according to the values of invariants can be carried out only "locally". Functions $\mathbf{B}$ and $\mathbf{E}$ are continuous, and with them the invariants are continuous; therefore their properties are conserved within a sufficiently small neighbourhood around a selected point in space-time.

We see that if $I_2 = 0$, the corresponding fields $\mathbf{B}$ and $\mathbf{E}$ are mutually orthogonal in all inertial reference frames. In addition, invariant types of the electromagnetic field are distinguished by the values which the other invariant, I_1, assumes. Let $I_1 > 0$, that is $\mathbf{B}^2 > \mathbf{E}^2$ in all reference frames. Then there exists such a reference frame K' in which $\mathbf{E}' = 0$. Indeed, we see from the first two equations of (7.13) that if in reference frame K fields $\mathbf{B}$ and $\mathbf{E}$ are fixed and mutually orthogonal, we can select velocity $\mathbf{v}$ in K' in such a manner that $\mathbf{E}'_{\perp} = \gamma\left(\mathbf{E}_{\perp} + \frac{1}{c}\,\mathbf{v}\times\mathbf{B}\right) = 0$ and $\mathbf{E}'_{\parallel} = \mathbf{E}_{\parallel} = 0$, that is $\mathbf{E}_\perp = \mathbf{E}$. This will hold if the absolute value of velocity is such that $v/c = E/B_\perp$. Direction of velocity $\mathbf{v}$ can always be chosen orthogonal to vector $\mathbf{B}$. Besides, $v/c = E/B < 1$, that is such a reference frame K' indeed exists.

On the contrary, if $I_1 < 0$, then similar arguments demonstrate that by directing vector $\mathbf{v}$ orthogonally to the plane passing through vectors $\mathbf{B}$ and $\mathbf{E}$ and by defining its magnitude by $v/c = B/E$, we can introduce such a reference frame K' in which the electromagnetic field becomes purely electric, that is $\mathbf{B}' = 0$. Here again $v < c$.

The above arguments show that if $\mathbf{E} = 0$ or $\mathbf{B} = 0$ in one inertial reference frame, then both fields are nonzero and mutually orthogonal in all other inertial reference frames; moreover, in the first case always $E' < B'$ while in the second case $B' < E'$.

Consider now a class of fields for which $I_2 \neq 0$. It can be shown that in this case there is always an inertial reference frame in which the electric and magnetic fields are parallel to one another. In other words, if fields $\mathbf{B}$ and $\mathbf{E}$ satisfy in one reference frame K the condition $\mathbf{B} \cdot \mathbf{E} \neq 0$, then there is also a reference frame K' in which a condition $\mathbf{E}' \times \mathbf{B}' = 0$ is also satisfied.

If equality $\mathbf{E} \times \mathbf{B} = 0$ is satisfied already in frame K, the problem becomes trivial. We assume therefore that $\mathbf{E} \times \mathbf{B} \neq 0$. We can define therefore a plane containing vectors $\mathbf{B}$ and $\mathbf{E}$. Let us choose velocity $\mathbf{v}$ of a new reference frame K' in the direction orthogonal to this plane. With this choice of velocity, $\mathbf{E}'_\parallel = \mathbf{E}_\parallel = 0$ and $\mathbf{B}'_\parallel = \mathbf{B}_\parallel = 0$, and therefore $\mathbf{E} = \mathbf{E}_\perp$, $\mathbf{B} = \mathbf{B}_\perp$, $\mathbf{E}' = \mathbf{E}'_\perp$, and $\mathbf{B}' = \mathbf{B}'_\perp$. As follows from transformation formulas (7.13) and condition $\mathbf{E}' \times \mathbf{B}' = 0$, for given $\mathbf{B}$ and $\mathbf{E}$ velocity $\mathbf{v}$ must satisfy the relation

$$\left[\mathbf{E} + \frac{1}{c}\mathbf{v} \times \mathbf{B}\right] \times \left[\mathbf{B} - \frac{1}{c}\mathbf{v} \times \mathbf{E}\right] = 0$$

Multiplying and taking into account that vectors $\mathbf{v}$ and $\mathbf{B}$, as well as $\mathbf{E}$ and $\mathbf{v}$, are orthogonal, we recast the above formula to

$$\mathbf{E} \times \mathbf{B} - \frac{1}{c}(\mathbf{E}^2 + \mathbf{B}^2)\,\mathbf{v} + \frac{1}{c^2}[(\mathbf{E} \times \mathbf{B}) \cdot \mathbf{v}]\,\mathbf{v} = 0$$

As we have seen above, vector $\mathbf{v}$ can be either parallel or antiparallel to vector $\mathbf{E} \times \mathbf{B}$. Let us choose the parallel arrangement. Then, by projecting the last equality on the direction of vector $\mathbf{v}$, we arrive at a quadratic equation for $\beta \equiv v/c$:

$$\beta^2 - \frac{E^2 + B^2}{|\mathbf{E} \times \mathbf{B}|}\,\beta + 1 = 0$$

For fixed $\mathbf{B}$ and $\mathbf{E}$ such an equation always has two positive roots β_1 and β_2, and $\beta_1\beta_2 = 1$. Hence, one of these roots is always less than unity. By choosing the absolute value of velocity corresponding to this root, we completely define the motion of reference frame K'

in which $\mathbf{E}' \times \mathbf{B}'=0$. If from this reference frame K' we now change to any other inertial reference frame K'' whose velocity in reference frame K' coincides in direction with parallel to one another vectors $\mathbf{B}'$ and $\mathbf{E}'$, then it follows from (7.14) that regardless of the magnitude of velocity of reference frame K'', vectors $\mathbf{B}''$ and $\mathbf{E}''$ remain parallel to one another. Velocity $\mathbf{u}'$ of any such reference frame K'' with respect to the initial reference frame K is given by the relativistic formula of addition of mutually orthogonal velocities of reference frame K' with respect to K, and of K'' with respect to K' (see p. 51).

7.3. Let us derive the equations for potentials of the electromagnetic field. It will be easy to verify by substituting the expression for field tensor (7.5), written in terms of four-dimensional potential Φ_i, into the left-hand side of (7.9) that the relativistic Maxwell equation becomes identity. This result is clear since (7.9) is an invariant form of precisely those two equations (M.2) and (M.3) which serve to define the relation between field strengths and potentials. Further, relations (7.3) make it possible to write in the invariant form the Lorentz gauge condition:

$$\frac{\partial \Phi^i}{\partial x^i}=0 \tag{7.15}$$

This relation was derived directly from (2.5). Clearly, the condition of the Coulomb gauge (2.8) is not invariant. But the gauge transformations (2.3) and (2.4) take the form

$$\tilde{\Phi}^i=\Phi^i+\frac{\partial \psi}{\partial x_i} \tag{7.15'}$$

Therefore, if function ψ is invariant, then the gauge transformation does not affect the vector character of potential Φ^i. And finally, substitution of (7.5) into (7.7) yields, with (7.15) taken into account, a second-order equation for four-dimensional potential:

$$\frac{\partial^2}{\partial x^k \partial x_k}\,\Phi^i=-\frac{1}{c}\,s^i \tag{7.16}$$

In order to find the electromagnetic field, we have therefore to solve wave equation (7.16) for given boundary and initial conditions, taking into account the constraint (7.15). Equations (7.16) are completely identical to the derived earlier (2.6_1) and (2.6_2) for any choice of inertial reference frame.

Let us study in more detail the properties of current vector s^i. Usually we can write $\mathbf{j}=\rho\mathbf{v}$, so that it follows from definition (5.17_1) of four-dimensional velocity u^i that

$$s^i=\rho_0 u^i \tag{7.17}$$

where $\rho_0=\rho\sqrt{1-\beta^2}$. Equality (7.17) can be satisfied only if ρ_0 is assumed to be a scalar (known as the *invariant charge density*).

The physical meaning of the invariant charge density is readily understood if we consider a charge filling volume dV' with density ρ_0 in a reference frame in which all points of volume dV' are at rest. Then in a different reference frame in which volume dV' moves at a constant velocity $\mathbf{v}$, equation (5.10′) yields $\rho\, dV = \rho_0\, dV'$. In other words, the total amount of charge remains invariant under the Lorentz transformation.

Let us imagine now an observer in a reference frame K. As s^i is a vector in space-time, its components are transformed according to formulas (5.12), with x^α replaced by $s^\alpha = j^\alpha$ and x^0 by $s^0 = c\rho$, that is

$$\mathbf{j} = \mathbf{j}' + (\gamma - 1)\, \frac{1}{v^2}\, (\mathbf{j}' \cdot \mathbf{v})\, \mathbf{v} + \gamma\rho'\mathbf{v}$$

$$\rho = \gamma\left(\rho' + \frac{1}{c^2}\, \mathbf{j}' \cdot \mathbf{v}\right) \tag{7.18}$$

According to this, we have written the formulas realizing a transformation from K' to K. They show, in particular, that density ρ' of a charge which is at rest in reference frame K' determines in reference frame K a fraction of the conduction current proportional to $\rho'\,\mathbf{v}$ and known as the *convective current*. On the other hand, an additional relativistic term $\frac{1}{c^2}\,\mathbf{j}' \cdot \mathbf{v}$ appears in the formula for the current density; as a result, an observer in reference frame K must notice a certain distribution of charge in a body moving with respect to the observer, and measure the electric field corresponding to this distribution, even in the case when in the moving reference frame $\rho' = 0$ but $\mathbf{j}' \neq 0$.

7.4. Our next point is the construction of the relativistic theory of electromagnetic field in *material media*. First, we have to write the fundamental formulas of relativistic electrodynamics for vacuum in the International System of Units, instead of the Gaussian system we were using so far. As a result, we can distinguish between vectors $\mathbf{D}$ and $\mathbf{H}$, on the one hand, and vectors $\mathbf{B}$ and $\mathbf{H}$, on the other. These vectors are related by formulas $\mathbf{D} = \varepsilon_0 \mathbf{E}$ and $\mathbf{B} = \mu_0 \mathbf{H}$. In the SI system of units, coefficient α in equations (M) is assumed equal to unity. Components of current s^i are defined as before, but components of potential are usually redefined. Namely, let us replace (7.4) by a vector $\overline{\Phi}_i$ having components

$$\overline{\Phi}_0 = -\frac{1}{c}\,\varphi, \quad \overline{\Phi}_\alpha = A_\alpha \tag{7.19}$$

The Lorentz condition (2.5) in vacuo has the form $\operatorname{div} \mathbf{A} + \varepsilon_0\mu_0 \frac{\partial\varphi}{\partial t} = 0$, that is, as follows from (1.25), $\operatorname{div} \mathbf{A} + \frac{1}{c^2}\,\frac{\partial\varphi}{\partial t} = 0$. Assuming differentiation to be carried out with respect to contravariant components x^α, x^0, and substituting contravariant components of $\overline{\Phi}^i$ defined by (7.19) for **A** and φ we find that the Lorentz condition takes on an invariant form (7.15) with Φ^i replaced by $\overline{\Phi}^i$.

Let us turn now to the definition of the field tensor. We saw in § 1 that in SI quantities $c\mathbf{B}$ and $\mathbf{E}$ have identical dimensions. Taking (7.19) into account, we can express the components of tensor

$$\overline{F}_{ik}=c\left(\frac{\partial\overline{\Phi}_k}{\partial x^i}-\frac{\partial\overline{\Phi}_i}{\partial x^k}\right) \tag{7.20}$$

as

$$\overline{F}_{\alpha 0}=E_\alpha,\quad \overline{F}_{12}=cB_3,\ \text{etc.} \tag{7.21}$$

Obviously, the formulas of transformation of field strengths (7.13) and their corollaries remain valid, but $c\mathbf{B}$ has to be substituted for **B**. The same is true for the relativistic Maxwell equation (7.9). Here F_{ik} must be replaced by $\overline{F}_{ik}$.

In order to obtain the relativistic form of the Maxwell equations with sources, we have to use not vectors **B** and **E** but vectors **H** and **D** with different dimensions. The arguments mentioned in § 1 regarding the dimensions show that **H** and $c\mathbf{D}$ have identical dimensions in SI. In addition,

$$\mathbf{H}=\frac{1}{\mu_0}\mathbf{B}=\sqrt{\frac{\varepsilon_0}{\mu_0}}\,c\mathbf{B},\quad c\mathbf{D}=c\varepsilon_0\mathbf{E}=\sqrt{\frac{\varepsilon_0}{\mu_0}}\,\mathbf{E} \tag{7.22}$$

As follows from definitions (7.20) and (7.21), the quantity

$$f_{ik}=\sqrt{\varepsilon_0/\mu_0}\,\overline{F}_{ik} \tag{7.23}$$

is a tensor, and

$$f_{\alpha 0}=cD_\alpha,\quad f_{12}=H_3,\ \text{etc.} \tag{7.24}$$

From this tensor we obtain that

$$\frac{\partial f^{ik}}{\partial x^k}=s^i \tag{7.25}$$

This equation replaces equation (7.7). Equations (7.25), (7.23), (7.20), and (7.15) yield the wave equation for potentials:

$$\frac{\partial^2}{\partial x^k\,\partial x_k}\overline{\Phi}^i=-\frac{1}{c}\sqrt{\frac{\mu_0}{\varepsilon_0}}\,s^i=-\mu_0 s^i \tag{7.26}$$

Field tensor $\overline{F}_{ik}$ and tensor f_{ik}, which we shall call the *induction tensor*, must be considered simultaneously when the relativistic electrodynamics of material media is developed. This development can be realized only with an additional postulate which characterizes the behaviour of fields **B** and **E**, as well as of fields **H** and **D**, in material media in a transition from one inertial reference frame to another. Such a postulate can be verified experimentally. In particular, it must satisfy a condition that the equations of electromagnetic field for an observer who is at rest with respect to the medium be identical with the Maxwell equations (M).

An assumption which appears to be most natural here is that vectors $c\mathbf{B}$ and $\mathbf{E}$ in the medium are represented, as in the case of vacuum, by tensor $\overline{F}_{ik}$, and vectors $\mathbf{H}$ and $c\mathbf{D}$, by tensor f_{ik}, although in the general case the relation between $\mathbf{B}$ and $\mathbf{H}$, as well as between $\mathbf{D}$ and $\mathbf{E}$, is nonlinear. The invariant form of the Maxwell equations will then be given by relativistic equations of already familiar types (7.9) and (7.25). When electromagnetic properties of media are studied in the relativistic limit, these equations are called the *Minkowski equations*. We must emphasize, in order to avoid confusion, that tensors f_{ik} and $\overline{F}_{ik}$ will not be related by a proportionality formula of the type (7.23) and tensor f_{ik} will be determined in any intertial reference frame only by relations (7.24).

The difference between induction tensor (7.23) in vacuo and induction tensor f_{ik} in a medium is also a tensor

$$\mathfrak{M}_{ik}=\sqrt{\frac{\varepsilon_0}{\mu_0}}\,\overline{F}_{ik}-f_{ik} \tag{7.27}$$

A comparison of definitions (7.21) and (7.24) with definitions of polarization (1.11) and magnetization (1.17) easily shows that components of $\mathfrak{M}_{ik}$ have the following physical meaning:

$$\mathfrak{M}_{\alpha 0}=-cP_{\alpha},\quad \mathfrak{M}_{12}=M_3,\ (\text{etc.} \tag{7.28}$$

Tensor $\mathfrak{M}_{ik}$ is known as the *tensor of moments*. Geometrically, the structure of this tensor is quite similar to the structure of field tensor F_{ik} discussed earlier. In particular, if $\mathbf{P}$ and $\mathbf{M}$ are the polarization and magnetization of a moving medium in a co-moving reference frame, then the transformation formulas for components of the tensor of moments can be immediately written by analogy with equations (7.13):

$$\mathbf{P}'_{\|}=\mathbf{P}_{\|},\quad \mathbf{P}'_{\perp}=\gamma\left(\mathbf{P}_{\perp}+\frac{1}{c^2}\,\mathbf{v}\times\mathbf{M}\right)$$
$$\mathbf{M}'_{\|}=\mathbf{M}_{\|},\quad \mathbf{M}'_{\perp}=\gamma\,(\mathbf{M}_{\perp}-\mathbf{v}\times\mathbf{P}) \tag{7.29}$$

An important physical corollary follows from (7.29). Namely, a physical object which possesses electric polarization and not magnetization in a co-moving inertial reference frame is found to be magnetized in any other inertial reference frame. Conversely, if an object possesses only magnetization in the reference frame at rest, then it also possesses an electric polarization in a reference frame moving with respect to this object. This "kinematic" (i.e. following from the Lorentz transformations) relationship between polarization and magnetization is confirmed experimentally. If $\mathbf{P} = 0$ but $\mathbf{M} \neq 0$, then in the nonrelativistic limit ($\gamma \approx 1$) equations (7.29) take the form

$$\mathbf{P}' \simeq \frac{1}{c^2}\,\mathbf{v}\times\mathbf{M}, \quad \mathbf{M}' \simeq \mathbf{M} \tag{7.29$_1$}$$

This effect of electric polarization of a moving magnet is called the *unipolar induction*. No sophisticated experimental techniques are required to detect this effect, it has already been used for generating electric currents for quite a long time.

But if $\mathbf{M} = 0$ and $\mathbf{P} \neq 0$, then

$$\mathbf{M}' \simeq -\mathbf{v}\times\mathbf{P}, \quad \mathbf{P}' \simeq \mathbf{P} \tag{7.29$_2$}$$

The effect predicted by these relations was experimentally confirmed by Eikhenwald and Roentgen.

Let us turn now to an important particular case in which the relation between inductions and field strengths in a reference frame in which the medium is at rest is given by (1.27). What will be the relations between these quantities in a moving medium? This question can be answered if one takes into account that transformation of inductions $\mathbf{D}$ and $\mathbf{H}$ is written in a form quite analogous to the transformation of field strengths (7.13):

$$\begin{aligned} \mathbf{D}'_{\|} &= \mathbf{D}_{\|}, \quad \mathbf{D}'_{\perp} = \gamma\left(\mathbf{D}_{\perp} + \frac{1}{c^2}\,\mathbf{v}\times\mathbf{H}\right) \\ \mathbf{H}'_{\|} &= \mathbf{H}_{\|}, \quad \mathbf{H}'_{\perp} = \gamma\left(\mathbf{H}_{\perp} - \mathbf{v}\times\mathbf{D}\right) \end{aligned} \tag{7.30}$$

Coefficients of vector products in these formulas correspond to the SI system of units; in the Gaussian system both these coefficients must equal $1/c$ as in the case of (7.13).

Assume now that a medium is at rest in reference frame K', so that $\mathbf{D}' = \varepsilon\mathbf{E}'$ and $\mathbf{B}' = \mu\mathbf{H}'$. In the left-hand sides of these relations we shall use the transformation equations (7.30), and in the right-hand side, equations (7.13) (we have already mentioned that in the SI system of units, $c\mathbf{B}$ in (7.13) must be substituted for $\mathbf{B}$). Finally, we obtain

$$\begin{aligned} \mathbf{D} + \frac{1}{c^2}\,\mathbf{v}\times\mathbf{H} &= \varepsilon\,(\mathbf{E} + \mathbf{v}\times\mathbf{B}) \\ \mathbf{B} - \frac{1}{c^2}\,\mathbf{v}\times\mathbf{E} &= \mu\,(\mathbf{H} - \mathbf{v}\times\mathbf{D}) \end{aligned} \tag{7.31}$$

It follows again for longitudinal and transverse components that

$$\begin{aligned} \mathbf{D}_{\|} &= \varepsilon\mathbf{E}_{\|}, \quad \left(1 - \frac{\varepsilon\mu}{\varepsilon_0\mu_0}\,\beta^2\right)\mathbf{D}_{\perp} = \varepsilon\,(1-\beta^2)\,\mathbf{E}_{\perp} + (\varepsilon\mu - \varepsilon_0\mu_0)\,\mathbf{v}\times\mathbf{H} \\ \mathbf{B}_{\|} &= \mu\mathbf{H}_{\|}, \quad \left(1 - \frac{\varepsilon\mu}{\varepsilon_0\mu_0}\,\beta^2\right)\mathbf{B}_{\perp} = \mu\,(1-\beta^2)\,\mathbf{H}_{\perp} + (\varepsilon\mu - \varepsilon_0\mu_0)\,\mathbf{v}\times\mathbf{E} \end{aligned} \tag{7.32}$$

Here we have taken into account that $c = (\varepsilon_0\mu_0)^{-1/2}$.

A problem of finding such conserved quantities as energy and momentum in the relativistic case of a field in the medium is far from trivial and requires careful analysis[11]. Nevertheless, we shall restrict the presentation to the discussed-above formal fundamentals of the electrodynamics of moving media[12]. They are sufficient to understand what problems can be expected and what is the distinction of this case from the relativistic electrodynamics in vacuo. From the experimental standpoint, the media moving at relativistic velocities were studied very insufficiently; consequently, the physical internipretation of the theory becomes difficult. The case of practical sigficance is the nonrelativistic limit for media moving at sufficiently low velocities with respect to the observer. A number of interesting effects are found in this limit (when only terms of the order not higher than v/c are taken into account). One of such cases (Faraday's induction in a moving loop) will be analyzed in details in connection with the fundamentals of magneto hydrodynamics (see § 35). Other effects (for instance, unipolar induction) can be found in: I. E. Tamm, *Fundamentals of the Theory of Electricity*, Mir Publishers, Moscow, 1979.

§ 8. Relativistic equations of charge motion

8.1. Relativistic equations of motion of a charge in a given electromagnetic field constitute a particular case of relativistic mechanics presented in § 6. We only have to find a new expression for force $\overleftarrow{F}$, namely, how it is related to the electromagnetic field tensor F^{lm}.

It was shown in § 3 that the effect of field on charges and currents is given by the volume density of the Lorentz force (3.13). It is this expression for force that we have now to analyze in order to derive its relativistic generalization.

Let us construct a 4-vector $\frac{1}{c} s_l F^{lk}$. By using definitions (7.1), (7.6), and (7.8), we can express the components of this vector in any inertial reference frame. These expressions are

$$\frac{1}{c} s_l F^{l\alpha} = \rho E^\alpha + \frac{1}{c} (\mathbf{j} \times \mathbf{B})^\alpha$$
$$s_l F^{l0} = \mathbf{j} \cdot \mathbf{E} \qquad (8.1)$$

Formulas (8.1) show that expression $\frac{1}{c} s_l F^{lk}$ can be regarded as the required relativistic generalization of the Lorentz force. Indeed, its

[11] See V. L. Ginzburg, *Theoretical Physics and Astrophysics*, Pergamon Press, Oxford, 1979.

[12] Additional information can be found in: A. Sommerfeld, *Electrodynamics*, Academic Press, New York, 1952 and V. A. Ugarov, *The Special Theory of Relativity*, Mir Publishers, Moscow, 1979.

spatial components are identical to (3.13), and, in accordance with § 6, the fourth (time) component is equal to the work of the Lorentz force required to displace the charges.

It is obvious from expressions (8.1) that the derived expression for the volume density of the force refers to one chosen reference frame. Hence, it must be set equal to the derivative, with respect to proper time, of the volume density of momentum of the medium moving under the action of the applied "external" electromagnetic field. We shall consider a "dustlike" medium characterized by a practically negligible interaction between the particles; in other words, the particles move in the electromagnetic field independently of one another. It is clear that at any rate volume density of momentum in the medium (not necessarily "dustlike") can be written as $\varkappa_0 u^k$, where $\varkappa_0$ is the invariant density of the rest mass (defined in complete analogy to the invariant charge density, see p. 64). Similarly to the charge conservation law, the *law of mass conservation* must hold. Obviously, this law can be expressed by setting the four-dimensional divergence equal to zero:

$$\frac{\partial}{\partial x^k}(\varkappa_0 u^k) = 0 \tag{8.2}$$

If we recall the mass-energy interrelation (see § 6), we come to a conclusion that the rest mass of each element of the medium depends on its interaction with all the elements of this medium. As we have neglected this interaction in a dustlike medium, $d(\varkappa_0 \delta V_0)/d\tau = 0$, where δV_0 is an element of volume of the medium in the corresponding reference frame (this element is invariant under the Lorentz transformation).

Let us recall the equations of motion derived in § 6. As follows from the above arguments, the relativistic equation of motion of an infinitesimally small element of the spatial volume of a dustlike medium can be written in the form

$$\frac{d}{d\tau}(\varkappa_0 u^k \delta V_0) = \frac{1}{c} s_l F^{lk} \delta V_0$$

that is,

$$\varkappa_0 \frac{du^k}{d\tau} = \frac{1}{c} s_l F^{lk} = \frac{1}{c} \rho_0 u_l F^{lk}$$

The second of these equalities is based on expression (7.17) for four-dimensional current s_l.

The introduced interaction with electromagnetic field is in agreement with the assumed conservation of the rest mass. Indeed, if the rest mass is constant, we must check whether relation (5.19) still holds. It is valid because the field tensor is antisymmetric. In other words, the four-dimensional force and velocity are mutually orthogonal.

8.2. The dustlike medium will be discussed again at the end of this chapter. Here we shall analyze in more detail the motion of a pointlike particle under the action of a given field. We have then to assume that the invariant densities of charge and rest mass are given by formulas

$$\varkappa_0 = m_0 \delta(\vec{x} - \vec{x}(\tau)), \quad \rho_0 = q\delta(\vec{x} - \vec{x}(\tau))$$

Here m_0 is the rest mass, and q is the charge of a pointlike particle. The delta function is defined by

$$\delta(\vec{x} - \vec{x}(\tau)) = \prod_{l=0}^{3} \delta(x^l - x^l(\tau))$$

where $x^l(\tau)$ are defined as coordinates of a particle, corresponding to its position on the appropriate world line in a point given by the value of the proper time τ.

By substituting these formulas into the equations of motion and integrating both sides of the equation over the entire space-time (using the basic property of delta function), we obtain

$$m_0 \frac{du^k}{d\tau} = \frac{q}{c} u_l F^{lk} \tag{8.3}$$

in which F^{lk} depends on τ via the coordinates of the particle in space-time.

We shall use in the right-hand side of (8.3) the expression (7.5) for the field tensor. As

$$\frac{\partial \Phi^k}{\partial x_l} u_l = \frac{\partial \Phi^k}{\partial x^l} \frac{dx_l}{d\tau} = \frac{d\Phi^k}{d\tau}$$

we can rewrite (8.3) in the form

$$\frac{d\pi^k}{d\tau} = -\frac{q}{c} u_l \frac{\partial \Phi^l}{\partial \Phi_k} \tag{8.4}$$

where

$$\pi^k = m_0 u^k - \frac{q}{c} \Phi^k \tag{8.5}$$

Equation (8.4) can be derived from the variational principle. Assume, in accordance with our initial assumptions, that a four-dimensional potential of electromagnetic field $\vec{\Phi}(\vec{x})$ can be considered fixed in a certain region of space-time. Within this region, we choose two fixed space-time points $\vec{x}_1$ and $\vec{x}_2$ (we shall specify that $\vec{x}_2$ is in the region of the absolute future with respect to $\vec{x}_1$). Consider all possible timelike curves connecting point $\vec{x}_1$ with point $\vec{x}_2$. The *variational principle* states that there exists a Lagrangian $\mathscr{L}(\vec{x}, \vec{u})$

with which the action integral

$$\mathcal{S}=\int\limits_{x_1}^{x_2}\mathcal{L}(\vec{x},\vec{u})\,d\tau$$

can be constructed; this integral takes on an extremal value on the world line along which the particle actually moves under the action of the given forces. In other words, the condition

$$\delta\mathcal{S}=\delta\int\limits_{\vec{x}_1}^{\vec{x}_2}\mathcal{L}(\vec{x},\vec{u})\,d\tau=0 \tag{8.6}$$

is satisfied on this world line. One direct corollary of this extremal condition is the *Euler-Lagrange equations*

$$\frac{d}{d\tau}\frac{\partial\mathcal{L}}{\partial u_k}=\frac{\partial\mathcal{L}}{\partial x_k} \tag{8.7}$$

In the case under consideration, that is for a pointlike particle moving in an external field, equations (8.7) are identical to the motion equations (8.4) if the Lagrangian is chosen in the form

$$\mathcal{L}=\frac{1}{2}m_0(u^ku_k-c^2)-\frac{q}{c}\Phi^ku_k \tag{8.8}$$

This is verified by direct substitution of (8.8) into (8.7)[13]. Consequently, formula (8.8) defines the Lagrangian of a pointlike charge interacting with a given electromagnetic field.

The following features are characteristic of the variational principle (8.6) with Lagrangian (8.8). First, the Lagrangian and the variational principle formulated with this function obviously satisfy the condition of the relativistic invariance. Here we apply the variational principle to determine the world line given parametrically by equations $x^h = x^h(\tau)$. Actually the parameter is chosen in such manner that equality (5.18) holds on the extremal, that is $u^ku_k = c^2$.

And finally, the gauge transformation of the type of (7.15′) leaves the equations of motion (8.4) unaltered. Indeed, as we see from (8.8), this transformation adds a derivative $\frac{\partial\psi}{\partial x_k}u_k=\frac{\partial\psi}{\partial\tau}$ to the Lagrangian (if function ψ does not depend explicitly on parameter τ); as follows from (8.6), this derivative does not affect the analysis of the extremal condition.

8.3. Equations of motion of a charged particle can also be written in the *Hamiltonian form*. First of all, a comparison of (8.5) and (8.8)

---

[13] Coefficient $m_0/2$ plays the role of the Lagrange multiplier which is assumed constant. The corresponding additional condition serves to specify parameter τ on the extremal.

yields

$$\partial\mathscr{L}/\partial u_k = \pi^k \tag{8.9}$$

Equations (8.5) can be used to find components of velocity u_k as functions of variables of $\vec{x}$ and $\vec{\pi}$. The Hamiltonian form of the equations of motion will be obtained if we assume that the motion of the particle is determined precisely by coordinates x_k and momenta π_k.

Hamiltonian $\mathscr{H}$ will be defined, similarly to what is done in the classical mechanics of the material pointlike mass, by the equality

$$\mathscr{H} = \pi^k u_k - \mathscr{L} \tag{8.10}$$

Then

$$\frac{\partial\mathscr{H}}{\partial x_k} = \pi^l \frac{\partial u_l}{\partial x_k} - \frac{\partial\mathscr{L}}{\partial x_k} - \frac{\partial\mathscr{L}}{\partial u_l}\frac{\partial u_l}{\partial x_k} = -\frac{\partial\mathscr{L}}{\partial x_k} = -\frac{d\pi^k}{d\tau}$$

$$\frac{\partial\mathscr{H}}{\partial \pi^k} = u_k + \pi^l \frac{\partial u_l}{\partial \pi_k} - \frac{\partial\mathscr{L}}{\partial u_l}\frac{\partial u_l}{\partial \pi^k} = u_k = \frac{dx_k}{d\tau} \tag{8.11}$$

These are the equations of motion in the Hamiltonian form. They have been derived from the Lagrange equations of motion (8.7) and relation (8.9).

The following expression for the Hamiltonian is obtained from definition (8.10) taken simultaneously with (8.8) and (8.5):

$$\mathscr{H} = \frac{1}{2m_0}\left(\vec{\pi} + \frac{q}{c}\vec{\Phi}\right)^2 + \frac{m_0 c^2}{2} \tag{8.12}$$

The first term on the right-hand side of this expression is equal, according to (8.5), to $\frac{1}{2} m_0 \vec{u}^2 = \frac{1}{2} m_0 c^2$[14]. We find, therefore, that momentum π_k of a particle on the world line which is actually realized in the given field of potential $\vec{\Phi}$, that is on the extermal of the action function $\mathscr{S}$, must satisfy the relation

$$\mathscr{H}(\vec{\pi}, \vec{x}) - \frac{m_0 c^2}{2} = \frac{1}{2m_0}\left(\vec{\pi} + \frac{q}{c}\vec{\Phi}\right)^2 = \frac{1}{2} m_0 c^2 \tag{8.13}$$

By using this relation we can express the time component π^0 of 4-momentum $\vec{\pi}$ as a function of the 4-dimensional radius vector $\vec{x}$ and of the remaining three components of the momentum, π^α: $\pi^0 = \pi^0(\vec{x}, \pi)$. If function π^0 is substituted into Hamiltonian $\mathscr{H}$, equation (8.13) becomes an identity:

$$\mathscr{H}(\pi, \pi^0(\vec{x}, \pi), \vec{x}) = m_0 c^2 \tag{8.14}$$

[14] This equality follows from the abovementioned condition determining the choice of parameter τ.

Differentiation of (8.14) yields

$$\frac{\partial \mathcal{H}}{\partial x_k}+\frac{\partial \mathcal{H}}{\partial \pi^0}\frac{\partial \pi^0}{\partial x_k}=0,\quad \frac{\partial \mathcal{H}}{\partial \pi^\alpha}+\frac{\partial \mathcal{H}}{\partial \pi^0}\frac{\partial \pi^0}{\partial \pi^\alpha}=0 \tag{8.15}$$

From the Hamiltonian equation (8.11) together with (8.15) we find

$$\frac{\partial x_\alpha}{\partial x_0}=\frac{\partial \mathcal{H}}{\partial \pi^\alpha}\Big/\frac{\partial \mathcal{H}}{\partial \pi^0}=-\frac{\partial \pi^0}{\partial \pi^\alpha},\quad \frac{\partial \pi^\alpha}{\partial x_0}=\frac{\partial \pi^0}{\partial x_\alpha}$$

Consequently, if we choose inertial reference frame, we can refer to function $\widetilde{\mathcal{H}} = c\pi^0(\vec{x}, \boldsymbol{\pi})$ as the "3-dimensional" Hamiltonian of the particle, with the last equations taking the form

$$\frac{\partial x^\alpha}{\partial t}=\frac{\partial \widetilde{\mathcal{H}}}{\partial \pi^\alpha},\quad \frac{\partial \pi^\alpha}{\partial t}=-\frac{\partial \widetilde{\mathcal{H}}}{\partial x^\alpha} \tag{8.16}$$

An explicit expression for function $\widetilde{\mathcal{H}}$ can be found if equality (8.14) is rewritten in the chosen inertial reference frame

$$-\left(\boldsymbol{\pi}-\frac{q}{c}\mathbf{A}\right)^2+\left(\frac{\widetilde{\mathcal{H}}}{c}-\frac{q}{c}\varphi\right)^2=m_0^2c^2$$

Here it is necessary to recall formulas (7.3) for potential and to use its contravariant components or its covariant components simultaneously for $\vec{\pi}$ and for $\vec{\Phi}$. Hence,

$$\widetilde{\mathcal{H}}=q\varphi+c\left[m_0^2c^2+\left(\boldsymbol{\pi}-\frac{q}{c}\mathbf{A}\right)^2\right]^{1/2} \tag{8.17}$$

In the nonrelativistic approximation, when $v \ll c$, that is $\left(\boldsymbol{\pi}-\frac{q}{c}\mathbf{A}\right)^2 \ll m_0^2c^2$,

$$\mathcal{H}\simeq q\varphi+\frac{1}{2m_0}\left(\boldsymbol{\pi}-\frac{q}{c}\mathbf{A}\right)^2+m_0c^2 \tag{8.18}$$

In this approximation the Galilean transformations can be used instead of the Lorentz transformations. In this case energy and momentum do not form a 4-vector and the term m_0c^2, that is the rest energy, in (8.18) can be ignored. Then this condition simply defines the choice of origin for the energy of the particle, set equal to $\mathcal{H} - m_0c^2$. Equation (8.18) also shows that $\boldsymbol{\pi} - \frac{q}{c}\mathbf{A} = m_0\mathbf{v}$ [15].

Since on the extremal $\vec{u}^2 = c^2$, we obtain, if we denote $-\widetilde{\mathcal{L}} = \mathcal{L} + m_0c^2$,

$$\delta\mathcal{S}=\delta\int_{t_1}^{t_2}\widetilde{\mathcal{L}}\,dt=0,\quad \widetilde{\mathcal{L}}=-m_0c^2\sqrt{1-\frac{v^2}{c^2}}-q\varphi+\frac{q}{c}\mathbf{A}\cdot\mathbf{v} \tag{8.19}$$

[15] In the SI system of units $\mathbf{A}$ must be replaced by $c\mathbf{A}$.

The Lagrangian is often used in this form. When $\beta \ll 1$, this function is approximately equal to the difference between kinetic energy and "potential function" (if the term $-m_0c^2$ is dropped).

§ 9*. Variational principle for electromagnetic field

9.1. Equations of motion of charged pointlike mass in an external electromagnetic field can be derived from the variational principle; similarly, it is possible to formulate the variational principle from which Maxwell equations can be derived. This is important since, on the one hand, variational principles are at the foundation of a number of computational methods, and on the other hand, the variational principle of electrodynamics is a prototype for all variational principles which are used to find field equations for microparticles in modern physics.

Application of the variational principle to the theory of electromagnetic field involves integration in the 4-dimensional space-time. Therefore, we first discuss the geometric aspect of this integration. Indeed, it can be carried out over objects of very different geometric nature: over a four-dimensional volume, over a three-dimensional hypersurface, over two-dimensional surfaces, and finally, along one-dimensional curves.

When integrating over a four-dimensional volume an infinitesimal element of this volume can be written in an arbitrary inertial reference frame as $d\Omega = dV\,dx^0$. This element of the four-dimensional volume is invariant under the Lorentz transformations without reflections, that is transformations conserving orientation of the initially chosen reference frame. Indeed, if $x^{i'} = A^{i'}_i x^i$, then

$$d\Omega' = d\Omega \frac{\partial (x^{0'}, x^{1'}, x^{2'}, x^{3'})}{\partial (x^0, x^1, x^2, x^3)} = d\Omega \cdot \det (A^{i'}_i) = d\Omega \qquad (9.1)$$

Consider now a three-dimensional hypersurface Σ. An element of volume of this hypersurface is given by an infinitesimal parallelepiped formed on 3 noncoplanar vectors $\vec{dx}$, $\vec{dy}$, $\vec{dz}$, coming from the same point and lying on the tangent hyperplane. The volume element can be measured by means of a pseudovector $\overset{*}{n}_m$ defined by

$$\overset{*}{n}_m \, d\Sigma = \varepsilon_{mikl} \, dx^i \, dy^k \, dz^l \qquad (9.2)$$

Here ε_{mikl} is a unit pseudoscalar of the 4-dimensional space-time introduced in (A.9), and $\overset{*}{n}_m$ is normalized by a condition $| \overset{*}{n}_m \overset{*}{n}{}^m | = 1$. Factor $d\Sigma$ (that is the value of the volume element) is assumed equal to the square root of the absolute value of the sum obtained in the right hand side after $\overset{*}{n}_m \, d\Sigma$ is scalarly multiplied by itself. The properties of pseudoscalar ε_{mikl} are such that this sum is equal to a determinant composed of all possible scalar products of vectors $\vec{dx}$, $\vec{dy}$, $\vec{dz}$ by one another[16]. Pseudovector $\overset{*}{n}_m$ is orthogonal to hypersurface Σ at the point where it is considered. Indeed, any vector $\vec{da}$ radiating from this point and belonging to hypersurface Σ can be expressed as a linear combination of vectors $\vec{dx}$, $\vec{dy}$ and $\vec{dz}$:

$$da^m = \alpha\, dx^m + \beta\, dy^m + \gamma\, dz^m$$

But this means that

$$(\overset{*}{n}_m\, da^m)\, d\Sigma = \alpha\, dx^m \varepsilon_{mikl}\, dx^i\, dy^k\, dz^l + \ldots \equiv 0$$

In particular, vector $\overset{*}{n}_m$ is timelike if Σ is a spacelike hypersurface. This definition of the volume of a three-dimensional hypersurface in the four-dimensional space is quite similar to the definition of surface area of a two-dimensional surface in the three-dimensional space (see, for example (B.7)).

Let us find the formulas for a particular case of a three-dimensional spacelike hyperplane defined in a chosen reference frame by equation $x^0 = \text{const}$. In this case $dx^0 = dy^0 = dz^0 = 0$ for vectors $\vec{dx}$, $\vec{dy}$, $\vec{dz}$ radiating from any point of this hyperplane. If, in addition, these vectors are directed along mutually orthogonal coordinate axes x^1, x^2, x^3, then

$$\overset{*}{n}_\alpha = 0 \ (\alpha = 1, 2, 3); \quad d\Sigma = dx^1\, dx^2\, dx^3 = dV \tag{9.3}$$

A closed four-dimensional volume Ω is bounded by a three-dimensional hypersurface Σ. In complete analogy with the standard derivation of the Gauss theorem, this theorem can be generalized for the above case in the form

$$\int_\Omega \frac{\partial A^i}{\partial x^i}\, d\Omega = \int_\Sigma A^i \overset{*}{n}_i\, d\Sigma \tag{9.4}$$

and also in the form

$$\int_\Omega \frac{\partial T^{ik}}{\partial x^k}\, d\Omega = \oint_\Sigma T^{ik} \overset{*}{n}_k\, d\Sigma \tag{9.5}$$

In what follows we shall not consider the reflection operations, and therefore shall not distinguish between vectors and pseudovectors.

---

[16] The proof of this statement follows from the relation

$$\varepsilon_{mk_1k_2k_3}\varepsilon^{mk_1'k_2'k_3'} = \varepsilon_{abc}\delta^{k_a'}_{k_1}\delta^{k_b'}_{k_2}\delta^{k_c'}_{k_3}$$

No further explanations will be required concerning the properties of integrals over two-dimensional surfaces and their relations to integrals over three-dimensional volume, since these operations will always be considered within a three-dimensional hypersurface chosen in advance on the basis of specific arguments. Hence, no additional complications should arise compared to the familiar theorems given in Appendix B.

The most important hereafter will be the integration over a four-dimensional volume defined as follows. Take two spacelike hypersurfaces Σ_1 and Σ_2 such that any point of hypersurface Σ_1 is within the region of the absolute future with respect to a certain point of hypersurface Σ_2 (Fig. 3). We first integrate over a four-dimensional region bounded by a cylindrical three-dimensional hypersurface whose bases lie on Σ_1 and Σ_2. We often shall have to find the limit of this integral when the lateral surface of the cylinder is moved to infinity in spacelike directions, so that its bases fill up the whole surface Σ_1 and the whole surface Σ_2. Note that the Gauss theorem in the form (9.4) or (9.5) is valid for the inner region of this cylinder.

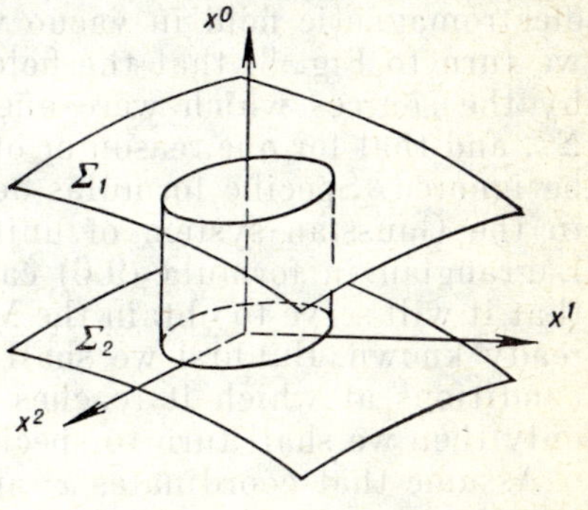

Fig. 3

9.2. The preliminary considerations completed, let us turn to the formulation of electromagnetic field equations on the basis of the *variational principle*.

The state of electromagnetic field in a certain region of space-time can be considered completely defined if four-dimensional potential Φ^l is known in this region. We define the action function of the field as an integral of a Lagrangian $\mathscr{L}$ over the appropriate space-time region:

$$\mathscr{S}\left[\Phi_l(x^i)\right]=\int_{\Omega}\mathscr{L}\left(x^i,\ \Phi_l(x^i),\ \frac{\partial\Phi_l(x^i)}{\partial x^m}\right)d\Omega \tag{9.6}$$

It is assumed here that indices i, l, m take on all values possible for them. The same condition must be kept in mind with respect to all other formulas of the present section. The integral of action $\mathscr{S}$ is regarded as a functional depending on electromagnetic potential Φ_l (this is indicated in the left-hand side of (9.6)). This integral is obviously a function of the type of region Ω over which we integrate.

The fundamental statement of the variational principle consists in assuming that equations of motion of electromagnetic field, that is the Maxwell equations, as well as the physical conservation laws which govern this field, can be derived as the necessary conditions for extremum of functional $\mathscr{S}$. This extremum must be reached with respect to infinitesimal variations of space-time coordinates x^i and functions $\Phi_l(x^i)$.

We shall apply the variational principle primarily to study the electromagnetic field in vacuo with no field sources. This means, if we turn to Fig. 3, that the field between Σ_1 and Σ_2 was generated by the sources which were effective during time interval prior to Σ_2, and that for one reason or other the sources within this layer can be ignored. Specific formulas dealing with this case will be written in the Gaussian system of units. We shall have to show that the Lagrangian in formula (9.6) can indeed be chosen in such manner that it will serve to obtain the Maxwell equations whose form is already known. But first we shall analyze the action integral and the conditions at which it reaches extremum in the general form, and only then we shall turn to specific results for electromagnetic field[17].

Assume that coordinates x^i and functions $\Phi_l(x^i)$ in (9.6) are subjected to a transformation in which x^i are replaced by new coordinates $\tilde{x}^i$, and $\Phi_l(x^i)$ by new functions $\tilde{\Phi}_l(x^i)$. We define variations δx^i and $\delta\Phi_l(x^i)$ (assumed infinitesimal) by equations

$$\tilde{x}^i - x^i = \delta x^i \tag{9.7$_1$}$$

$$\tilde{\Phi}_l(\tilde{x}^i) - \Phi_l(x^i) = \delta\Phi_l(x^i) \tag{9.7$_2$}$$

One important example of infinitesimal variation (9.7_2) of coordinates is formula (5.24) for the Lorentz transformation. It should also be noted that in (9.7_2) the variation of functions Φ_l can be performed, in the general case, independently of the variation of coordinates, but the new functions $\tilde{\Phi}_l$ can always be regarded dependent on $\tilde{x}^i$ since equation (9.7_1) gives coordinates x^i in terms of $\tilde{x}^i$. Indeed, variations δx^i in (9.7_1) must be considered as fixed functions.

Obviously, a transition from x^i to $\tilde{x}^i$ is a transition to a new range of integration, which we denote by $\tilde{\Omega}$. Therefore, the variation results in the new action integral having the form

$$\mathscr{S}[\tilde{\Phi}_l(\tilde{x}^i)] = \int\limits_{\tilde{\Omega}} \mathscr{L}\left(\tilde{x}^i, \tilde{\Phi}_l(\tilde{x}^i), \frac{\partial\tilde{\Phi}_l}{\partial\tilde{x}^m}(\tilde{x}^i)\right) d\tilde{\Omega} \tag{9.8}$$

The necessary condition for extremum of $\mathscr{S}$ consists in that the difference

$$\mathscr{S}[\tilde{\Phi}_l(\tilde{x}^i)] - \mathscr{S}[\Phi_l(x^i)] \tag{9.9}$$

[17] Our exposition of the variational principle follows I. M. Gelfand and S. V. Fomin, *Calculus of Variations*, Prentice-Hall, Engelwood Cliffs, N.Y., 1963 and N. N. Bogoliubov and D. V. Shirkov, *Introduction to the Theory of Quantized Fields*, Wiley, New York, 1959. Detailed calculation of variations is required for a correct interpretation of results.

expanded into the Taylor series in variations δx^i and $\delta\Phi_l$ must have the first-order term equal to zero for arbitrary infinitesimals δx^i and $\delta\Phi_l$.

9.3. In order to analyze formula (9.9), we need some new information on variations; it will be obtained below.

First, define an infinitely small difference

$$\tilde{\Phi}_l(x^i) - \Phi_l(x^i) = \bar{\delta}\Phi_l(x^i) \tag{9.10}$$

which is naturally referred to as the *variation in form* of function $\Phi_l(x^i)$. As follows from (9.7_2) and (9.10),

$$\delta\Phi_l(x^i) = \tilde{\Phi}_l(\tilde{x}^i) - \tilde{\Phi}_l(x^i) + \bar{\delta}\Phi_l(x^i)$$
$$\simeq \frac{\partial\tilde{\Phi}_l}{\partial x^i}\delta x^i + \bar{\delta}\Phi_l(x^i) \simeq \frac{\partial\Phi_l}{\partial x^i}\delta x^i + \bar{\delta}\Phi_l \tag{9.11}$$

The total variation of function $\delta\Phi_l$ is thus represented to within infinitesimals of the order above first as a sum of two terms the first of which is a function of the variation of coordinates and the second is independent of it. In what follows the symbol $\simeq$ always designates equalities valid to the same accuracy. Then

$$\frac{\partial}{\partial x^k} = \frac{\partial\tilde{x}^i}{\partial x^k}\frac{\partial}{\partial\tilde{x}^i} = \left(\delta_{ik} + \frac{\partial(\delta x^i)}{\partial x^k}\right)\frac{\partial}{\partial\tilde{x}^i} = \frac{\partial}{\partial\tilde{x}^k} + \frac{\partial(\delta x^i)}{\partial x^k}\frac{\partial}{\partial\tilde{x}^i}$$

that is,

$$\frac{\partial}{\partial x^k} - \frac{\partial}{\partial\tilde{x}^k} = \frac{\partial(\delta x^i)}{\partial x^k}\frac{\partial}{\partial\tilde{x}^i} \tag{9.12}$$

We now denote by $\delta\left(\frac{\partial\Phi_l}{\partial x^i}\right)$ the principal part of the difference

$$\frac{\partial\tilde{\Phi}_l(\tilde{x}^k)}{\partial\tilde{x}^i} - \frac{\partial\Phi_l(x^k)}{\partial x^i}$$

We can write

$$\frac{\partial\tilde{\Phi}_l(\tilde{x}^k)}{\partial\tilde{x}^i} - \frac{\partial\Phi_l(x^k)}{\partial x^i} = \frac{\partial}{\partial\tilde{x}^i}[\tilde{\Phi}_l(\tilde{x}^k) - \Phi_l(\tilde{x}^k)]$$
$$+ \frac{\partial}{\partial x^i}[\Phi_l(\tilde{x}^k) - \Phi_l(x^k)] + \left(\frac{\partial\Phi_l(\tilde{x}^k)}{\partial\tilde{x}^i} - \frac{\partial\Phi_l(\tilde{x}^k)}{\partial x^i}\right) \tag{9.13}$$

The differentiable function in the first term of the right-hand side of (9.13) is a quantity of the first order of smallness; therefore, by virtue of (9.12) $\partial/\partial\tilde{x}^k$ can be replaced by $\partial/\partial x^k$. To within the terms of higher order of smallness the first term can thus be written in the form

$$\frac{\partial}{\partial x^i}(\bar{\delta}\Phi_l(\tilde{x}^k)) \simeq \frac{\partial}{\partial x^i}(\bar{\delta}\Phi_l(x^k))$$

After expanding the differentiable function into the Taylor series, the second term becomes equal, to the same accuracy, to

$$\frac{\partial}{\partial x^i}\left(\frac{\partial \Phi_l}{\partial x^k}\,\delta x^k\right)$$

Finally,

$$\left(\frac{\partial}{\partial \widetilde{x}^i}-\frac{\partial}{\partial x^i}\right)\Phi_l\,(\widetilde{x}^k)\simeq\left(\frac{\partial}{\partial \widetilde{x}^i}-\frac{\partial}{\partial x^i}\right)\Phi_l\,(x^k)\simeq-\frac{\partial\,(\delta x^m)}{\partial x^i}\,\frac{\partial \Phi_l\,(x^k)}{\partial x^m}$$

and ultimately we obtain

$$\delta\left(\frac{\partial \Phi_l}{\partial x^i}\right)\simeq\frac{\partial}{\partial x^i}\,(\bar{\delta}\Phi_l)+\frac{\partial^2\Phi_l}{\partial x^i\,\partial x^m}\,\delta x^m \tag{9.14}$$

As could be expected, the integration element $d\widetilde{\Omega}$ in integral (9.8) is related to $d\Omega$ by

$$d\widetilde{\Omega}=\frac{\partial\,(\widetilde{x}^0,\,\widetilde{x}^1,\,\widetilde{x}^2,\,\widetilde{x}^3)}{\partial\,(x^0,\,x^1,\,x^2,\,x^3)}\,d\Omega$$

Formula (9.7_1) shows that

$$\frac{\partial \widetilde{x}^i}{\partial x^k}=\delta_{ik}+\frac{\partial\,(\delta x^i)}{\partial x^k}$$

so that the principal part of the Jacobian is

$$\frac{\partial\,(\widetilde{x}^0,\,\widetilde{x}^1,\,\widetilde{x}^2,\,\widetilde{x}^3)}{\partial\,(x^0,\,x^1,\,x^2,\,x^3)}\simeq 1+\frac{\partial\,(\delta x^i)}{\partial x^i} \tag{9.15}$$

If we denote by $\delta\mathcal{S}$ the sum of terms in (9.9), linear in δx^i and $\delta\Phi_l$ then we obtain, taking into account (9.15),

$$\delta\mathcal{S}=\int\limits_{\Omega}\left[\frac{\partial\mathcal{L}}{\partial x^i}\delta x^i+\frac{\partial\mathcal{L}}{\partial\Phi_l}\delta\Phi_l+\frac{\partial\mathcal{L}}{\partial\,(\partial\Phi_l/\partial x_i)}\,\delta\left(\frac{\partial\Phi_l}{\partial x^i}\right)+\mathcal{L}\,\frac{\partial\,(\delta x^i)}{\partial x^i}\right]d\Omega \tag{9.16}$$

If we express $\delta\Phi_l$ in terms of $\bar{\delta}\Phi_l$ via (9.11) and also use formula (9.14), then

$$\delta\mathcal{S}=\int\limits_{\Omega}\left[\frac{\partial\mathcal{L}}{\partial x^i}\delta x^i+\frac{\partial\mathcal{L}}{\partial\Phi_l}\,\frac{\partial\Phi_l}{\partial x^k}\,\delta x^k+\frac{\partial\mathcal{L}}{\partial\,(\partial\Phi_l/\partial x_i)}\,\frac{\partial^2\Phi_l}{\partial x^k\,\partial x^i}\,\delta x^k\right.$$
$$\left.+\mathcal{L}\,\frac{\partial\,(\delta x^i)}{\partial x^i}+\frac{\partial\mathcal{L}}{\partial\,(\partial\Phi_l/\partial x^k)}\,\frac{\partial}{\partial x^k}\,(\bar{\delta}\Phi_l)+\frac{\partial\mathcal{L}}{\partial\Phi_l}\,\bar{\delta}\Phi_l\right]d\Omega$$

The first four terms in the integrand can be written in the form

$$\frac{\partial}{\partial x^k}\,(\mathscr{L}\delta x^k)$$

and the fifth term, in the form

$$\frac{\partial}{\partial x^k}\left(\frac{\partial \mathscr{L}}{\partial\,(\partial\Phi_l/\partial x^k)}\,\bar{\delta}\Phi_l\right)-\left(\frac{\partial}{\partial x^k}\,\frac{\partial \mathscr{L}}{\partial\,(\partial\Phi_l/\partial x^k)}\right)\bar{\delta}\Phi_l$$

As a result, we obtain

$$\delta\mathscr{S}=\int\limits_{\Omega}\left\{\left[\frac{\partial\mathscr{L}}{\partial\Phi_l}-\frac{\partial}{\partial x^k}\,\frac{\partial\mathscr{L}}{\partial\,(\partial\Phi_l/\partial\Phi^k)}\right]\bar{\delta}\Phi_l\right.$$

$$\left.+\frac{\partial}{\partial x^k}\left[\frac{\partial\mathscr{L}}{\partial\,(\partial\Phi_l/\partial x^k)}\,\bar{\delta}\Phi_l+\mathscr{L}\delta x^k\right]\right\}d\Omega \qquad (9.17)$$

Vector Φ_l is known as the *extremal vector of functional* $\mathscr{S}$ if it represents a solution of the Euler-Lagrange equations

$$\frac{\partial\mathscr{L}}{\partial\Phi_l}=\frac{\partial}{\partial x^k}\,\frac{\partial\mathscr{L}}{\partial\,(\partial\Phi_l/\partial x^k)} \qquad (9.18)$$

Assume now that the action function $\mathscr{S}$ is invariant under the realized variation of coordinates x^i and potentials Φ_l, that is, difference (9.9) vanishes. Then, as the range of integration Ω is arbitrary, it follows from (9.17), (9.18) and (9.11) that any extremal vector Φ_l also satisfies the equation

$$\frac{\partial}{\partial x^k}\left[\frac{\partial\mathscr{L}}{\partial\,(\partial\Phi_l/\partial x^k)}\,\bar{\delta}\Phi_l+\mathscr{L}\,\delta x^k\right]$$

$$=\frac{\partial}{\partial x^k}\left[\frac{\partial\mathscr{L}}{\partial\,(\partial\Phi_l/\partial x^k)}\left(\delta\quad{}_l-\frac{\partial\Phi_l}{\partial x^m}\,\delta x^m\right)+\mathscr{L}\,\delta x^k\right]=0 \qquad (9.19)$$

The relation is fundamental for the derivation of conservation laws.

§ 10*. The Noether theorem. Relativistic differential and integral conservation laws for electromagnetic fields

10.1. Let an infinitesimal variation of coordinates be the inhomogeneous Lorentz transformation (5.24), that is

$$\delta x^k=g^{kk}\omega_{kj}x^j+\omega^k \qquad (10.1)$$

All parameters ω_{kj} and ω^k are linearly independent. Function Φ^l is known to be transformed as a vector under proper Lorentz transformations and to be invariant under translations. Then, according to the definition of the vector, infinitesimal transformations must satisfy the relation

$$\delta\Phi^k=g^{kk}\omega_{kj}\Phi^j \qquad (10.2)$$

similar to (10.1). When (10.1) and (10.2) are substituted into (9.19), coefficients of each parameter ω_{kj} and ω^k must vanish independently of one another.

Let us define the following tensors:

$$T^k_{\cdot m} = \frac{\partial \mathscr{L}}{\partial (\partial \Phi_l / \partial x^k)} \frac{\partial \Phi_l}{\partial x^m} - \delta_{km} \mathscr{L} \tag{10.3}$$

$$\mu_{kmj} = T_{km} x_j - T_{kj} x_m \tag{10.4}$$

and

$$\sigma_{kmj} = \frac{\partial \mathscr{L}}{\partial (\partial \Phi^j / \partial x^k)} \Phi_m - \frac{\partial \mathscr{L}}{\partial (\partial \Phi^m / \partial x^k)} \Phi_j \tag{10.5}$$

If we take into account the antisymmetric character of parameters ω_{kj}, the result of the abovementioned substitution will take the form of equations

$$\frac{\partial T^{km}}{\partial x^k} = 0 \tag{10.6}$$

and

$$\frac{\partial}{\partial x^k} (\mu^{kmj} + \sigma^{kmj}) = 0 \tag{10.7}$$

For reasons which will be presently clear, T^{km} is called the *energy-momentum tensor*, μ^{kmj}, the *angular momentum tensor*, and σ^{kmj}, the *spin tensor* of electromagnetic field. To be precise, these tensors describe densities of the abovementioned physical quantities. We see that (10.6) follows from the invariance of the action function $\mathscr{S}$ with respect to the group of translations in the space-time, and (10.7) follows from the invariance under a group of proper Lorentz transformations. And both these equalities hold simultaneously as a result of the invariance of the action function $\mathscr{S}$ with respect to the inhomogeneous Lorentz group which includes the indicated transformations as its subgroups. Relations (10.6) and (10.7) are known as the differential laws of energy-momentum conservation and of the 4-momentum conservation. In our particular case we have proved the so-called *first Noether theorem*. This theorem states that invariance of action functions with respect to any group of transformation with a finite number of parameters corresponds to a differential conservation law. It must be kept in mind that, as is clear from the above, these conservation laws are satisfied only with functions Φ^l which have the extremal properties, that is with functions representing solutions of the Lagrange equations.

We see that the differential conservation law for a certain quantity is defined as the statement that the four-dimensional divergence of this quantity vanishes. The meaning of this terminology becomes clear if one recalls formula (7.2) which expresses charge conservation in differential form, that is expresses the continuity equation as the equality of four-dimensional divergence of current vector to zero. Formulas (10.6) and (10.7) differ only in that a similar condition is applied to quantities represented by tensors. Later we shall give a more detailed interpretation of these formulas.

10.2. Let us analyze some important features of tensors (10.3)-(10.5), in connection with the formulation of conservation laws for tensors.

First of all, let ψ^{lkm} be an arbitrary tensor of rank 3 antisymmetric with respect to indices k and l, that is $\psi^{klm} = -\psi^{lkm}$. Consider a quantity

$$T'^{km} = T^{km} + \partial\psi^{lkm}/\partial x^l \tag{10.8}$$

As $\partial^2\psi^{lkm}/\partial x^k\,\partial x^l \equiv 0$ by virtue of the assumed antisymmetric character of tensor ψ^{klm}, tensor T'^{km} satisfies the conservation law (10.6) if T^{km} satisfies this law. In other words, transformation of type (10.8) modifies the form of the energy-momentum tensor without violating the conservation law.

Let us consider now angular momentum (10.4) and calculate its divergence $\partial\mu^{kmj}/\partial x^k$ which is found in (10.7). It follows directly from (10.4) that

$$\partial\mu^{kmj}/\partial x^k = T^{jm} - T^{mj} \tag{10.9}$$

if T^{km} satisfies (10.6). Therefore if T^{km} is a symmetric tensor, the right-hand side of (10.9) vanishes. Hence, taking account of (10.7), we arrive at

$$\frac{\partial\mu^{kmj}}{\partial x^k} = 0, \quad \frac{\partial\sigma^{kmj}}{\partial x^k} = 0 \tag{10.10}$$

In other words, if the energy-momentum tensor is symmetric, then angular momentum and spin of the field are conserved independently. But if the energy-momentum tensor is not symmetric, then in principle we can try to use transformation (10.8) choosing function ψ^{lkm} in such a manner that the tensor T'^{km} be symmetric. Therefore we shall assume hereafter that $T^{km} = T^{mk}$.

Integral conservation laws can be derived from differential conservation laws (10.6) and (10.7). For this purpose it is sufficient to integrate (10.6) and (10.7) first over the cylindrical four-dimensional region shown in Fig. 3. If Σ is the total surface bounding this region, then the Gauss theorem of the type (9.5) yields

$$\oint_\Sigma T^{ik} n_k\,d\Sigma = 0, \quad \oint_\Sigma (\mu^{kmj} + \sigma^{kmj})\,n_k\,d\Sigma = 0$$

We then eliminate the lateral surface of the cylinder by infinitely expanding it between fixed spacelike surfaces Σ_1 and Σ_2. We shall restrict our analysis to the fields for which integrals of the relevant tensors over this lateral surface vanish in the mentioned limit transition. This means that the field must diminish sufficiently rapidly at the spacelike infinity. Then an integral over Σ reduces to integrals over Σ_1 and Σ_2. As usual, the application of the Gauss theorem requires a choice of the positive direction of normal $\overset{*}{n}_k$; we choose, for instance, the direction external with respect to the volume over which integration is carried out. If instead we define a vector n_k with positive direction toward the absolute future region on both hypersurfaces, we obtain

$$\int_{\Sigma_1} T^{ik} n_k \, d\Sigma_1 = \int_{\Sigma_2} T^{ik} n_k \, d\Sigma_2 \tag{10.11}$$

and a similar equality for the second integral. We introduce two definitions

$$P^i [\Sigma] = A \int_{\Sigma} T^{ik} n_k \, d\Sigma$$

$$M^{mj} [\Sigma] = B \int_{\Sigma} (\mu^{kmj} + \sigma^{kmj}) \, n_k \, d\Sigma \tag{10.12}$$

where Σ is an arbitrary spacelike hypersurface and A and B are scalar coefficients. As follows from the above arguments, if T^{ik} and $\mu^{kmj} + \sigma^{kmj}$ satisfy the differential conservation laws, then P^i and M^{mj} are independent of the choice of hypersurface Σ. This means that the *integral conservation laws* hold for P^i and M^{mj}.

10.3. The general theory presented above must be applied to a study of electromagnetic field. Clearly this application must be based on a correct choice of the Lagrangian. Invariance of the action function (9.6) under the Lorentz transformation will be realised if the Lagrangian is an invariant under these transformations since $d\Omega$ in itself already possesses this quality (see p. 75). Hence, we shall take for the Lagrangian an invariant of the Lorentz transformations formed by potentials and their derivatives. The relativistic theory presented in § 7 shows that invariant I_1 defined by (7.10_1) is directly associated with the description of electromagnetic field. Let us choose Lagrangian $\mathscr{L}$ in the form

$$\mathscr{L} = -(1/2)\, I_1 = -(1/4)\, F_{ik}F^{ik} = (1/2)\,(\mathbf{E}^2 - \mathbf{B}^2) \tag{10.13}$$

and show that this choice yields results correct from the physical standpoint. Obviously, multiplication of the Lagrangian by a constant factor, and, in particular, the reversal of its sign, does not significantly affect these results. It will be more convenient to rewrite function $\mathscr{L}$ in the form

$$\mathscr{L} = -(1/4)\, g^{ii} g^{kk} F_{ik}^2 = -(1/4)\, g^{ii} g^{kk} \left(\frac{\partial \Phi_k}{\partial x^i} - \frac{\partial \Phi_i}{\partial x^k} \right)^2 \tag{10.14}$$

in accordance with formula (7.5). This Lagrangian $\mathscr{L}$ is gauge invariant. This is obvious since tensor F_{ik} is gauge invariant. First of all, we find derivative $\partial\mathcal{L}/\partial(\partial\Phi_l/\partial x^m)$. As summation in (10.14) was carried out over all values of i and k, derivative $\partial\Phi_l/\partial x^m$ will be encountered in this sum twice for each pair of fixed values of l and m: in term F_{lm}^2 and in F_{ml}^2. As a result,

$$\frac{\partial\mathcal{L}}{\partial(\partial\Phi_l/\partial x^m)}=-g^{ll}g^{mm}\left(\frac{\partial\Phi_l}{\partial x^m}-\frac{\partial\Phi_m}{\partial x^l}\right)=-F_{ml}g^{ll}g^{mm}=-F^{ml}$$

In this formula there is no summation over l and m.

In our case the Euler-Lagrange equations (9.18) take the form

$$\frac{\partial}{\partial x^i}\frac{\partial\mathcal{L}}{\partial(\partial\Phi_k/\partial x^i)}=0 \qquad (10.15)$$

that is,

$$\frac{\partial^2\Phi_k}{\partial x_i\,\partial x^i}=\frac{\partial}{\partial x^k}\left(\frac{\partial\Phi_i}{\partial x_i}\right) \qquad (10.16)$$

Owing to the gauge invariance of function $\mathcal{L}$, we can always consider that potentials are subjected to the Lorentz condition (7.15). Then the above equality becomes a homogeneous wave equation

$$\frac{\partial^2\Phi_k}{\partial x_i\,\partial x^i}=0 \qquad (10.17)$$

for electromagnetic field in the absence of sources.

We turn now to conserved quantities. The given above calculation of the derivative shows that expression (10.3) for the energy-momentum tensor includes a sum $F^{kl}\partial\Phi_l/\partial x^m$. We transform it as follows:

$$F^{kl}\frac{\partial\Phi_l}{\partial x^m}=F^{kl}F_{ml}+\frac{\partial}{\partial x^l}(F^{kl}\Phi_m)-\Phi_m\frac{\partial F^{kl}}{\partial x^l} \qquad (10.18)$$

But the last term vanishes owing to (10.16), and the one preceding it has the form of an addition to the tensor, as we have discussed in connection with transformation (10.18). We drop this term and thus, without violating the conservation law, switch to a new tensor T^{km}. This new energy-momentum tensor is found to be symmetric:

$$T_{km}=-g^{ll}F_{kl}F_{ml}-g_{km}\mathcal{L}=-g^{ll}F_{kl}F_{ml}+(1/4)\,g_{km}F_{pq}F^{pq} \qquad (10.19)$$

Superscript k of tensor $T^{k}_{\cdot m}$ was transformed into a subscript by multiplying both sides of the above equation by g_{kk}. It follows from (10.19) that

$$T_{00}=-g^{\alpha\alpha}F_{0\alpha}F_{0\alpha}+(1/2)(\mathbf{B}^2-\mathbf{E}^2)=(1/2)(\mathbf{E}^2+\mathbf{B}^2) \qquad (10.20_1)$$

In § 3 we have formulated the law of field energy conservation; recalling it, we find that component T_{00} has the physical meaning of the density of the field energy.

Similarly, a comparison of (10.19) with definitions (7.6) and formulas of § 3 directly yields[18]

$$T_{\alpha 0}=T_{0\alpha}=-\mathbf{E}\times\mathbf{B}|_{\alpha}=-\frac{1}{c}S_{\alpha} \qquad (10.20_2)$$

$$-T_{\alpha\beta}=T^{(e)}_{\alpha\beta}+T^{(m)}_{\alpha\beta}\equiv T^{(M)}_{\alpha\beta} \qquad (10.20_3)$$

We have used above definitions (3.6), (3.9) and (3.11) of the Poynting vector **S** and the Maxwell stress tensor, which here we denote by $T^{(\mathrm{M})}_{\alpha\beta}$.

10.4. We turn now to integral conserved quantities (10.12). We shall choose hypersurface Σ as a hyperplane defined by the condition t = const. Then vector n_k is given by (9.3). Besides,

$$P^0[\Sigma]=A\int\limits_V T^{00}\,dV$$

$$P^{\alpha}[\Sigma]=A\int\limits_V T^{\alpha 0}\,dV=\frac{1}{c}A\int\limits_V S_{\alpha}\,dV \qquad (10.21)$$

Consequently, if we choose $A=c^{-1}$, then the time component of vector $P^i[\Sigma]$ is equal to the total energy of the field divided by c, and its spatial components are equal to the corresponding components of the total momentum of the field. Vector $P^i[\Sigma]$ is therefore called the *energy-momentum vector*; for a hypersurface Σ of the general type it is

$$P^i[\Sigma]=\frac{1}{c}\int\limits_{\Sigma} T^{ik}n_k\,d\Sigma \qquad (10.22)$$

If, in addition, we choose $B=c^{-1}$, then it follows from (10.22) and (10.4) that in the second formula of (10.12)[19] we can write

$$dM^{mj}=\frac{1}{c}\mu^{kmj}n_k\,d\Sigma=x^j\,dP^m-x^m\,dP^j \qquad (10.23)$$

This expression is similar to the definition of angular momentum in classical mechanics, introduced as a vector product of radius vector by momentum.

As tensor M^{mj} is antisymmetric, it has six linearly independent components. Formula (10.23) immediately shows that spatial components $dM^{\alpha\beta}$ may be represented in three-dimensional notations by vector product $\mathbf{r}\times d\mathbf{P}$, that is, they coincide with the classic definition of angular momentum density. The other three components are written as $dM^{0\alpha}=x^{\alpha}\frac{dw}{c}-ct\,dP^{\alpha}$. A comparison with classical me-

[18] Rules of operations with covariant and contravariant indices show that $T^{\alpha 0}=-T_{\alpha 0}$, $T^{\alpha\beta}=T_{\alpha\beta}=-T^{\alpha}_{\cdot\beta}$.

[19] Only density μ^{kmj} is taken into account since (10.10) are valid here.

chanics leads to a conclusion that conservation of these three quantities is an equation of motion of the center of inertia of the considered volume element of the field. Indeed, $dw/c^2 \equiv dm$ is the mass corresponding to the energy within this volume element. In a particular case when the law of conservation of $dM^{0\alpha}$ is applied to two infinitely close hypersurfaces $t =$ const, we obtain $dM^{0\alpha}/dt = 0$, that is $(dm)\,\mathbf{v} = d\mathbf{P}$, where $v^\alpha = dx^\alpha/dt$.

In three-dimensional notations, the law of conservation (10.6) for energy-momentum tensor is rewritten in the form

$$\frac{\partial T^{00}}{\partial x^0} + \frac{\partial T^{\alpha 0}}{\partial x^\alpha} = 0, \quad \frac{\partial T^{0\beta}}{\partial x^0} + \frac{\partial T^{\alpha\beta}}{\partial x^\alpha} = 0 \tag{10.24}$$

If (10.20) is taken into account, the first equation of (10.24) becomes identical to the energy conservation law (3.3), $\partial w/\partial t + \operatorname{div} \mathbf{S} = 0$, for the field in the region in space where sources $\mathbf{j}$ are absent. The second of these equations must be compared with the momentum conservation law (3.13), since $\frac{1}{c} T^{0\alpha}$ is interpreted as the field momentum density. Here again the coincidence is complete.

From the standpoint of the Noether theorem mentioned on p. 82, the above result must be interpreted as follows. If the action integral of the field is invariant with respect to transformations under the inhomogeneous Lorentz group, then different subgroups which are included into this group correspond to the following conservation laws: energy conservation corresponds to the subgroup of translations by a timelike vector; momentum conservation corresponds to the subgroup of translations in the three-dimensional space; angular momentum conservation corresponds to the subgroup of three-dimensional rotations; and finally, the integral of motion of the center of inertia corresponds to the subgroup of rotations in planes of the type $(0, \alpha)$.

We saw that there is one additional conserved quantity, namoly the spin angular momentum given by formulas (10.5) and (10.14). The physical interpretation of this quantity is supplied by the quantum field theory. In this theory the electromagnetic field is described as an ensemble of elementary particles, photons, each of photons possessing an intrinsic angular momentum called spin (in addition to the orbital angular momentum mentioned above). The classical approximation (10.5) describes the total spin momentum (cf. § 5 of the monograph by Bogoliubov and Shirkov cited on p. 78). It should be recalled again that according to formula (10.10) the spin and orbital angular momenta are conserved independently. Therefore, the spin angular momentum can be ignored if we restrict the analysis to classical electrodynamics.

Definition (10.13) of the Lagrangian used above is not the only one possible in the theory of electromagnetic field. For example, a frequently used form of

Lagrangian is

$$\mathscr{L}'+\mathscr{L}-\frac{1}{2}\left(\frac{\partial\Phi^m}{\partial x^m}\right)^2 \tag{10.25}$$

where $\mathscr{L}$ is defined, as before, by (10.13). We easily find that

$$\mathscr{L}'=-\frac{1}{2}\left(\frac{\partial\Phi^m}{\partial x^l}\right)^2 g^{mm}g^{ll}+\frac{1}{2}\frac{\partial}{\partial x^m}\left(\Phi^k\frac{\partial\Phi^m}{\partial x^k}-\Phi^m\frac{\partial\Phi^k}{\partial x^k}\right)$$

As the second term has the form of divergence, the first term alone can be used as the Lagrangian, without changing the physical content of the variational principle. If this Lagrangian is used, equations of motion immediately take the form (10.17); Lagrangian $\mathscr{L}'$ thus incorporates the Lorentz condition. Physical quantities obtained by means of $\mathscr{L}'$ will not satisfy gauge invariance. It can be shown, however, that the differences between these quantities and the gauge invariant obtained by means of $\mathscr{L}$ make no contribution into integral dynamic characteristics of the field satisfying conservation laws. This fact will not be discussed here in more detail although the Lagrangian representation in the form (10.25) proved to be convenient in quantum electrodynamics.

10.5. Our next step is the formulation of the variational principle for electromagnetic field interacting with sources in vacuo. Obviously, this formulation depends on the assumptions concerning the properties of sources. If sources are distributed in space as a "dustlike" medium discussed in § 8, the Lagrangian can be taken in the form

$$\widetilde{\mathscr{L}}=\mathscr{L}+\frac{1}{c}\Phi_i s^i-\frac{1}{2}\varkappa_0 u_i u^i$$

where $s^i=\rho_0 u^i$ is the current vector. Variation of the action function $\delta\widetilde{\mathscr{F}}=\delta\int\widetilde{\mathscr{L}}\,d\Omega$ can be calculated in two ways. If this variation is made under an assumption that vector u^i, and consequently current s^i, is not varied but potentials are varied (as was the case earlier in she analysis of the free field), then the Euler-Lagrange equations, as could be expected, take the form

$$\frac{\partial F^{ik}}{\partial x^k}=\frac{1}{c}s^i$$

instead of (10.16).

Assume now that vector Φ^i can be considered fixed and should not be varied. The integral in the formula for variation of action function can be transformed as follows:

$$\frac{1}{c}\int\Phi^i s_i\,d\Omega=\int dq\int\Phi_i\frac{dx^i}{d\tau}\,d\tau$$

Here $dq=\rho_0\,dV$ and dV is the volume of an element of medium in the co-moving reference frame. Assume that variation is carried out for world lines of the elements of medium with fixed initial and final points. With respect to this variation the extremum condition takes the form of the Euler-Lagrange equations

$$\frac{d}{d\tau}\frac{\partial\widetilde{\mathscr{L}}}{\partial u^i}=\frac{\partial\widetilde{\mathscr{L}}}{\partial x^i}$$

Substitution of the Lagrangian $\widetilde{\mathscr{L}}$ yields

$$-\varkappa_0\frac{du^i}{d\tau}+\frac{\rho_0}{c}\frac{d\Phi^i}{d\tau}=\frac{1}{c}s_k\frac{\partial\Phi^k}{\partial x^i}$$

that is,

$$\varkappa_0\frac{du^i}{d\tau}=\frac{1}{c}s_kF^{ki}$$

since

$$\frac{d\Phi^i}{d\tau}=\frac{\partial\Phi^i}{\partial x^k}u^k$$

In fact we have repeated above the arguments of § 8 for the case of Lagrangian $\widetilde{\mathscr{L}}$ taking into account additional conditions with respect to the mode of variation. Lagrangian $\widetilde{\mathscr{L}}$ can be used to derive the conservation laws for a system of fields and charges. This derivation will not be discussed here since in principle it does not involve anything new. Note only that, for instance, conservation of the energy-momentum tensor takes in this case the form

$$\frac{\partial}{\partial x^k}(T^{ik}+T^{ik}_{\text{source}})=0$$

where $T^{ik}_{\text{source}}=\varkappa_0u^iu^k$ is the energy-momentum tensor of the charges, that is of the "dustlike" medium.

No doubt, we could analyze also distributions of sources with more complicated properties than those of a "dustlike" medium, for example, the properties of the ideal liquid. The rest mass $\varkappa_0$ then should be considered as a function of proper time, because of the interaction between elements of the medium. However, we shall not be able to use this theory in the later chapters, and so need not elaborate it here[20].

---

[20] Presentation of the variational principle for different properties of the medium and the definition of the energy-momentum tensor see in §§ 32, 46-49 of the monograph by V. A. Fock, cited on p. 46.

CHAPTER 3

STATIC FIELDS. SOLUTION OF THE WAVE EQUATION. RADIATION FIELD

§ 11. Electrostatic field

11.1. We have demonstrated in § 2 that wave equations (2.6_1) and (2.6_2) for Lorentz-gauged potentials can be considered as the basic equations describing behavior of electromagnetic field in homogeneous isotropic media. These equations become resolvable if we demand that the initial and boundary conditions correspond to a physically meaningful situation. Functions $\rho(\mathbf{r}, t)$ and $\mathbf{j}(\mathbf{r}, t)$ which determine spatial distribution of field sources and evolution of this distribution, will be considered known to an extent required for the solution.

Assume that $\mathbf{j} = 0$, magnetic field is absent completely, and charge density ρ and scalar potential φ are independent of time. Equation (2.6_2) then takes the form of the Poisson equation

$$\Delta\varphi = -\frac{1}{\varepsilon}\rho(\mathbf{r}) \tag{11.1}$$

which is the fundamental equation of electrostatic field. Indeed, if function $\varphi(\mathbf{r})$ is known, it is possible to calculate electric field strength from the formula

$$\mathbf{E} = -\text{grad}\,\varphi \tag{11.2}$$

We shall remind the reader some facts from the theory of equations of mathematical physics dealing with the solution of the Poisson equation (11.1).

The essential step is to find the so-called fundamental solution of the Poisson equation. By definition, it is a solution corresponding to a pointlike source in infinite space (when boundary conditions reduce to the requirement that the solutions diminish sufficiently rapidly at infinity). The distribution density of a pointlike source in space is given by the delta function $\delta(\mathbf{r} - \mathbf{r}')$, where $\mathbf{r}'$ is the radius vector of the point at which the source is located. In other words, the fundamental solution of equation (11.1) is a particular solution $f(\mathbf{r}, \mathbf{r}')$ of the nonhomogeneous equation

$$\Delta f(\mathbf{r}, \mathbf{r}') = -\frac{1}{\varepsilon}\delta(\mathbf{r} - \mathbf{r}') \tag{11.1'}$$

which diminishes rapidly enough when $|\mathbf{r}-\mathbf{r}'|\to\infty$. It can be shown that

$$f(\mathbf{r}, \mathbf{r}')=\frac{1}{4\pi\varepsilon\,|\mathbf{r}-\mathbf{r}'|} \tag{11.3}$$

Indeed, denote $\mathbf{R}=\mathbf{r}-\mathbf{r}'$. If $\mathbf{R}\neq 0$, we always have $\Delta\left(\frac{1}{R}\right)=0$, which is readily verified by direct calculations. However, $\int\Delta\left(\frac{1}{R}\right)dV'\neq 0$, owing to a singularity of the integrand at $R=0$ (we differentiate with respect to variable $\mathbf{r}$ and integrate over variable $\mathbf{r}'$). This integral can be evaluated by using the following (far from rigorous) arguments. Let us integrate over a three-dimensional volume incorporating point $\mathbf{r}'=\mathbf{r}$ and bounded by a closed sphere σ with the center at this point. According to the Gauss theorem, $\int\Delta\left(\frac{1}{R}\right)dV'=\oint\frac{\partial}{\partial n}\left(\frac{1}{R}\right)d\sigma$. But since $\frac{\partial}{\partial n}\left(\frac{1}{R}\right)=-\frac{1}{R^2}$ and $d\sigma=R^2\,d\Omega$ (here $d\Omega$ is an element of solid angle on the sphere), we obtain $\int\Delta\left(\frac{1}{R}\right)dV'=-4\pi$, that is $\Delta\left(\frac{1}{R}\right)=-4\pi\delta(\mathbf{R})$. We shall not try to substantiate this result more rigorously.

It immediately follows from (11.3) that the solution of equation (11.1) in infinite space is of the form

$$\varphi(\mathbf{r})=\int f(\mathbf{r}, \mathbf{r}')\,\rho(\mathbf{r}')\,dV'=\frac{1}{4\pi\varepsilon}\int\frac{\rho(\mathbf{r}')\,dV'}{|\mathbf{r}-\mathbf{r}'|} \tag{11.4}$$

Here and below $\mathbf{r}$ denotes the radius vector of the observation point, and $\mathbf{r}'$ is the radius vector of the point where the source is located[1]. If we apply operator Δ to both sides of (11.4) and take into account (11.1′) and properties of delta function, then (11.4) indeed proves to be a solution of equation (11.1). In our notation differential operations with respect to $\mathbf{r}'$ are primed. It is important to remember that

$$\operatorname{grad} f(\mathbf{r}-\mathbf{r}')=-\operatorname{grad}' f(\mathbf{r}-\mathbf{r}') \tag{11.5}$$

This is a frequently used relation.

The general solution of the nonhomogeneous equation (11.1′) can be rewritten in the form

$$G(\mathbf{r}, \mathbf{r}')=f(\mathbf{r}, \mathbf{r}')+F(\mathbf{r}, \mathbf{r}') \tag{11.6}$$

where F is the general solution of a homogeneous equation $\Delta' F=0$ (here and later it will be more convenient to use operator Δ' and consider functions G and F to be symmetric).

Function $G(\mathbf{r}, \mathbf{r}')$ is called *Green's function* for the Poisson equation. Using Green's function, we can obtain a solution of this equation in a finite volume for specific boundary conditions on the surface σ bounding this volume. The Dirichlet conditions, when potential φ

[1] A brief summary of formulas expressing elementary properties of delta function is given in Appendix C.

is fixed on σ, and the Neumann conditions, when the normal derivative of potential $\partial\varphi/\partial n$, that is the normal component of electric strength (11.2), is given on σ, are often used as such boundary conditions.

The Dirichlet and Neumann conditions provide unique solution of the boundary problem. Let us assume that two solutions, φ_1 and φ_2, of equation (11.1) are found, satisfying the same boundary conditions on surface σ.

We denote $u = \varphi_1 - \varphi_2$. We use formula (B.27), assuming $\psi = \varphi = u$. This formula now becomes

$$\int_V (u\Delta' u + |\text{grad}' u|^2)\, dV' = \oint_\sigma u \frac{\partial u}{\partial n'}\, d\sigma' \tag{11.7}$$

But within V we have $\Delta' u = 0$, and at the boundary σ of this region either $u = 0$ (if φ_1 and φ_2 satisfy the same Dirichlet conditions), or $\partial u/\partial n' = 0$ (in the case of Neumann conditions). In both cases $\int_V |\text{grad}' u|^2\, dV' = 0$ should hold, which is only possible if $\text{grad}' u = 0$ within V, that is if u within this volume is constant. In the case of Dirichlet conditions we conclude that $u = 0$, that is $\varphi_1 = \varphi_2$, while for the Neumann conditions φ_1 may differ from φ_2 by an additive constant, which is unimportant.

Let us apply now to our region V Green's formula (B.28) assuming that φ is an arbitrary solution of equation (11.1) and ψ is an arbitrary Green's function of type (11.6), satisfying equation (11.1′). By using the properties of delta function, we arrive at the integral equation for function φ:

$$\varphi(\mathbf{r}) = \int_V \rho(\mathbf{r}')\, G(\mathbf{r}, \mathbf{r}')\, dV' + \varepsilon \oint \left[G(\mathbf{r}, \mathbf{r}') \frac{\partial \varphi}{\partial n} - \varphi(\mathbf{r}') \frac{\partial G(\mathbf{r}, \mathbf{r}')}{\partial n'} \right] d\sigma' \tag{11.8}$$

With a proper choice of Green's function G, we can satisfy specific boundary conditions. Thus, if Green's function G_D has a property $G_D(\mathbf{r}, \mathbf{r}') = 0$ for $\mathbf{r}' \in \sigma$ then the first term in the integral over σ will vanish and we shall have a solution corresponding to Dirichlet boundary conditions. Derivation of the integral formula for the Neumann boundary conditions is more complicated. Finally, if we assume $G = f$, formula (11.8) becomes

$$\varphi(\mathbf{r}) = \frac{1}{4\pi\varepsilon} \int_V \frac{\rho(\mathbf{r}')\, dV'}{|\mathbf{r} - \mathbf{r}'|} + \frac{1}{4\pi} \oint_\sigma \left[\frac{1}{|\mathbf{r} - \mathbf{r}'|} \frac{\partial \varphi}{\partial n'} - \varphi(\mathbf{r}') \frac{\partial}{\partial n'} \frac{1}{|\mathbf{r} - \mathbf{r}'|} \right] d\sigma' \tag{11.9}$$

It is clear from the above arguments that, since solution φ of equation (11.1) is uniquely specified by setting on surface σ the value of only function φ or only of derivative $\partial\varphi/\partial n'$, equation (11.9) must be considered as an integral relation which must be satisfied by the solution corresponding to the chosen value of $G = f$.

If the charge is distributed not in a three-dimensional volume as has been assumed until now, but over a two-dimensional surface σ with surface density $\lambda(\mathbf{r}')$, then, by analogy with the first term in (11.9) or to formula (11.4) for infinite space, we can write

$$\varphi(\mathbf{r}) = \frac{1}{4\pi\varepsilon} \int_\sigma \frac{\lambda(\mathbf{r}')\,d\sigma'}{|\mathbf{r}-\mathbf{r}'|} \tag{11.10}$$

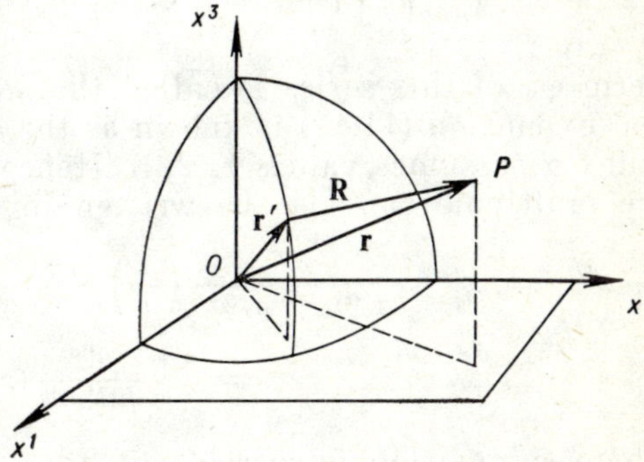

Fig. 4

This is the so-called *potential of the elementary layer* (that is, of a layer of charges distributed over surface σ). Terms of the type (11.10) must be added to the right-hand side of formula (11.8) when volume V contains two-dimensional elementary layers. At the same time, by analogy with (11.10), the first term in the surface integral of (11.9) can be referred to as the potential of an elementary layer deposited on the boundary surface σ with the surface charge density numerically equal to $\partial\varphi/\partial n'$. The second term in this integral also can be given a physical interpretation which we shall discuss later.

11.2. Expansion of potential in multipole potentials. Let us apply formula (11.4) to calculation of potential in one important particular case when the whole charge generating the field is located inside a sphere with finite radius A (bounded charge distribution); we are interested in the potential outside of this sphere. For the origin of coordinates we take any point O inside this sphere (it will be convenient, however, to choose the center of this sphere as the origin), and we shall determine the values of the field at the observation point P outside of this sphere (Fig. 4). If quantity $1/|\mathbf{r}-\mathbf{r}'|$ is considered a function of components x'^α of vector $\mathbf{r}'$, we can expand this function into the Taylor series in the neighbourhood of point $\mathbf{r}' = 0$:

$$f(\mathbf{r}') = \frac{1}{r} - x'^\alpha \frac{\partial}{\partial x^\alpha}\left(\frac{1}{r}\right) + \frac{x'^\alpha x'^\beta}{2!}\frac{\partial^2}{\partial x^\alpha \partial x^\beta}\left(\frac{1}{r}\right) - \cdots + \frac{(-1)^n}{n!} x'^{\alpha_1} \ldots x'^{\alpha_n} \frac{\partial^n}{\partial x^{\alpha_1} \ldots \partial x^{\alpha_n}}\left(\frac{1}{r}\right) + \cdots \tag{11.11}$$

Note that the chosen constraints $r > A$, $r' < A$ guarantee uniform convergence of the above series. Substitution of the series into the initial equation (11.4) yields an infinite series for potential

$$\varphi = \sum_{n=0}^{\infty} \varphi_n \tag{11.12}$$

where

$$\varphi_n = \frac{(-1)^n}{4\pi\varepsilon n!} \int \rho(\mathbf{r}') x'^{\alpha_1} \ldots x'^{\alpha_n} dV' \frac{\partial^n}{\partial x^{\alpha_1} \ldots \partial x^{\alpha_n}} \left(\frac{1}{r}\right) \tag{11.12'}$$

Term φ_n of this series is called the *multipole potential of order* $2n$, and expansion (11.12) is known as the *expansion in multipoles*. Each index α_k assumes values 1, 2, 3. Hence, the sum in the definition of the multipole can also be written in the form

$$\frac{1}{n!} x'^{\alpha_1} \ldots x'^{\alpha_n} \frac{\partial^n}{\partial x^{\alpha_1} \ldots \partial x^{\alpha_n}} \left(\frac{1}{r}\right) = \sum \frac{1}{k!l!m!} (x_1')^k (x_2')^l (x_3')^m \frac{\partial^n}{\partial x_1^k \partial x_2^l \partial x_3^m} \left(\frac{1}{r}\right)$$

where summation is carried out over all positive integers k, l, m, for which $k + l + m = n$. The theory of spheric functions, when applied to an analysis of multipoles, is most efficient in revealing their characteristic properties. Here, however, we restrict the analysis to several simplest cases.

Let us discuss in more detail the first two terms in expansion (11.12). First, the term

$$\varphi_0 = \frac{1}{4\pi\varepsilon r} \int \rho(\mathbf{r}') dV'$$

has the same form as the potential of a pointlike charge $q = \int \rho(\mathbf{r}') dV'$ located at the origin. The expression for φ_0 can be derived if $\rho(\mathbf{r}') = q\delta(\mathbf{r}')$ is substituted into (11.4).

In the vector notation, the next term takes the form

$$\varphi_1 = -\frac{1}{4\pi\varepsilon} \left(\mathbf{p} \operatorname{grad} \frac{1}{r}\right) \tag{11.13}$$

where vector $\mathbf{p}$ is defined by

$$\mathbf{p} = \int \rho(\mathbf{r}') \mathbf{r}' dV' \tag{11.14}$$

and is referred to as the *dipole moment* of charge distribution.

An expression of the form of (11.13) can also be derived from the general formula (11.4). Assume that a pointlike negative charge $-q$ is placed at the origin of coordinates, and a pointlike charge $+q$ at a point with radius vector $\mathbf{l}$. The density of such distribution of charges is written in the form

$$\rho(\mathbf{r}') = q\,[\delta(\mathbf{r}' - \mathbf{l}) - \delta(\mathbf{r}')]$$

By assuming the magnitude of vector $\mathbf{l}$ to be small, we can formally expand the first term into a Taylor series in $\mathbf{l}$ and retain only the first two terms of this expansion, so that

$$\rho(\mathbf{r}') \simeq -q\,(\mathbf{l}, \text{grad}'\,\delta(\mathbf{r}')) \tag{11.15}$$

Now let the magnitude of vector $\mathbf{l}$ tend to zero, and that of charge q to infinity in such a manner that the limit of their product, $\lim_{q\to\infty,\, l\to 0} q\mathbf{l} \equiv \mathbf{p}$, exists. Vector $\mathbf{p}$ is then called the dipole moment of the *pointlike dipole* (in our case it is located at the origin). After this limiting transition we substitute (11.15) into (11.4) and use the properties of a derivative of delta function (see Appendix C) and obtain, having taken into account (11.5),

$$\begin{aligned}\varphi(\mathbf{r}) &= -\frac{1}{4\pi\varepsilon}\int \frac{(\mathbf{p}, \text{grad}'\,\delta(\mathbf{r}'))}{|\mathbf{r}-\mathbf{r}'|}\,dV' = \frac{1}{4\pi\varepsilon}\left(\mathbf{p}, \text{grad}'\,\frac{1}{|\mathbf{r}-\mathbf{r}'|}\bigg|_{\mathbf{r}'=0}\right)\\ &= -\frac{1}{4\pi\varepsilon}\left(\mathbf{p}, \text{grad}\,\frac{1}{|\mathbf{r}-\mathbf{r}'|}\bigg|_{\mathbf{r}'=0}\right) = -\frac{1}{4\pi\varepsilon}\left(\mathbf{p}, \text{grad}\,\frac{1}{r}\right)\end{aligned} \tag{11.16}$$

We immediately note that formulas (11.13) and (11.16) formally are identical. Therefore, term φ_1 of the expansion of the potential of bounded charge distribution in multipoles can be regarded as the potential of a pointlike dipole located at the point chosen for the origin of coordinates. Numerically, the dipole moment of this dipole is given by formula (11.14). Notice that equation (11.14) does not unambiguously determine the dipole moment, since it depends on the choice of the origin. It is readily found that if the origin is displaced to point $\mathbf{r}_0$, so that $\mathbf{r}' = \mathbf{r}_0 + \mathbf{r}''$, the dipole moment is modified to

$$\mathbf{p}' = \int \rho(\mathbf{r}'')\,\mathbf{r}''\,dV'' = \mathbf{p} - q\mathbf{r}_0$$

where q is the net charge within sphere σ. The dipole moment is independent of the choice of the origin only for a system with zero net charge ($q = 0$). The same is true for multipoles of higher orders (see below). A multipole of order n is determined unambiguously only if all the multipoles of order below n are equal to zero.

Expression (11.16) can be used to obtain directly some generalizations. If, for example, vector $\mathbf{p}$ is considered a function of spatial coordinates, we can operate in terms of distributions of pointlike dipoles in three-dimensional space or on a two-dimensional surface. In the case of a volume distribution of dipoles, electric potentials produced by such distributions in the point of observation $\mathbf{r}$ will be given by the formula

$$\varphi(\mathbf{r}) = -\frac{1}{4\pi\varepsilon} \int_V \left(\mathbf{p}(\mathbf{r}'), \operatorname{grad} \frac{1}{|\mathbf{r}-\mathbf{r}'|}\right) dV' \qquad (11.17)$$

If dipoles are distributed over surface σ, we only have to replace integration over the volume by integration over the surface. This surface on which dipoles are distributed is called the *double layer*. It can be noticed that the formal structure of the double layer potential is found in the second term of the surface integral in formula (11.9). We also have to assume that dipoles are directed normally to surface σ and that $p_{n'}(\mathbf{r}') = p(\mathbf{r}') = -\varphi(\mathbf{r}')$. On the whole, therefore, the surface integral is a physical picture of an elementary and a double layer placed simultaneously on surface σ.

The electrostatic field strength $\mathbf{E}$, calculated, as we have mentioned above, by means of formula (11.2), can be written in the discussed cases as

$$\mathbf{E}(\mathbf{r}) = -\frac{1}{4\pi\varepsilon} \int \rho(\mathbf{r}') \operatorname{grad} \frac{1}{|\mathbf{r}-\mathbf{r}'|} dV' \qquad (11.18_1)$$

$$\mathbf{E}(\mathbf{r}) = \frac{1}{4\pi\varepsilon} \operatorname{grad}\left(\mathbf{p}, \operatorname{grad}\frac{1}{r}\right) = -\frac{1}{4\pi\varepsilon} \operatorname{grad} \frac{(\mathbf{pr})}{r^3}$$

$$= -\frac{1}{4\pi\varepsilon} (\mathbf{p}, \operatorname{grad}) \frac{\mathbf{r}}{r^3} \qquad (11.18_2)$$

$$\mathbf{E}(\mathbf{r}) = -\frac{1}{4\pi\varepsilon} \int_V (\mathbf{p}(\mathbf{r}'), \operatorname{grad}) \frac{(\mathbf{r}-\mathbf{r}')}{|\mathbf{r}-\mathbf{r}'|^3} dV' \qquad (11.18_3)$$

The first of these formulas is obtained from (11.4), the second, which gives the field strength produced by a pointlike dipole, is obtained from (11.16) via (B.18); and finally, the third of them is a corollary of (11.17).

We shall use one important theorem which states that if charge density $\rho(\mathbf{r}')$ is a bounded and piecewise-continuous function of spatial coordinates[2], then potential $\varphi(\mathbf{r})$ and field strength $\mathbf{E}(\mathbf{r})$, given by formulas (11.4) and (11.18_1), are finite continuous functions of $\mathbf{r}$. A similar property is characteristic of the potential produced by a distribution of dipoles (see formulas (11.17) and (11.18_3)), but the conditions of the theorem must now hold for the distribution density of dipoles, $\mathbf{p}(\mathbf{r}')$. If function $\lambda(\mathbf{r}')$ determining the potential of elementary layer in expression (11.10) is bounded and piecewise-continuous on surface σ, then this potential is bounded and continuous everywhere, and therefore, has no discontinuity along the lines crossing surface σ. On the opposite, field strength produced by an elementary layer has a finite discontinuity on such lines. This

[2] That is, there is a finite number of regions in each of which function $\rho(\mathbf{r}')$ is continuous.

directly follows from boundary conditions for the Maxwell equations discussed in § 4. And finally, the double layer has the following property: its potential has a finite discontinuity across this layer.

11.3. In analogy to our treatment of the field of two charges with opposite signs on p. 94, let us turn to two pointlike dipoles with oppositely directed dipole moments with equal magnitudes, located at points $\mathbf{r}' = 0$ and $\mathbf{r}' = \mathbf{l}$. The integrand in (11.17) can then be transformed via an expansion

$$p^{\alpha}(\mathbf{r}') = p^{\alpha}[\delta(\mathbf{r}'-\mathbf{l}) - \delta(\mathbf{r}')] \simeq -p^{\alpha}l^{\beta}\frac{\partial}{\partial x'^{\beta}}\delta(\mathbf{r}')$$

Let us assume that there exists a limit $d^{\alpha\beta} = \lim\limits_{p\to\infty,\, l\to 0} p^{\alpha}l^{\beta}$ when all considered dipoles are placed at one and the same point, and the magnitude of the dipole moment rises infinitely. Tensor $d^{\alpha\beta}$ is called the *quadrupole moment* of the obtained charge distribution known as the *quadrupole*[3]. In this limit

$$\varphi(\mathbf{r}) = \frac{1}{4\pi\varepsilon} d^{\alpha\beta}\int_V \frac{\partial}{\partial x'^{\beta}}\delta(\mathbf{r}')\frac{\partial}{\partial x^{\alpha}}\frac{1}{|\mathbf{r}-\mathbf{r}'|}\,dV'$$

$$= -\frac{1}{4\pi\varepsilon} d^{\alpha\beta}\frac{\partial^2}{\partial x'^{\beta}\partial x^{\alpha}}\frac{1}{|\mathbf{r}-\mathbf{r}'|}\bigg|_{\mathbf{r}'=0} = \frac{1}{4\pi\varepsilon} d^{\alpha\beta}\frac{\partial^2}{\partial x^{\alpha}\partial x^{\beta}}\frac{1}{r}$$

A comparison with formula (11.12) shows that a term in the expansion of potential over multipoles can be considered as the potential of a quadrupole. This process of formation of multipoles of higher and higher order can be carried on indefinitely.

Tensor $d^{\alpha\beta}$ can be considered symmetric since its antisymmetric part is cancelled out in the summation; in other words, the product $p^{\alpha}l^{\beta}$ can be symmetrized before carrying out the limit operation. In the case of continuous distribution of charges we have to assume

$$d_{\alpha\beta} = \frac{1}{2}\int x'_{\alpha}x'_{\beta}\rho(\mathbf{r}')\,dV' \qquad (11.19_1)$$

Consider a tensor

$$Q_{\alpha\beta} = Q_{\beta\alpha} = \int \rho(\mathbf{r}')(3x'_{\alpha}x'_{\beta} - r'^2\delta_{\alpha\beta})\,dV' \qquad (11.19_2)$$

Using (11.19_1) and (11.19_2) it can be readily shown that

$$\varphi_2(\mathbf{r}) = \frac{d_{\alpha\beta}}{4\pi\varepsilon}\frac{\partial^2}{\partial x_{\alpha}\,\partial x_{\beta}}\frac{1}{r} = \frac{1}{6}\frac{Q_{\alpha\beta}}{4\pi\varepsilon}\frac{\partial^2}{\partial x_{\alpha}\,\partial x_{\beta}}\frac{1}{r}$$

Hence, tensor $Q_{\alpha\beta}$ can be chosen as a definition of the quadrupole moment of continuous charge distribution. Relation (11.19_2) shows that this tensor has the following property: $\sum_{\alpha} Q_{\alpha\alpha} = 0$, that is it has only five linearly independent components.

[3] Note that a quantity usually referred to as the quadrupole moment is $2d_{\alpha\beta}$.

To conclude, we shall find force **F** with which a given electrostatic field **E** acts on a dipole. It must be clear from the foregoing that this force is given by the relation

$$\mathbf{F}(\mathbf{r}) = \lim_{q\to\infty,\, l\to\infty} q\,[\mathbf{E}(\mathbf{r}+\mathbf{l}) - \mathbf{E}(\mathbf{r})]$$

if the dipole is located at point **r**. By expanding **F** into a Taylor series and recalling the definition of dipole moment **p**, we obtain

$$\mathbf{F} = (\mathbf{p}\cdot\text{grad})\,\mathbf{E} \tag{11.20}$$

For a static field formula (11.20) can be written in the form $\mathbf{F} = -\text{grad}\, U$, where

$$U = -\mathbf{p}\cdot\mathbf{E} \tag{11.21}$$

is the potential energy of the dipole in field **E**. The torque acting on the dipole by field **E** can be calculated by the formula

$$\mathbf{N} = \lim_{q\to\infty,\, l\to 0} \mathbf{l}\times(q\mathbf{E}) = \mathbf{p}\times\mathbf{E} \tag{11.22}$$

because the forces $+\,|q|\mathbf{E}$ and $-\,|q|\mathbf{E}$, having equal magnitudes to within infinitesimal corrections, applied in the opposite directions to the positive and the negative charges, constitute a mechanical couple.

Later we shall return to the study of electrostatic field, and in particular to its energy properties, the Maxwell stress tensor, and so on (see Chapter 7).

§ 12. Magnetostatic field generated by currents

12.1. Let us turn now to the static magnetic field. The source of this field is a distribution of currents with time-independent density **j** (**r**) considered in an inertial reference frame. If the medium is homogeneous and isotropic, that is $\mathbf{B} = \mu\mathbf{H}$ and the magnetic permeability μ is constant, the fundamental equations (1.19) and (1.20) of static magnetic field take the form

$$\text{div}\,\mathbf{B} = 0, \quad \text{curl}\,\mathbf{B} = \frac{\mu}{\alpha}\,\mathbf{j} \tag{12.1}$$

If $\partial\mathbf{A}/\partial t = 0$ (this condition ensures the absence of the electric field of external sources in the chosen reference frame), the vector potential **A** satisfies, in Cartesian coordinates, equation (2.6_1) in the form

$$\Delta\mathbf{A} = -\frac{\mu}{\alpha}\,\mathbf{j} \tag{12.2}$$

Note that charge density ρ is assumed to be zero[4].

[4] As far as the definition of vector potential is concerned, compare the end of Subsection 2.2, and in particular, equations (2.12′). Note that condition $\partial\mathbf{B}/\partial t = 0$ does not lead to an immediate conclusion that $\partial\mathbf{A}/\partial t = 0$. If, how-

In the infinite space the solution of equation (12.2) satisfying the condition of sufficiently rapid decrease at infinity, can be constructed for each component of vector **A** independently, in exactly the same way as this was done in § 11 for scalar potential φ. The result takes the form

$$\mathbf{A}(\mathbf{r}) = \frac{\mu}{4\pi\alpha} \int \frac{\mathbf{j}(\mathbf{r}')}{|\mathbf{r}-\mathbf{r}'|} dV' \tag{12.3}$$

Magnetic induction vector $\mathbf{B}(\mathbf{r})$ is calculated by formula (2.1), $\mathbf{B} = \text{curl}\,\mathbf{A}$. The differential operation curl is carried out with respect to radius vector **r** of the observation point. By using formula (B.14_3), we obtain

$$\mathbf{B}(\mathbf{r}) = \frac{\mu}{4\pi\alpha} \int \text{grad} \frac{1}{|\mathbf{r}-\mathbf{r}'|} \times \mathbf{j}(\mathbf{r}')\, dV' \tag{12.4}$$

Consider a case in which current distribution can be represented by a certain number of closed linear contours. An element of volume of each of these contours can be written in the form $dV' = d\sigma'\, ds'$, where $d\sigma'$ is the area of cross section, and ds' is an element of tangent at a given point. Assuming that the directions of current density vector **j** and tangent vector **s**$'$ coincide, and defining the total current intensity I by $I = j\, d\sigma$, we obtain a relation which will be useful later:

$$\mathbf{j}(\mathbf{r}')\, dV' = I\, d\mathbf{s}' \tag{12.5}$$

It is clear from the continuity equation that $I = \text{const}$ along the whole contour. Taking this into account, we can derive a particular form of (12.4) which gives strength $\mathbf{B}(\mathbf{r})$ of the magnetic field produced at the point of observation **r** by a given closed current loop:

$$\mathbf{B}(\mathbf{r}) = \frac{\mu I}{4\pi\alpha} \oint \left(\text{grad} \frac{1}{|\mathbf{r}-\mathbf{r}'|}\right) \times d\mathbf{s}' \tag{12.6}$$

This is the mathematical form of the *Ampère law*, also called Laplace's law, or Biot Savart relation. Often this law is written in the differential form:

$$d\mathbf{B}(\mathbf{r}) = \frac{\mu I}{4\pi\alpha} \frac{d\mathbf{s}' \times (\mathbf{r}-\mathbf{r}')}{|\mathbf{r}-\mathbf{r}'|^3} \tag{12.7}$$

Replace now in (12.7) I by I_1, $d\mathbf{s}'$ by $d\mathbf{s}_1$, and $\mathbf{r}'$ by $\mathbf{r}_1$, and assume that at point $\mathbf{r}_2$ there is an electric current with density $\mathbf{j}_2(\mathbf{r}_2)$. Then field strength $\mathbf{B}(\mathbf{r}_2)$ determines, by formula (3.13), the bulk

---

ever, we decide at the start that the fields arriving from the "outside" of the chosen reference frame are ignored (this is possible because equations are linear), then (12.2) is a direct corollary of (12.1) under Coulomb gauge $\text{div}\,\mathbf{A} = 0$.

density of force $\mathbf{f}_{12}(\mathbf{r}_2)$ applied to this current:

$$\mathbf{f}_{12}(\mathbf{r}_2)=\frac{1}{\alpha}\,\mathbf{j}_2(\mathbf{r}_2)\times\mathbf{B}(\mathbf{r}_2)$$

$$=\frac{\mu I_1}{4\pi\alpha^2}\oint \mathbf{j}_2(\mathbf{r}_2)\times\left(d\mathbf{s}_1\times\operatorname{grad}\frac{1}{|\mathbf{r}_2-\mathbf{r}_1|}\right) \quad (12.8)$$

The total force applied by the first loop to the second can be found by integrating $\mathbf{f}_{12}(\mathbf{r}_2)$ along the entire second loop. By using the relation $\mathbf{j}_2(\mathbf{r}_2)\,dV_2=I_2\,d\mathbf{s}_2$, which is an analog of (12.5), we obtain

$$\mathbf{F}_{12}=\int \mathbf{f}_{12}\,dV_2=\frac{\mu I_1}{4\pi\alpha^2}\int \mathbf{j}_2\,dV_2\times\oint d\mathbf{s}_1\times\frac{\mathbf{R}_{12}}{R_{12}^3}$$

$$=\frac{\mu I_1 I_2}{4\pi\alpha^2}\oint_1\oint_2 d\mathbf{s}_2\times\left(d\mathbf{s}_1\times\frac{\mathbf{R}_{12}}{R_{12}^3}\right) \quad (12.9)$$

where $\mathbf{R}_{12}\equiv\mathbf{r}_2-\mathbf{r}_1$. By using (B.6) we can rewrite integrals in (12.9) in the form

$$\oint_1\oint_2 \frac{d\mathbf{s}_1\,(d\mathbf{s}_2\cdot\mathbf{R}_{12})}{R_{12}^3}-\oint_1\oint_2\frac{\mathbf{R}_{12}}{R_{12}^3}\,(d\mathbf{s}_1\cdot d\mathbf{s}_2)$$

However,

$$d\mathbf{s}_2\cdot\operatorname{grad}\frac{1}{R_{12}}=\frac{d}{ds_2}\left(\frac{1}{R_{12}}\right)ds_2$$

so that the first term contains an integral of a complete differential over a closed contour and thus vanishes. Hence,

$$\mathbf{F}_{12}=-\frac{\mu I_1 I_2}{4\pi\alpha^2}\oint_1\oint_2\frac{\mathbf{R}_{12}}{R_{12}^3}\,(d\mathbf{s}_1\cdot d\mathbf{s}_2)=-\mathbf{F}_{21} \quad (12.10)$$

where $\mathbf{F}_{21}$ is the force applied by the second loop to the first; it is obtained by permuting indices 1 and 2 and taking into account that $\mathbf{R}_{21}=-\mathbf{R}_{12}$. Hence, forces of interaction between two closed loops satisfy Newton's third law. In particular, we find from (12.10) that parallel currents are attracted, while antiparallel currents are repulsed.

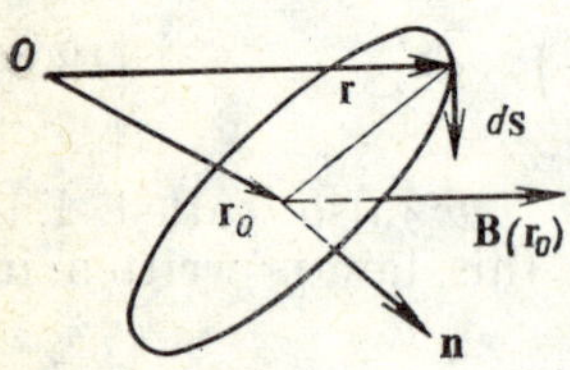

Fig. 5

12.2. Consider a sufficiently small closed current loop (Fig. 5). By analogy with the preceding derivation, we shall write for the force applied to element $d\mathbf{s}$ of the contour in magnetic field the expression

$$d\mathbf{F}=\frac{1}{\alpha}\,\mathbf{j}\,dV\times\mathbf{B}=\frac{1}{\alpha}\,I\,d\mathbf{s}\times\mathbf{B}$$

Force $d\mathbf{F}$ corresponds to torque

$$d\mathbf{N}=\mathbf{r}\times d\mathbf{F}=\frac{I}{\alpha}\,\mathbf{r}\times(d\mathbf{r}\times\mathbf{B})$$

since $d\mathbf{s} = d\mathbf{r}$. By using (B.6), we find that the total torque applied to the loop is

$$\mathbf{N} = \frac{1}{\alpha} I \oint \mathbf{r} \times (d\mathbf{r} \times \mathbf{B}) = \frac{I}{\alpha} \oint (\mathbf{r} \cdot \mathbf{B})\, d\mathbf{r} - \frac{I}{\alpha} \oint \mathbf{B}\, (\mathbf{r} \cdot d\mathbf{r})$$

If the loop is infinitely small for any point $\mathbf{r}_0$ inside the loop, we can assume that $\mathbf{B}(\mathbf{r}) \simeq \mathbf{B}(\mathbf{r}_0)$ everywhere on the loop. The second term is then transformed to $\mathbf{B}(\mathbf{r}_0) \oint d(\mathbf{r}^2/2) = 0$. The integrand in the first integral can be recast, again via (B.6), to the following form:

$$\mathbf{r} \cdot \mathbf{B}\, d\mathbf{r} = (1/2)\, [(\mathbf{r} \times d\mathbf{r}) \times \mathbf{B}] + (1/2)\, d\, ((\mathbf{r} \cdot \mathbf{B})\, \mathbf{r}) \tag{12.11}$$

Here we have taken into account that $d\mathbf{B} = 0$. An integral of a total differential over a closed contour is zero, as that finally we obtain

$$\mathbf{N} = \mathbf{m} \times \mathbf{B} \tag{12.12}$$

where the quantity

$$\mathbf{m} = \frac{I}{2\alpha} \oint \mathbf{r} \times d\mathbf{r} = \frac{I}{\alpha} \int \mathbf{n}\, d\sigma = \frac{I}{\alpha} S\mathbf{n} \tag{12.13}$$

is called the *magnetic moment* of the considered small current loop. Vector $\mathbf{n}$ in (12.13) is the normal to the plane containing the loop, and $d\sigma$ is an element of area. In other words, magnetic moment is proportional to the area enclosed by the loop[5].

Let us compare formulas (12.12) and (11.22). The comparison shows that the effect of magnetic field $\mathbf{B}$ on an elementary current loop possessing magnetic moment $\mathbf{m}$ is analogous to the effect of electric field $\mathbf{E}$ on a pointlike dipole possessing a dipole moment $\mathbf{p}$. So far this analogy has been proved in this text only with respect to the "passive" behavior of currents in a field generated by external sources. In § 36 we shall demonstrate that this is also true for the currents when they are considered as the sources of magnetic field.

12.3. Let us investigate in more detail the form of formula (12.3) in the case when all currents generating magnetic field are within a bounded three-dimensional volume. We can make use of expansion (11.11). Substituting it into (12.3), we obtain, in vector notations,

$$\mathbf{A}(\mathbf{r}) = \sum_{n=0}^{\infty} \mathbf{A}^{(n)}(\mathbf{r}) = \frac{\mu}{4\pi\alpha} \int \mathbf{j}(\mathbf{r}')\, dV' - \frac{\mu}{4\pi\alpha} \int \left(\mathbf{r}' \cdot \operatorname{grad} \frac{1}{\mathbf{r}}\right) \mathbf{j}(\mathbf{r}')\, dV'$$

$$+ \frac{\mu}{8\pi\alpha} \int \left[\mathbf{r}' \cdot \operatorname{grad}\left(\mathbf{r}' \cdot \operatorname{grad} \frac{1}{r}\right)\right] \mathbf{j}(\mathbf{r}')\, dV' + \ldots \tag{12.14}$$

[5] Magnetic moment is independent of the choice of origin. If, for example, $\mathbf{r} = \tilde{\mathbf{r}} + \mathbf{r}_0$, where $\mathbf{r}_0$ is a constant vector, then

$$\oint \mathbf{r} \times d\mathbf{r} = \oint \tilde{\mathbf{r}} \times d\tilde{\mathbf{r}} + \mathbf{r}_0 \times \oint d\tilde{\mathbf{r}}$$

but the integral in the second term is equal to zero.

Consider the first two terms. We again assume that the distribution of currents is such that it can be decomposed into a certain number of closed loops. As the final formula can be obtained by summation over these loops, let us analyze one of them separately. The first term is then equal to zero since $\int \mathbf{j}\, dV' = I \oint d\mathbf{s}' = 0$. The term which differs from zero and decreases at the minimum rate with distance is the second term. The integrand in this term can be transformed by a formula similar to (12.11). We shall only need to replace $\mathbf{r}$ by $\mathbf{r}'$, $d\mathbf{r}$ by $d\mathbf{r}'$ and $\mathbf{B}$ by grad $\frac{1}{r}$, and to apply differentiation in the second term of the right-hand side to $\mathbf{r}'$, assuming the observation point to be fixed, so that $d\left(\text{grad}\,\frac{1}{r}\right) = 0$. Therefore

$$\left(\mathbf{r}'\cdot\text{grad}\,\frac{1}{r}\right) d\mathbf{r}'$$

$$= \frac{1}{2}(\mathbf{r}' \times d\mathbf{r}') \times \text{grad}\,\frac{1}{r} + \frac{1}{2}\, d\left[\left(\mathbf{r}'\cdot\text{grad}\,\frac{1}{r}\right)\mathbf{r}'\right] \tag{12.15}$$

Integration of the total differential along a closed contour yields zero; at sufficiently large distances from the volume occupied by currents we can ignore the terms diminishing more rapidly and thus arrive at the formula

$$\mathbf{A}(\mathbf{r}) \simeq \mathbf{A}^{(1)}(\mathbf{r}) = -\frac{\mu}{4\pi}\,\mathbf{m} \times \text{grad}\,\frac{1}{r} \tag{12.16}$$

in which the magnetic moment of the loop generating the magnetic field is again given by (12.13).

Let us return to the fundamental equations (12.1); note that at all points of space where $\mathbf{j} = 0$ the equality curl $\mathbf{B} = 0$, as well as curl $\mathbf{H} = 0$, holds. Let us try, on the basis of these equalities, to introduce a scalar magnetic potential ψ instead of the vector potential $\mathbf{A}$, which we used so far, by a definition

$$\mathbf{H}(\mathbf{r}) = -\text{grad}\,\psi \tag{12.17}$$

For the sake of simplification, assume that, as before, the field-generating currents form several closed loops. The integral form (1.18) of the basic equation of magnetostatics can be written in our case in the following form:

$$\oint \mathbf{H}\, d\mathbf{s} = \frac{1}{\alpha}\, I \tag{12.18}$$

where I is the total current flowing in the loop along which we integrate the left-hand side of (12.18). Now, if $\psi_0(\mathbf{r})$ is a value of the magnetic potential established at the start of integrating along the loop, then at the moment of return to point $\mathbf{r}$ potential ψ must assume a new value $\psi(\mathbf{r})$ such that

$$|\psi(\mathbf{r})-\psi_0(\mathbf{r})|=\frac{1}{\alpha}I \tag{12.19}$$

this result can be checked by substituting (12.17) into (12.18). Consequently, the magnetic scalar potential cannot be found as a single-valued function of $\mathbf{r}$, in contrast to the electrostatic scalar potential; indeed, curl $\mathbf{E}=0$ everywhere. It should be emphasized that the above discussion deals with a magnetic field generated by electric currents. However, magnetic field can also be generated by permanent magnets, that is by ferromagnetics, as we have already mentioned in § 1. As far as this case is concerned, the concept of the magnetic scalar potential will be useful, and we shall discuss it again in § 36[6]. Further problems concerning the structure of the magnetic field of currents, as well as the energy properties of magnetic fields in various media, will be discussed in Chapter 8.

§ 13. Solution of the nonhomogeneous wave equation. The Liénard-Wiechert potentials

13.1. In a number of important cases solutions of the Maxwell equations satisfy wave equations with constant coefficients. Thus, in § 1 we have derived the homogeneous wave equation (1.24) for electromagnetic field strength in vacuo in the absence of sources, and in § 2, the wave equations for potentials for the same situation. These last equations assume especially symmetric form (2.6_1) and (2.6_2) if the Lorentz gauge is used. Let us recall in this connection that in § 7 we have established the relativistic invariance of equations (2.6) and of the Lorentz condition for fields in vacuo. Properties of solutions of such wave equations can be analyzed for specific boundary and initial conditions using as a model a wave equation of the type

$$\left(\Delta-\frac{1}{c^2}\frac{\partial^2}{\partial t^2}\right)\psi(\mathbf{r},t)=-g(\mathbf{r},t) \tag{13.1}$$

for a scalar function ψ, assuming function g to be known.

If $g\equiv 0$, we obtain a homogeneous equation whose spherically symmetric solutions can be written (this can be verified by direct differentiation) in the form

$$\chi(r,t)=\frac{f(r-ct)}{r}+\frac{h(r+ct)}{r} \tag{13.2}$$

[6] It should be noted here that the source of this potential is the distribution of magnetic dipoles. In contrast to the electric dipole, magnetic dipole need not be treated as a pair of a positive and a negative charge. Individual magnetic charges do not exist. The important fact is that the simplest (ideally, pointlike) sources of a magnetic field possess orientation which is characterized by the direction of their magnetic moments.

where f and h are arbitrary twice-differentiable functions. This solution has a singularity at point $r = 0$. The first term describes waves diverging from this point at moment $t = 0$, and the second term describes waves converging to this point. Solution (13.2) plays with respect to equation (13.1) a role which is similar to the role played by function $1/R$ in the Poisson equation analyzed in § 11.

Consider a region V of the three-dimensional space, bounded by surface σ. As the origin of spatial coordinates we choose an arbitrary point; as in the preceding sections, we denote by $\mathbf{r}$ the fixed radius vector of the observation point within region V, and by $\mathbf{r}'$, the varying radius vector corresponding to all points in this region, and define $\mathbf{R} = \mathbf{r} - \mathbf{r}'$. Therefore, the observation point is given by the condition $\mathbf{R} = 0$. In order to find a solution of the equation (13.1) at moment $t' = 0$, we shall make use of Green's formula (B.28), assuming that function ψ in (B.28) satisfies equation (13.1), and that function φ is a solution of a homogeneous wave equation. With these assumptions, we obtain

$$\int_V \varphi(\mathbf{r}', t')\, g(\mathbf{r}', t')\, dV' + \frac{1}{c^2}\int_V \left(\psi \frac{\partial^2 \varphi}{\partial t'^2} - \varphi \frac{\partial^2 \psi}{\partial t'^2}\right) dV' = \oint_\sigma \left(\psi \frac{\partial \varphi}{\partial n'} - \varphi \frac{\partial \psi}{\partial n'}\right) d\sigma \qquad (13.3)$$

Let us integrate both sides of (13.3) over time. Integration limits t'_1 and t'_2 will be chosen so that the inequalities

$$t'_1 + \frac{R}{c} < 0, \quad t'_2 + \frac{R}{c} > 0 \qquad (13.4)$$

be valid everywhere on the boundary surface σ. Assuming now that surface σ is at a finite distance from the observation point, we can always satisfy conditions (13.4). As

$$\psi \frac{\partial^2 \varphi}{\partial t'^2} - \varphi \frac{\partial^2 \psi}{\partial t'^2} = \frac{\partial}{\partial t'}\left(\psi \frac{\partial \varphi}{\partial t'} - \varphi \frac{\partial \psi}{\partial t'}\right)$$

we obtain

$$\int_{t'_1}^{t'_2} dt' \int_V dV'\, \varphi(\mathbf{r}', t')\, g(\mathbf{r}', t') + \frac{1}{c^2}\int_V dV' \left(\psi \frac{\partial \varphi}{\partial t'} - \varphi \frac{\partial \psi}{\partial t'}\right)\Big|_{t'_1}^{t'_2} = \int_{t'_1}^{t'_2} dt' \oint d\sigma \left(\psi \frac{\partial \varphi}{\partial n'} - \varphi \frac{\partial \psi}{\partial n'}\right)$$

It is clear from the physical point of view that for function φ we shall take that solution of the homogeneous wave equation of type (13.2) which describes waves converging to the observation point

$\mathbf{R} = 0$. In addition, we shall assume that these waves are "instantaneous pulses", that is, we assume $h(t' + R/c) = \delta(t' + R/c)$. Finally, since the chosen function has a singularity at point $R = 0$, we surround this point by a sphere σ_1 whose radius will then tend to zero. By virtue of condition (13.4), delta function vanishes within the integration ranges. Therefore, the second term in the left-hand side vanishes. As a result,

$$\int_{t_1'}^{t_2'} dt \int dV' \, \delta\left(t' + \frac{R}{c}\right) \frac{g(\mathbf{r}', t')}{r'}$$

$$= \int_{t_1'}^{t_2'} dt' \oint_{\sigma} \left(\psi \, \mathrm{grad}' \frac{\delta(t' + R/c)}{R} - \frac{\delta(t' + R/c)}{R} \mathrm{grad}' \psi\right) \cdot \mathbf{n}' \, d\sigma \quad (13.5)$$

We have used here the definition $\partial/\partial n' = (\mathbf{n}' \cdot \mathrm{grad}')$. We must keep in mind that the surface integral in the right-hand side consists of two terms: the integral over the former ("external") boundary surface and the integral over sphere σ_1 introduced above. In addition,

$$\mathrm{grad}' \frac{\delta(t' + R/c)}{r'} = \delta\left(t' + \frac{R}{c}\right) \mathrm{grad}' \frac{1}{R} + \frac{1}{R} \mathrm{grad}' \, \delta\left(t' + \frac{R}{c}\right)$$

$$= \delta\left(t' + \frac{R}{c}\right) \mathrm{grad}' \frac{1}{R} + \frac{1}{cR}\left[\frac{\partial}{\partial t'} \delta\left(t' + \frac{R}{c}\right)\right] \mathrm{grad}' R \quad (13.5')$$

and

$$\int_{t_1'}^{t_2'} dt' \, \psi(\mathbf{r}', t') \frac{\partial}{\partial t'} \delta\left(t' + \frac{R}{c}\right) = -\left.\frac{\partial \psi}{\partial t'}\right|_{t' = -R/c}$$

Using the properties of delta function to transform the remaining terms as well, we can rewrite the results for the external boundary in the following form:

$$\oint \left(\psi \, \mathrm{grad}' \frac{1}{R} - \frac{1}{cR} \frac{\partial \psi}{\partial t'} \mathrm{grad}' R - \frac{1}{R} \mathrm{grad}' \psi\right) \cdot \mathbf{n}' \Big|_{t' = -R/c} d\sigma' \quad (13.6)$$

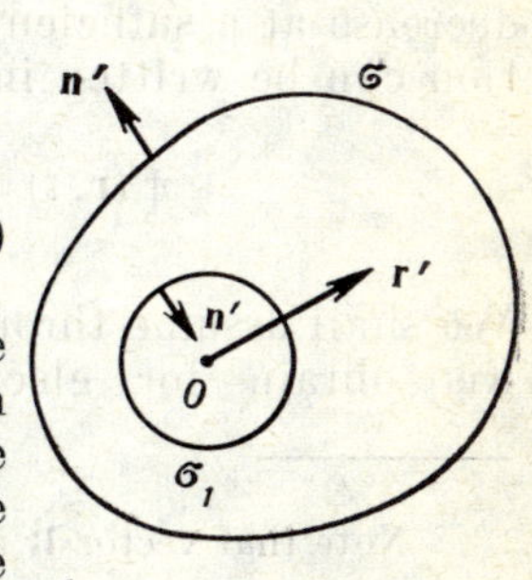

Fig. 6

Let us turn now to the integral over the internal sphere. To facilitate the evaluation of this integral, we move the origin for the time being into the observation point inside this sphere (see Fig. 6). We shall seek the solution ψ in a class of such functions which are regular together with their derivatives at the observation point $r' = 0$. Note that $d\sigma' = r'^2 \, d\Omega'$ where $d\Omega'$ is an element of a solid angle; an analysis of an integral of the type (13.5) in which R is replaced by r' and $r' \to 0$ shows that only the addend determined by the first term in (13.5′) remains finite in this limiting transition. We also have to take into account that the direction of the positive normal $\mathbf{n}'$ to the internal sphere σ_1 is opposite to the direction of vector $\mathbf{r}'$ (Fig. 6). As a result,

$$\oint_{\sigma_1} \psi \left(\text{grad}' \frac{1}{r'} \cdot \mathbf{n}' \right) d\sigma' = \oint \psi \frac{1}{r'^2} r'^2 \, d\Omega' \to 4\pi\psi|_{r'=0} \qquad (13.7)$$

Integration over time t' in the limit $r' \to 0$ reduces to the operation $\delta(t')$, that is to the substitution of $t'=0$ for the argument of function ψ. Returning to the initially chosen origin, we have to replace condition $r' = 0$ by $R = 0$ in the right-hand side of (13.7)[7]. And finally, we can replace the integration variable t' by $t' - t$, which corresponds to the time of observation t. Collecting the results (13.5), (13.6), and (13.7), we obtain

$$\psi(\mathbf{r}, t) = \frac{1}{4\pi} \int_{-\infty}^{+\infty} dt' \int_V dV' \frac{\delta(t'-t+|\mathbf{r}-\mathbf{r}'|/c)}{|\mathbf{r}-\mathbf{r}'|} g(\mathbf{r}', t') + I_K \qquad (13.8)$$

where the integral over the outer boundary surface of the region under consideration,

$$I_K = \frac{1}{4\pi} \oint_{\sigma} \left(\frac{1}{R} \text{grad}' \psi + \frac{1}{cR} \frac{\partial \psi}{\partial t'} \text{grad}' R - \psi \, \text{grad}' \frac{1}{R} \right)_{t'=t-R/c} \cdot \mathbf{n}' d\sigma \qquad (13.9)$$

is called the *Kirchhoff integral*[8]. This integral will be analyzed in more detail in § 20 in connection with the theory of diffraction of electromagnetic waves. At the moment we assume that surface σ expands to infinity and that this expansion is accompanied by $I_K \to 0$. This condition singles out a class of such solutions ψ which decrease at a sufficiently high rate at infinity. In this case the solution can be written in the form

$$\psi(\mathbf{r}, t) = \frac{1}{4\pi} \int_V dV' \frac{g(\mathbf{r}', \mathbf{r}')}{|\mathbf{r}-\mathbf{r}'|} \Bigg|_{t'=t-|\mathbf{r}-\mathbf{r}'|/c} \qquad (13.10)$$

We shall assume throughout this chapter that $I_K = 0$. In particular we obtain for electromagnetic potentials in Cartesian coor

---

[7] Note that vector $\mathbf{R}$ is directed from point $\mathbf{r}'$ to point $\mathbf{r}$, so that $\text{grad}' \frac{1}{R} = \frac{\mathbf{R}}{R^3}$ has the same direction as $\mathbf{n}'$ on the smaller sphere; the result of evaluation of (13.7) is therefore the same if r' is substituted by R. But it appears sufficiently obvious that this result must be independent of the choice of origin.

[8] For simplification, we tend t_1' to $-\infty$ and t_2' to $+\infty$; as seen from the above arguments, this does not affect the final result.

dinates

$$\varphi(\mathbf{r}, t) = \frac{1}{4\pi\varepsilon} \int_{-\infty}^{+\infty} dt' \int_V dV' \frac{\delta(t'-t+|\mathbf{r}-\mathbf{r}'|/c)}{|\mathbf{r}-\mathbf{r}'|} \rho(\mathbf{r}', t') \qquad (13.11_1)$$

$$\mathbf{A}(\mathbf{r}, t) = \frac{\mu}{4\pi\alpha} \int_{-\infty}^{+\infty} dt' \int_V dV' \frac{\delta(t'-t+|\mathbf{r}-\mathbf{r}'|/c)}{|\mathbf{r}-\mathbf{r}'|} \mathbf{j}(\mathbf{r}', t') \qquad (13.11_2)$$

Such potentials are known as the *retarded potentials.*

13.2. The physical meaning of formulas (13.11) is quite obvious. The potential observed at point **r** at time t is the total effect of instantaneous pulses which at this moment have arrived from sources to the observation point, propagating at a finite velocity c. Clearly, the solution for an arbitrary homogeneous isotropic medium is simply obtained by replacing c by the velocity v of propagation of electromagnetic pulses in this medium. With this is mind, in the remaining part of this section we shall give the formulas valid in vacuo, and use the Gaussian system of units.

If we apply the operator $\Delta - \frac{1}{c^2}\frac{\partial^2}{\partial t^2}$ to both sides of formula (13.8), we obtain the corollary of the fact that this formula is a solution of equation (13.1) (in the infinite space, because $I_K = 0$):

$$\left(\Delta - \frac{1}{c^2}\frac{\partial^2}{\partial t^2}\right) G(\mathbf{r}-\mathbf{r}', t-t') = -\delta(\mathbf{r}'-\mathbf{r})\,\delta(t'-t) \qquad (13.12)$$

Function G has the form

$$G(\mathbf{r}-\mathbf{r}', t-t') = \frac{\delta(t'-t+|\mathbf{r}-\mathbf{r}'|/c)}{4\pi|\mathbf{r}-\mathbf{r}'|} \qquad (13.12')$$

This function is called the *fundamental solution* of the nonhomogeneous wave equation. It can be seen that formula (13.8) can also be written in the form

$$\psi(\mathbf{r}, t) = \int_{-\infty}^{+\infty} dt' \int_V dV' \, G(\mathbf{r}-\mathbf{r}', t-t')\, g(\mathbf{r}', t') \qquad (13.13)$$

Note that the fundamental solution (13.12′) ensures that the causality condition holds; this condition coincides with the condition of retardation analyzed above[9].

Function (13.12′) as a relativistic invariant. In order to prove it, let us contract the notation by introducing $\tau \equiv t - t'$ and rewrite formula (13.12), taking into account property (C.7) of delta function, in the form $G = \frac{c}{4\pi} \frac{\delta(c\tau - R)}{R}$, for $\tau > 0$. The main property of delta function corresponds to equality $R = c\tau$; using this, we can trans-

[9] Formula (13.13) is additionally analyzed in Appendix E.

form function G in the integral over dt' in the following manner:

$$G=\frac{2c}{4\pi}\frac{\delta(c\tau-R)}{2R}=\frac{2c}{4\pi}\frac{\delta(c\tau-R)}{c\tau+R}=\frac{c}{2\pi}\delta[(c\tau)^2-R^2]=\frac{c}{2\pi}\delta(\vec{R}^2)$$

Here $\vec{R}$ is a vector in space-time with components $R^0=c(t-t')$ and $R^\alpha=r^\alpha-r'^\alpha$. If we introduce a symbol $d\Omega'\equiv c\,dt'\,dV'$ for four-dimensional volume, we transform formula (13.13) to the form

$$\psi(\mathbf{r},t)=\frac{1}{2\pi}\int\limits_{t>t'} d\Omega'\,\delta(\vec{R}^2)\,g(\mathbf{r}',t')$$

Integration is carried out for fixed t over that half of the light cone, having the apex at the observation point $t'=t$, $\mathbf{r}'=\mathbf{r}$, which is directed into the past. Using the relativistic notations for current (7.1) and potentials (7.3), we recast formulas (13.11) in an explicitly invariant form:

$$\Phi^i(\mathbf{r},t)=-\frac{1}{2\pi c}\int\limits_{t>t'} d\Omega'\,\delta(\vec{R}^2)\,s^i(\mathbf{r}',t') \qquad (13.14)$$

An analysis of the arguments involved in the above solution will demonstrate that the use of both terms of (13.2) (and not only of the second term, as we have done) with delta functions h and f would lead to a representation of potentials in the unbounded space as a sum of the considered above retarded potentials and the advanced potentials for which the inverse condition $t=t'-|\mathbf{r}-\mathbf{r}'|/c$ must hold and which therefore violate the causality condition. In the four-dimensional notation (13.14) this corresponds to integration over the whole light cone with the apex at the observation point ct, $\mathbf{r}$, that is, to rejection of the condition $t>t'$.

13.3. One important particular case of solution (13.11) for potentials is obtained when the electromagnetic field is generated by a single pointlike charge. In this case it will be convenient to use formulas (13.11) including delta function. We assume that the charge motion equation $\mathbf{r}'=\mathbf{r}'(t')$ is known. The charge density is given in the form $\rho(\mathbf{r}',t')=q\delta(\mathbf{r}'-\mathbf{r}'(t'))$, and the electric current corresponding to the motion of the charge is equal to $\mathbf{j}(\mathbf{r}',t')=\rho\mathbf{v}$. Velocity $\mathbf{v}$ can also be considered a known function of t'. Therefore,

$$\varphi(\mathbf{r},t)=\frac{q}{4\pi}\int\limits_{-\infty}^{+\infty}dt'\int dV'\,\frac{\delta(t'-t+|\mathbf{r}'-\mathbf{r}|/c)}{|\mathbf{r}-\mathbf{r}'|}\,\delta(\mathbf{r}'-\mathbf{r}'(t'))$$

$$=\frac{q}{4\pi}\int\limits_{-\infty}^{+\infty}dt'\,\frac{\delta(t'-t+|\mathbf{r}'(t')-\mathbf{r}|/c)}{|\mathbf{r}-\mathbf{r}'(t')|} \qquad (13.15)$$

A formula for $\mathbf{A}$ differs from (13.15) only by a factor $\boldsymbol{\beta} = \mathbf{v}(t')/c$ in the integrand.

Here the argument of delta function is $T = t' - t + |\mathbf{r}'(t') - \mathbf{r}|/c$, and the integration variable is t'. The integral is found by using formula (C.8). In addition,

$$\varkappa \equiv dT/dt' = 1 - \boldsymbol{\beta}\cdot\mathbf{n} \tag{13.16}$$

where $\mathbf{n} = \mathbf{R}/R$, $\mathbf{R} = \mathbf{r} - \mathbf{r}'(t)$ and $\boldsymbol{\beta} = \mathbf{v}/c = d\mathbf{r}'/c\,dt'$. Finally,

$$\varphi(\mathbf{r}, t) = \frac{q}{4\pi\varkappa R}\bigg|_{t'=t-R/c} \tag{13.17}$$

A completely similar calculation yields the following result:

$$\mathbf{A}(\mathbf{r}, t) = \frac{q\boldsymbol{\beta}}{4\pi\varkappa R}\bigg|_{t'=t-R/c} \tag{13.18}$$

By analogy with (13.14), we introduce a four-dimensional vector $\vec{R}$ which lies on the light cone in the past (for which $R^0 = |\mathbf{R}|$); it can be readily shown that this yields the relativistically invariant form to expressions (13.17) and (13.18):

$$\Phi^i(\mathbf{r}, t) = -\frac{q}{4\pi}\frac{u^i}{u^k R_k}\bigg|_{R^0=R} \tag{13.19}$$

Potentials (13.17) and (13.18) of a pointlike source are known as the *Liénard-Wiechert potentials*. Using these potentials, it is possible to find the field strengths generated by such sources; this will be demonstrated in § 14.

13.4. We should remind again that the wave equation assumes the explicitly relativistic invariant form (13.1) for all Cartesian components of four-dimensional potential only if the Lorentz gauge is used. If the Coulomb gauge is used (see § 2), equations for potentials take the form (2.9). Equation (2.9_2) for scalar potential is the Poisson equation, and its solution can be written in the form (11.4):

$$\varphi(\mathbf{r}, t) = \frac{1}{4\pi}\int \frac{\rho(\mathbf{r}', t)}{|\mathbf{r} - \mathbf{r}'|}\,dV' \tag{13.20}$$

The reader should pay attention to the fact that (13.20) describes the instantaneous effect of the source, because argument t must be identical on both sides of the equation. Besides, we may assume now that term $\frac{1}{c}\,\text{grad}\,\frac{\partial\varphi}{\partial t}$ in (2.9_1) for the vector potential is a known function of coordinates and time.

Using the continuity equation (1.13), we can write

$$\text{grad}\,\frac{\partial\varphi}{\partial t} = -\text{grad}\int \frac{\text{div}'\,\mathbf{j}(\mathbf{r}', t)}{4\pi\,|\mathbf{r} - \mathbf{r}'|}\,dV' \tag{13.21}$$

On the other hand, $\mathbf{j}$ can be written in the form

$$\mathbf{j} = \mathbf{j}_t + \mathbf{j}_l \tag{13.22}$$

where, by definition,

$$\mathbf{j}_l = -\operatorname{grad} \int \frac{\operatorname{div}' \mathbf{j}(\mathbf{r}', t)}{4\pi |\mathbf{r}-\mathbf{r}'|} dV' \tag{13.23}$$

Clearly, curl $\mathbf{j}_l = 0$. In addition.

$$\operatorname{div} \mathbf{j}_t = \operatorname{div} \mathbf{j} + \Delta \int \frac{\operatorname{div}' \mathbf{j}}{4\pi |\mathbf{r}-\mathbf{r}'|} dV' = \operatorname{div} \mathbf{j} - \int \operatorname{div} \mathbf{j} \delta(\mathbf{r}-\mathbf{r}') dV' = 0$$

This result is obtained by making use of (11.1′) and (11.3). Formula (13.22) is a decomposition of current $\mathbf{j}$ into a "transverse" component $\mathbf{j}_t$ and a "longitudinal" component $\mathbf{j}_l$. After substitution of this decomposition into the right-hand side of equation (2.9_2), and having taken into account that by virtue of (13.21) and (13.23) $\mathbf{j}_l = \operatorname{grad} \frac{\partial \varphi}{\partial t}$, we obtain

$$\left(\Delta - \frac{1}{c^2} \frac{\partial^2}{\partial t^2} \right) \mathbf{A} = -\frac{1}{c} \mathbf{j}_t \tag{13.24}$$

But the solution of the last equation is given by formula (13.11_2) if $\mathbf{j}$ in this formula is replaced by the transverse component $\mathbf{j}_t$.

If the Coulomb gauge is used, the physical meaning of the solutions of the Maxwell equations remains unaltered although it loses the explicit relativistic invariance. If, for example, the source of electromagnetic field is located in the same reference frame as the observer, the investigation of this field by using the Coulomb gauge is often simplified.

§ 14. Field strength around a pointlike charge. Radiation field. Uniform linear motion of a charge

14.1. Field strength generated by a pointlike charge moving arbitrarily can be calculated by formulas (2.1) and (2.2) if φ and $\mathbf{A}$ are the Liénard-Wiechert potentials given by (13.17) and (13.18). It is easier to take into account the conditions of retardation if these formulas are used in the form of (13.5). To perform differentiation with respect to time t and coordinates $\mathbf{r}$ of the observation, we first of all note that

$$\operatorname{grad} \delta\left(t' - t + \frac{R}{c}\right) = \frac{\partial}{\partial R} \delta\left(t' - t + \frac{R}{c}\right) \operatorname{grad} R$$
$$= \frac{\partial}{\partial T} \delta\left(t' - t + \frac{R}{c}\right) \frac{\partial T}{\partial R} \operatorname{grad} R = \frac{1}{c} \frac{\partial \delta(T)}{\partial T} \mathbf{n}$$

Here, as in § 13, $\mathbf{R} = \mathbf{r} - \mathbf{r}'(t')$, grad $= \partial/\partial\mathbf{r}$, $T \equiv t' - t + R/c$. We also obtain

$$\frac{1}{c}\frac{\partial\delta}{\partial t} = \frac{1}{c}\frac{\partial\delta}{\partial T}\frac{\partial T}{\partial t} = -\frac{1}{c}\frac{d\delta}{dT}$$

$$\operatorname{grad}\frac{\delta(T)}{R} = -\frac{1}{R^2}\mathbf{n}\delta(T) + \frac{1}{cR}\mathbf{n}\frac{d\delta(T)}{dT}$$

$$\operatorname{curl}\left(\boldsymbol{\beta}\frac{\delta(T)}{R}\right) = \left(\operatorname{grad}\frac{\delta(T)}{R}\right)\times\boldsymbol{\beta}$$

This yields then that

$$4\pi\mathbf{E}(\mathbf{r}, t) = q\int\left[\frac{\mathbf{n}}{R^2}\delta(T) + \frac{1}{cR}(\boldsymbol{\beta} - \mathbf{n})\frac{d\delta(T)}{dT}\right]dt'$$

$$4\pi\mathbf{B}(\mathbf{r}, t) = q\int\mathbf{n}\times\boldsymbol{\beta}\left\{\frac{1}{cR}\frac{d\delta(T)}{dT} - \frac{\delta(T)}{R^2}\right\}dt' \qquad (14.1)$$

Making use of formulas (C.3) and (C.8), as well of notation (13.16) for derivatives, we can obtain the following expressions for integrals:

$$\int\frac{\mathbf{n}}{R^2}\delta(T)\,dt' = \frac{\mathbf{n}}{\varkappa R^2}\Big|_{T=0}$$

$$\int\frac{\mathbf{n}\times\boldsymbol{\beta}}{R}\frac{d\delta(T)}{dT}dt' = -\frac{d}{dT}\left(\frac{\mathbf{n}\times\boldsymbol{\beta}}{\varkappa R}\right)_{T=0} = -\frac{1}{\varkappa}\frac{d}{dt'}\left(\frac{\mathbf{n}\times\boldsymbol{\beta}}{\varkappa R}\right)_{T=0}$$

Condition $T = 0$ is simply the condition of retardation. Equations (14.1) are therefore transformed to

$$4\pi\mathbf{E}(\mathbf{r}, t) = q\left[\frac{\mathbf{n}}{\varkappa R^2} + \frac{1}{c\varkappa}\frac{d}{dt'}\left(\frac{\mathbf{n} - \boldsymbol{\beta}}{\varkappa R}\right)\right]_{T=0}$$

$$4\pi\mathbf{B}(\mathbf{r}, t) = q\left[\frac{\boldsymbol{\beta}\times\mathbf{n}}{\varkappa R^2} + \frac{1}{c\varkappa}\frac{1}{dt'}\left(\frac{\boldsymbol{\beta}\times\mathbf{n}}{\varkappa R}\right)\right]_{T=0} \qquad (14.2)$$

It remains to carry out differentiation with respect to time t' in the obtained formulas. Hereafter this operation is marked by a dot over a variable. We have

$$\frac{1}{c}\dot{\mathbf{n}} = \frac{1}{cR}\dot{\mathbf{R}} - \frac{\mathbf{R}}{cR^2}\dot{R} = -\frac{\boldsymbol{\beta}}{R} + \frac{\mathbf{n}(\mathbf{n}\cdot\boldsymbol{\beta})}{R} = \frac{\mathbf{n}\times(\mathbf{n}\times\boldsymbol{\beta})}{R}$$

In the last formula we made use of (B.6), taking into account that $\mathbf{n}^2 = 1$. Differentiation first gives us equalities

$$4\pi\mathbf{E}(\mathbf{r}, t) = q\left[\frac{\mathbf{n} - \boldsymbol{\beta}}{\varkappa^2 R^2} + \frac{\mathbf{n}}{c\varkappa}\frac{d}{dt'}\left(\frac{1}{\varkappa R}\right) - \frac{1}{c\varkappa}\frac{d}{dt'}\left(\frac{\boldsymbol{\beta}}{\varkappa R}\right)\right]_{T=0}$$

$$4\pi\mathbf{B}(\mathbf{r}, t) = q\left[\frac{\boldsymbol{\beta}\times\mathbf{n}}{\varkappa^2 R^2} + \frac{1}{c\varkappa}\frac{d}{dt'}\left(\frac{\boldsymbol{\beta}}{\varkappa R}\right)\times\mathbf{n}\right]_{T=0}$$

whence

$$\mathbf{B} = \mathbf{n}\times\mathbf{E} \qquad (14.3)$$

Completing the differentiation indicated in the formula for $\mathbf{E}$, we obtain after an elementary though cumbersome collection of like terms

$$4\pi\mathbf{E}(\mathbf{r}, t) = q\,\frac{(\mathbf{n}-\boldsymbol{\beta})(1-\boldsymbol{\beta}^2)}{\varkappa^3 R^2}\Bigg|_{T=0} + \frac{q}{c}\,\frac{\mathbf{n}\times[(\mathbf{n}-\boldsymbol{\beta})\times\dot{\boldsymbol{\beta}}]}{\varkappa^3 R}\Bigg|_{T=0} \qquad (14.4)$$

The first term is a function of only the velocity of the source, and decreases in proportion to R^{-2} as the distance $\mathbf{R}$ from the source increases; the second term is the sum of all term which are functions of acceleration $\dot{\beta}$ and diminishes in proportion to R^{-1}. We denote the first term by $\mathbf{E}^0$, and the second by $\mathbf{E}^{(1)}$. Similarly, vector $\mathbf{B}$ given by (14.3) can also be decomposed into two components.

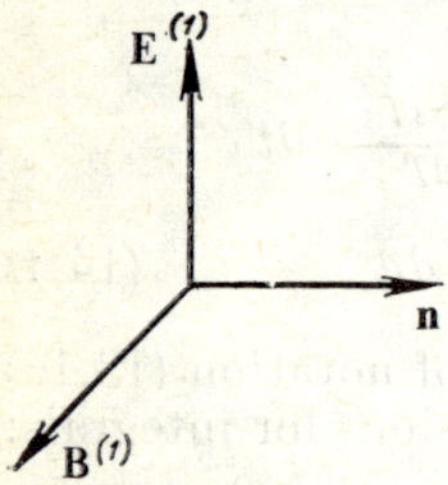

Fig. 7

14.2. Consider the energy flux across a two-dimensional sphere $R(t') = \text{const}$ (for a fixed t') which is found from (3.6) and (3.4). As an element of surface area of this sphere is given by $R^2\, d\Omega$, where $d\Omega$ is an element of solid angle, and as the product $\mathbf{E}\times\mathbf{B}$ determining the energy flux density contains terms proportional to R^{-2}, R^{-3} and R^{-4} [see (14.3) and (14.4)], a nonzero contribution to the total energy flux across the sphere at sufficiently large values of R is connected only to terms proportional to R^{-2}. But these are precisely the terms which are given by the vector product $\mathbf{E}^{(1)} \times \mathbf{B}^{(1)}$. Hence, an observer sufficiently remote from the emitting charge will detect an energy flux determined only by vectors $\mathbf{E}^{(1)}$ and $\mathbf{B}^{(1)}$. Consequently, $\mathbf{E}^{(1)}$ and $\mathbf{B}^{(1)} = \mathbf{n} \times \mathbf{E}^{(1)}$ are called the *radiation fields*. Fields $\mathbf{E}^0$ and $\mathbf{B}^0$, which fall off with distance more rapidly, can be called *quasistationary*. Note that (14.4) yields

$$\mathbf{n}\cdot\mathbf{E}^{(1)} = 0 \qquad (14.5)$$

Then it follows that the component of vector $\mathbf{S}$ which determines the energy flux at large distances from the charge is given, after formula (B.6) is applied, in the form

$$\mathbf{S} = c\mathbf{E}^{(1)} \times (\mathbf{n} \times \mathbf{E}^{(1)}) = c\mathbf{n}\, |\, \mathbf{E}^{(1)}\, |^2 \qquad (14.6)$$

On the other hand, as follows from (14.5) and (B.6),

$$\mathbf{E}^{(1)} = \mathbf{B}^{(1)} \times \mathbf{n} \qquad (14.7)$$

All the above formulas involve the retardation condition. Equations (14.3) and (14.7) show that vectors $\mathbf{n}$, $\mathbf{E}^{(1)}$ and $\mathbf{B}^{(1)}$ taken at time t at point $\mathbf{R}(t')$ form a right-handed trihedral (Fig. 7). As can be seen from (14.6), vector $\mathbf{n}$ points in the direction of propagation of the radiation energy. Hence, the radiation field is transverse (with respect to $\mathbf{n}$). It also follows from (14.3) or (14.7) that

$$|\, \mathbf{E}^{(1)}\, | = |\, \mathbf{B}^{(1)}\, | \qquad (14.8)$$

As a result, equation (14.6) for energy flux can also be used in the form

$$\mathbf{S} = c\mathbf{n} \mid \mathbf{B}^{(1)} \mid^2 = c\mathbf{n}\cdot(1/2)\,(\mid \mathbf{E}^{(1)} \mid^2 + \mid \mathbf{B}^{(1)} \mid^2) = c\mathbf{n}w^{(1)} \tag{14.9}$$

where $w^{(1)}$ is, according to (3.5), the radiation field energy density at a given point. The radiation phenomenon therefore represents transfer of energy $w^{(1)}$ at velocity c (we treat radiation in vacuo).

From the standpoint of a relativistic description of electromagnetic field, both invariants (7.10) for the radiation field vanish as a result of (14.3). Hence, the mutual orthogonality of vectors **E** and **B** of the radiation field and equality (14.8) are relativistic invariant properties of this field, that is, they hold in any inertial reference frame.

Properties of radiation will be later considered in more detail.

14.3. Let us analyze here the case of $\dot{\beta} = 0$, that is, the case when **E** and **B** completely reduce to their quasistationary components $\mathbf{E}^0$ and $\mathbf{B}^0$. Having in mind only these components, we shall drop superscript 0 throughout the remaining part of this section. Thus, formulas

$$4\pi\mathbf{E}(\mathbf{r}, t) = q\left[\frac{(\mathbf{n}-\boldsymbol{\beta})(1-\beta^2)}{\varkappa^3 R^2}\right]_{T=0},$$

$$\mathbf{B} = \mathbf{n}\times\mathbf{E} = \frac{q}{4\pi}\left[\frac{(\boldsymbol{\beta}\times\mathbf{n})(1-\beta^2)}{\varkappa^3 R^2}\right]_{T=0} \tag{14.10}$$

describe the electromagnetic field of a pointlike charge moving uniformly and linearly with respect to an observer. We shall compare these formulas to the result of relativistic transformations (7.13) of electromagnetic field.

In a reference frame (denoted by $\overline{K}$) in which the charge is fixed, that is for $\beta = 0$, we obtain first of all

$$\overline{\mathbf{E}}(\mathbf{r}, t) = \mathbf{E}\mid_{\beta=0} = \frac{q\mathbf{n}}{4\pi R^2}, \quad \overline{\mathbf{B}} = \mathbf{B}\mid_{\beta=0} = 0 \tag{14.11}$$

As could be expected, this is a standard expression for electrostatic field of a pointlike charge (cf. § 11). Here we need not take account of retardation condition $T = 0$ simply because in the case in question **R** is independent of time.

Invariant I_2 given by (7.10_2) equals zero, since there is a reference frame $\overline{K}$ in which $\overline{\mathbf{B}} = 0$. From this it follows immediately that fields **E** and **B** are mutually orthogonal in any inertial reference frame. Note that in reference frame $\overline{K}$ field **E** is pureley longitudinal with respect to vector **n**, while the radiation field, as we have shown above, is always transverse. And finally, we see from (7.10_1) that $I_1 < 0$.

If formulas (14.11) give the field generated by a pointlike charge in a reference frame moving uniformly and linearly with this charge, then the field observed in any other inertial reference frame can be obtained by the Lorentz transformations (7.13) for field strengths. It must be kept in mind that vector **R** in (14.11) must be understood as the distance between the locations of the charge and the observation point, which are simultaneous in the reference frame associated with the charge. In order to emphasize this fact, we introduce for this distance a new symbol **R** and rewrite formula (14.11) in the form

$$\overline{\mathbf{E}}=q\frac{\overline{\mathbf{R}}}{4\pi\overline{R}^{3}},\quad \overline{\mathbf{B}}=0 \tag{14.11'}$$

because $\mathbf{n}=\overline{\mathbf{R}}/R$. Consider now electric field **E** in reference frame K (observer's reference frame) with respect to which the charge moves at velocity **v**. Straightforward application of formulas inverse with respect to (7.13) yields

$$\mathbf{E}=\overline{\mathbf{E}}_{\parallel}+\gamma\overline{\mathbf{E}}_{\perp}=\frac{q}{4\pi\overline{R}^{3}}(\overline{\mathbf{R}}_{\parallel}+\gamma\overline{\mathbf{R}}_{\perp})=\gamma\frac{q}{4\pi\overline{R}^{3}}\mathbf{R} \tag{14.12}$$

where **R** denotes a radius vector connecting the position of the charge and the observer, which are simultaneous with respect to reference frame K[10]. Indeed, formulas $\overline{\mathbf{R}}_{\parallel}=\gamma\mathbf{R}_{\parallel}$ and $\overline{\mathbf{R}}_{\perp}=\mathbf{R}_{\perp}$, used in the derivation of (14.12), express the contraction of a moving scale, discussed in § 5.

It is not yet obvious that formulas (14.12) and (14.10) are identical, but this is really the case, as we could have expected. In order to verify this, let us recast formula (14.12) to a somewhat different form, defining a new vector $\mathbf{R}_*$ by equalities $\mathbf{R}_{*\parallel}=\mathbf{R}_{\parallel}$, $\mathbf{R}_{*\perp}=\gamma^{-1}\mathbf{R}_{\perp}$. Then

$$\overline{\mathbf{R}}^{2}=(\gamma\mathbf{R}_{\parallel}+\mathbf{R}_{\perp})^{2}=\gamma^{2}\mathbf{R}_{\parallel}^{2}+\mathbf{R}_{\perp}^{2}=\gamma^{2}\mathbf{R}_{*}^{2}$$

that is $\overline{R}=\gamma R$ and

$$\mathbf{E}=\frac{1}{4\pi}\frac{q}{\gamma^{2}R_{*}^{3}}\mathbf{R} \tag{14.13}$$

Besides, $R_{*}^{2}=R_{\parallel}^{2}+\gamma^{-2}R_{\perp}^{2}=R^{2}+(\gamma^{-2}-1)R_{\perp}^{2}=R^{2}-\beta^{2}R_{\perp}^{2}=R^{2}-(\boldsymbol{\beta}\times\mathbf{R})^{2}$.

Now we have to take into account that vector **R** used in formula (14.10) refers to reference frame K of the observer, but relates nonsimultaneous positions of the source and the observer and takes into account the condition of retardation. We can write $\mathbf{R}(t')=\mathbf{R}(t)+(t'-t)\frac{d\mathbf{R}}{dt'}$ because $\frac{1}{c}\frac{d\mathbf{R}}{dt'}=-\boldsymbol{\beta}=\text{const}$ $\left(\text{recall that } \boldsymbol{\beta}=\frac{1}{c}\frac{d\mathbf{r}'}{dt},\text{ and } \mathbf{R}=\mathbf{r}-\mathbf{r}'\right)$. If the retardation condition is satisfied, then $t'=t-R_{\mathrm{ret}}/c$ and $\mathbf{R}_{\mathrm{ret}}=\mathbf{R}-\frac{R_{\mathrm{ret}}}{c}\frac{1}{c}\frac{d\mathbf{R}}{dt'}=\mathbf{R}+\boldsymbol{\beta}R_{\mathrm{ret}}$. From

[10] We remind that subscripts $\parallel$ and $\perp$ denote components parallel and orthogonal, respectively, to relative velocity **v**.

this it follows immediately that $\beta \times R_{ret} = \beta \times R$, and that vector $\mathbf{R}$ connecting simultaneous positions has the same meaning as $\mathbf{R}$ in formula (14.13).

By using (13.16) and the above formulas, one readily finds that

$$(\varkappa R)^2_{ret} = R^2 - (\beta \times R_{ret})^2 = R^2 - (\beta \times R)^2 = R^2_*$$

and $\mathbf{n} - \beta = (\mathbf{R}_{ret} - \beta R_{ret})/R_{ret} = \mathbf{R}/R_{ret} = \varkappa \mathbf{R}/R_*$. Substitution of these results into formula (14.10) transforms it to (14.13).

Magnetic field generated around a uniformly and linearly moving charge can be written, by using similar arguments, in the form

$$\mathbf{B} = \frac{q}{4\pi} \frac{\beta \times \mathbf{R}}{\gamma^2 R^3_*} \tag{14.14}$$

In the nonrelativistic limit $\gamma \approx 1$, and therefore $R_* \approx R$. If we introduce additionally the symbol $\mathbf{j} = \rho \mathbf{v}$ for current produced by motion of the charge, then formally equation (14.14) coincides with the Ampère law (12.7) if we set $I\, d\mathbf{s}' = \mathbf{j}$. The physical meaning of this coincidence is nevertheless very limited. First of all, differential law (12.7) can be correctly interpreted only in relation to the integral formula (12.6) which gives magnetic field generated at a given point of space by a closed loop of direct current considered as a whole. Moreover, the magnetic field vector of a pointlike charge given by formula (14.14) varies in magnitude and direction as this charge moves, in contrast to the constant field given by (12.7). Asa result, the analogy between the two fields is limited to an infinitesimal interval of time. To complete the analogy, one can imagine that the pointlike charge leaving a given point of space is immediately replaced at this point by another identical charge moving at the same velocity. But even in this case the analogy remains limited because no reference frame can be found in which the field of a pointlike charge would be purely magnetic; we have already demonstrated that a reference frame can be found in which the fields is purely electrostatic.

We have seen above that from the standpoint of the relativistic theory of transformations of electromagnetic field (see § 7), the field of a pointlike charge is classified as a field satisfying inequality $I_1 < 0$, where invariant I_1 is given by formula (7.10_1). On the contrary, magnetostatic field discussed in § 12 is an example of the case of $I_1 > 0$. Indeed, this field can be produced, for example, by a current flowing in a closed loop (conductor) of an arbitrary shape; all parts of this loop are at rest in an inertial reference frame in which the formulas derived in § 12 are valid. The conductor can always be considered neutral, so that in this particular reference frame no electric field will be generated. The neutrality condition can be considered satisfied if the densities of the positive and negative charges are distributed uniformly over the conductor volume from the macroscopic viewpoint, and they balance each other at all points of this volume. A distribution of pointlike charges would result in a dipole moment, and thus we have to assume that charge density has no singularities.

In any other inertial reference frame K' the equality of the positive and negative charge densities is violated (obviously, the net charge will remain equal to zero)[11]. As a result, an electric field will be added to the magnetic field in reference frame K'; this electric field can be calculated by formulas (7.13) if field **B** in reference frame K is known. Of course, the field produced by a positive charge located in volume dV' of a conductor and moving at a velocity $\mathbf{v}$ in reference frame K with respect to the negative charge (for the sake of simplification, we assume this charge fixed) can be calculated from formula (12.7) by substituting into it $I\,d\mathbf{s}' = \mathbf{j}\,dV' = \rho\mathbf{v}\,dV'$.

Note in conclusion that a pointlike charge moving along a closed loop will be accelerated and in principle will radiate, while direct current generates no radiation.

§ 15*. Relativistic law of energy-momentum conservation for the electromagnetic field of a pointlike charge

15.1. The formulas derived in the preceding section make it possible to calculate directly the radiation energy flux produced by a moving pointlike charge, the angular distribution of this flux, and so on (see § 16). It is not easy, however, to grasp from these formulas the relationship between the phenomenon of emission and the special theory of relativity. This aspect will be elucidated if the study of the radiation field is conducted in relativistic terms, on the basis of the relativistically invariant form of Liènard-Wiechert potentials (13.19). This will be done below. The results given in this section have been obtained relatively recently by a number of authors.[12]

First, we want to modify the field strength generated by a pointlike charge and the energy-momentum tensor of this field to a form convenient for further manipulations.

Let us denote by $\vec{x}$ the four-dimensional radius vector of point P in the Minkowski space in which radiation is observed, and by $\vec{z}$, the radius vector of the world line of the pointlike charge (Fig. 8). We assume that the proper time parameter τ on this world line is fixed, so that $\vec{z} = \vec{z}\,(\tau)$. In further formulas differentiation with respect to τ is marked by a dot. Let us recall formulas (5.16), (5.18)

[11] See transformation formulas (7.18) on p. 65.

[12] Our presentation here and partially in § 23 is based on the paper by P. A. Hogan, *Nuovo Cimento* **15B**, 136 (1973) which in its turn is based on the results due to J. L. Synge and F. Rohrlich.

and (5.19):

$$\vec{u} = \dot{\vec{z}}, \quad \vec{w} = \dot{\vec{u}}, \quad \vec{u}^2 = c^2, \quad \vec{u}\vec{w} = 0 \tag{15.1}$$

Vector $\vec{u}/c$ is a unit timelike vector tangent to the world line of the charge at point $\vec{z}$, while the vector of four-dimensional acceleration $\vec{w}$ is spacelike.

Let us define

$$\vec{R} = \vec{x} - \vec{z}(\tau) \tag{15.2}$$

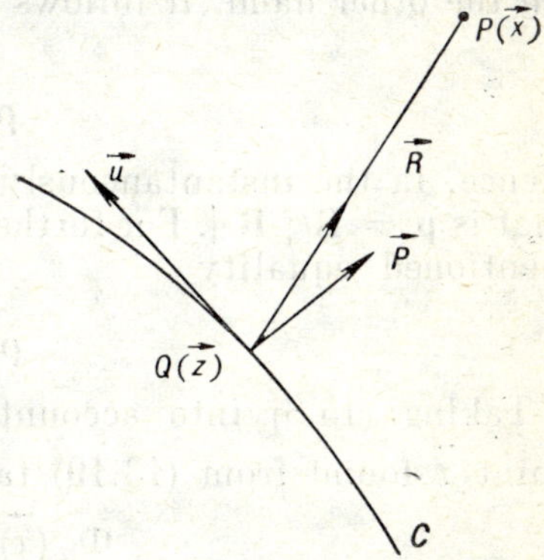

Fig. 8

We have shown in § 13 that electromagnetic radiation emitted at point Q can be observed in point P if

$$\vec{R}^2 = 0 \tag{15.3}$$

that is $R^0 = x^0 - z^0 = \pm |\mathbf{R}|$, where $|\mathbf{R}|^2 = \Sigma_{\alpha=1}^{3} (x^\alpha - z^\alpha)^2$. Observation at P takes place at a later moment than emission at Q if $R^0 > 0$, that is $R^0 = + |\mathbf{R}|$. For this reason we refer to zero vector $\vec{R} = \{+|\mathbf{R}|, \mathbf{R}\}$ as the retarded vector.

Let us introduce now at point Q a spacelike vector $\vec{p}$[13] defined so that

$$\vec{p}^2 = -1, \quad \vec{u}\vec{p} = 0 \tag{15.4}$$

We also demand that a decomposition

$$\vec{R} = \rho' \vec{u} + \rho \vec{p} \tag{15.5}$$

be possible, in which ρ and ρ' are certain numbers. In other words, $\vec{p}$ is a unit vector of projection of $\vec{R}$ onto a spacelike hyperplane orthogonal to vector $\vec{u}$. Taking into account (15.1), we obtain from (15.3) and (15.5) that $\rho' = \pm\rho/c$, that is $\vec{R} = \rho\,(\vec{p} \pm \vec{u}/c)$. This yields, via (15.4) and (15.1), $\rho = \pm\,(1/c)\,(\vec{u}\vec{R})$.

Now we want to take into account the condition of retardation. As the expression for ρ is relativistically invariant, we can find it in the instantaneously co-moving reference frame where $\vec{u} = \{c, 0\}$ and therefore $\rho = \pm R^0$; in other words, $\rho = \pm |\mathbf{R}|$ if $\vec{R}$ satisfies the condition of retardation. We can therefore drop the lower sign, assume that ρ for the "retarded" vector $\vec{R}$ is an arbitrary non-nega-

[13] Of course, this vector has nothing in common with 4-momentum.

tive number, and write a decomposition of this vector in the form

$$\vec{R}=\rho(\vec{p}+\vec{u}/c) \tag{15.6}$$

The second of conditions (15.4) can be written in the form $p^0=(1/c)\mathbf{v}\cdot\mathbf{p}$, so that in the instantaneously co-moving reference frame $p^0=0$. On the other hand, it follows from (15.6) that

$$\rho=-\vec{p}\vec{R} \tag{15.7}$$

Hence, in the instantaneously co-moving reference frame $\rho=\mathbf{p}\cdot\mathbf{R}$, that is $\mathbf{p}=\mathbf{R}/|\mathbf{R}|$. For further reference let us single out the above-mentioned equality

$$\rho=\frac{1}{c}\vec{u}\vec{R} \tag{15.8}$$

Taking (15.8) into account, the Liénard-Wiechert potential at point $\vec{x}$ found from (13.19) takes the form

$$\Phi_n(\vec{x})=-\frac{q}{4\pi c}\frac{u_n}{\rho} \tag{15.9}$$

15.2. Tensor $F_{mn}(\vec{x})$ of field strength is to be calculated by differentiating equations (15.9) with respect to coordinates x^r of the observation point P. In this operation we must consider virtual displacements dx^r of this point. But these displacements must be such that point P remains a possible point of observation. In order to satisfy this requirement, differentiation must include also a displacement of point Q along the world line of the particle so as to keep points P and Q connected by the zero retarded vector $\vec{R}$. Hence, differentiation at point P has to be carried out under a constraint $\vec{R}\,d\vec{R}/d\tau=0$, that is $\vec{R}\,(d\vec{x}/d\tau-\vec{u})=0$.

The derivative $d\vec{x}/d\tau$ appears here precisely because, for a fixed world line, point P cannot be displaced "no matter how", because of a risk of creating a situation when radiation emitted from this world line cannot be observed at P. From formulas (15.6), (15.1), and (15.4) we obtain

$$\frac{1}{c\rho}R_m\frac{dx^m}{d\tau}=1 \tag{15.10}$$

However, when points P and Q are related in this single-valued fashion, parameter τ can also be considered as a function of coordinates x^m. Consequently, $d\tau=\frac{\partial\tau}{\partial x^m}dx^m$. By comparing this expression with (15.10), we find that

$$\frac{\partial\tau}{\partial x^m}=\frac{1}{c\rho}R_m \tag{15.11}$$

Whence

$$\frac{\partial u_r}{\partial x^p}=\frac{\partial u_r}{d\tau}\frac{\partial \tau}{\partial x^p}=\frac{1}{c\rho}w_r R_p \tag{15.12}$$

As follows from definition (15.2),

$$\frac{\partial R^r}{\partial x^p}=\delta_p^r-\frac{dz^r}{d\tau}\frac{\partial \tau}{\partial x^p}=\delta_p^r-\frac{1}{c\rho}u^r R_p \tag{15.13}$$

and we find from (15.8), (15.12), and (15.13) that

$$c\frac{\partial \rho}{\partial x^r}=\frac{\partial}{\partial x^r}(u^l R_l)=u_r-\frac{c}{\rho}R_r\left(1-\frac{\vec{w}\vec{R}}{c^2}\right)$$

Let us introduce new notations:

$$W\equiv\frac{\vec{w}\vec{R}}{c^2}=\frac{\rho}{c^2}(\vec{w}\vec{p}) \tag{15.14}$$

and

$$B\equiv\frac{1}{\rho}(1-W) \tag{15.15}$$

From this we obtain

$$\frac{\partial \rho}{\partial x^r}=\frac{u_r}{c}-BR_r=W\frac{u_r}{c}-p_r(1-W) \tag{15.16}$$

Equations (15.9), (15.12), and (15.16) now yield

$$-\frac{4\pi c}{q}\frac{\partial \Phi_n}{\partial x^m}=\frac{\partial}{\partial x^m}\left(\frac{u_n}{\rho}\right)=\frac{1}{c\rho^2}w_n R_m+\frac{B}{\rho^2}u_n R_m-\frac{u_n u_m}{c\rho^2} \tag{15.17}$$

We define a vector

$$\vec{V}\equiv\frac{1}{c}\vec{w}+B\vec{u} \tag{15.18}$$

Then

$$F_{mn}=\frac{\partial \Phi_n}{\partial x^m}-\frac{\partial \Phi_m}{\partial x^n}=\frac{q}{4\pi c}\frac{V_{[m}R_{n]}}{\rho^2} \tag{15.19}$$

Here $A_{[m}B_{n]}\equiv A_m B_n-A_n B_m$. After antisymmetrization the last term in (15.17) vanishes. When we switch to three-dimensional notations, formula (15.19) becomes identical to expressions for field strengths derived in § 14. We suggest that the reader verify this statement as a useful exercise.

The auxiliary vector $\vec{V}$ has the following properties:

$$\vec{V}^2=\frac{1}{c^2}\vec{w}^2+c^2B^2,\quad \vec{V}\vec{R}=c,\quad \vec{V}\vec{u}=c^2B \tag{15.20}$$

Derivation of the second of them requires the use of (15.6) and of definitions (15.14) and (15.15).

15.3. Let us substitute formula (15.19) into definition (10.19) of the energy-momentum tensor of electromagnetic field. Taking account of the second relation of (15.20) as well as of (15.3), we obtain

$$\rho^4 \left(\frac{4\pi c}{q}\right)^2 F_{ab}F^{ab} = (V_a R_b - V_b R_a)(V^a R^b - V^b R^a)$$
$$= 2\vec{V}^2\vec{R}^2 - 2(\vec{V}\vec{R})^2 = -2c^2$$

The term $F_{kl}F^{l}{}_{\cdot m}$ is calculated in a similar way. As a result,

$$\left(\frac{4\pi c}{q}\right)^2 T_{km} = \frac{1}{\rho^4}\left[c(V_k R_m + V_m R_k) - R_k R_m \vec{V}^2 - \frac{c^2}{2} g_{km}\right] \quad (15.21)$$

Formula (15.21) readily gives the "projections" of the energy-momentum tensor onto vectors $\vec{p}$, $\vec{R}$, $\vec{u}/c$, these projections will be required in further calculations. Namely, we can use decomposition (15.6) and notations (15.14) and (15.15) to write

$$\left(\frac{4\pi}{q}\right)^2 T^{km} p_m = \frac{1}{\rho^4}\left[\left(\frac{W}{\rho} + \rho\frac{\vec{V}^2}{c^2}\right) R^k - \frac{\rho}{c} V^k - \frac{1}{2} p^k\right] \quad (15.22)$$

$$\left(\frac{4\pi}{q}\right)^2 T^{km} R_m = \frac{1}{2\rho^3}\left(p^k + \frac{u^k}{c}\right) = \frac{1}{2\rho^4} R^k \quad (15.23)$$

$$\left(\frac{4\pi}{q}\right)^2 T^{km}\frac{u_m}{c} = \frac{1}{c\rho^4}\left[\rho V^k + \left(cB - \frac{\rho\vec{V}^2}{c}\right) R^k - \frac{1}{2} u^k\right] \quad (15.24)$$

With the energy-momentum tensor known we can consider conservation laws. Let us drawn in space-time around world line C of the pointlike charge two cylindrical hypersurfaces described by the equations $\rho = \varepsilon$ and $\rho = R$ (Fig. 9). By Ω we denote a four-dimensional volume bounded by these surfaces and by two light cones issuing from world line C at points $\tau = \tau_1$ and $\tau = \tau_2$ and intersecting these hypersurfaces (see the figure). Integration over volume Ω of the equation $\partial T^{rs}/\partial x^s = 0$, whose solution is the energy-momentum tensor in this volume, and application of the Gauss theorem yield

$$\oint T^{rs}\, d\Sigma_s = 0 \quad (15.25)$$

Choosing the direction of normal to the boundary hypersurface, external with respect to volume Ω as the positive direction of normal, we can rewrite equation (15.25) in the form

$$P_r(\tau_1) - P_r(\tau_2) = Q_r(R) - Q_r(\varepsilon) \quad (15.26)$$

The notations used in the above equation are clear from the drawing. In particular

$$Q^r(R) = \frac{1}{c}\int\limits_{\substack{\tau=\tau_1 \\ (\rho=R)}}^{\tau=\tau_2} T^{rl}\, d\Sigma_l, \quad Q^r(\varepsilon) = -\frac{1}{c}\int\limits_{\substack{\tau=\tau_1 \\ (\rho=\varepsilon)}}^{\tau=\tau_2} T^{rl}\, d\Sigma_l \quad (15.27)$$

We shall be interested in equations (15.26) and (15.27) in the limit of $\varepsilon \to 0$ and $R \to \infty$. In order to calculate vectors $Q_r(\varepsilon)$ and $Q_r(R)$, we have first to clarify the geometric properties of hypersurfaces $\rho = \varepsilon$ and $\rho = R$ in the indicated limiting transition.

The direction of a normal to hypersurface $\rho=$const coincides with the direction of grad ρ[14]. This last can be found from (15.16) and (15.14). As $W \to 0$ for $\rho \to 0$, we find from (15.16) that in this limit

$$\frac{\partial \rho}{\partial x^r} \simeq - p_r \tag{15.28}$$

that is the normal is spacelike. For brevity, we shall consider such hypersurfaces as timelike.

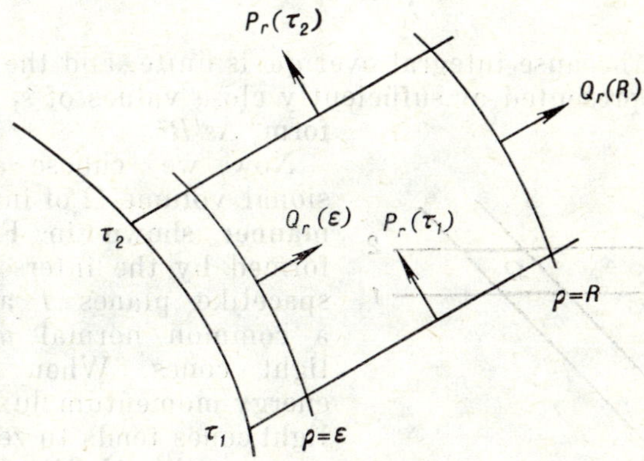

Fig. 9

In the limit of $R \to \infty$ inequality $|W| \gg 1$ holds, therefore (15.16) yields that

$$\frac{\partial \rho}{\partial x^r} \simeq W\left(\frac{u_r}{c} + p_r\right) = \frac{W}{R} R_r \tag{15.29}$$

In other words,

$$\left| \frac{\partial \rho}{\partial x^r} \frac{\partial \rho}{\partial x_r} \right| \to \frac{1}{c^4} (\overrightarrow{w}\overrightarrow{p})^2 \vec{R}^2 = 0$$

in the limit of $R \to \infty$ and the hypersurface $\rho = R$ acquires the properties of a light cone. In the present section we are interested in precisely this limit, that is in $Q_r(R)$ for $R \to \infty$[15].

The light cone volume element can be calculated by formula (D.8) for an arbitrary direction of the normal. We assume $\vec{n} = \vec{p}$; then $d\Sigma$ in (D.8) is an element of a timelike plane which in turn can be calculated from (D.4). This arbitrary timelike plane can be regarded locally coinciding with a segment of the timelike surface $\rho = \varepsilon$; we

[14] Obviously, here grad is the four-dimensional gradient in the Minkowski space.

[15] We shall calculate $Q_r(\varepsilon)$ for $\varepsilon \to 0$ and give it a physical interpretation in § 23.

have just established that a normal to this surface is vector $-\vec{p}$ so that we obtain from (D.9), (D.4), and (15.28) for integration element $d\Sigma_l$ in integral $Q_r(R)$

$$d\Sigma_l = \frac{R_l\, d\Sigma}{|\overrightarrow{pR}|} = \frac{R_l}{R}\left|\frac{\partial \rho}{\partial x^r}\, p^r\right| R^2\, d\omega\, ds = R_l R\, d\omega\, ds \qquad (15.30)$$

It can be found from (15.27) and (15.23) at $\rho \simeq R$ that

$$Q_r(R) \sim \frac{1}{2}\int\limits_{s_1}^{s_2} \frac{ds}{R^2} \int \left(p_r + \frac{u_r}{c}\right) d\omega \to 0 \qquad (15.31)$$

for $R \to \infty$, because integral over $d\omega$ is finite, and the integral over ds can be represented at sufficiently close values of s_1 and s_2 in the form $\Delta s/R^2$.

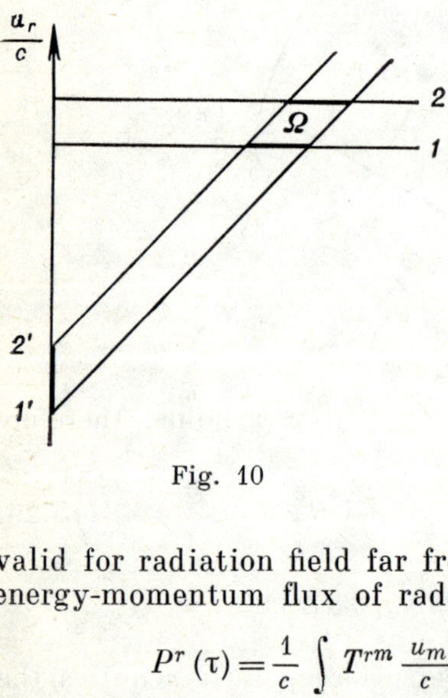

Fig. 10

Now we choose a four-dimensional volume Ω of integration in a manner shown in Fig. 10. It is formed by the intersection of two spacelike planes *1* and *2* having a common normal u_r/c with two light cones. When $R \to \infty$, the energy-momentum flux through both light cones tends to zero, in accordance with (15.31). Consequently, the energy-momentum flux across the part of hypersurface *1* cut out by these light cones is equal to the flux across a similar portion of hypersurface *2*; in this sense the energy-momentum conservation is valid for radiation field far from the source. For calculation of the energy-momentum flux of radiation we must evaluate the integral

$$P^r(\tau) = \frac{1}{c}\int T^{rm}\frac{u_m}{c}\, d\Sigma' \simeq \frac{1}{c^2}\int T^{rm} u_m R^2\, d\rho\, d\omega$$

(cf. formula (D.2)) over a segment of spacelike hyperplane with unit normal u_m/c. Let us calculate the radiation flux across a spherical layer with thickness $d\rho = c d\tau$ in the hyperplane. We obtain

$$\frac{dP^r}{d\tau} \simeq \frac{1}{c}\int T^{rm} u_m R^2\, d\omega$$

Now we can make use of formula (15.24). When $\rho = R \to \infty$, the final expression has the only term $-\frac{1}{c^2 R^3}\vec{V}^2 R^k = -\frac{1}{c^2R^2}\vec{V}^2\left(p^k + \frac{u^k}{c}\right)$. Substituting $\vec{V}^2$ from (15.20) and B from (15.15), we obtain for $R \to \infty$ that

$$\left(\frac{4\pi}{q}\right)^2 T^{rm}\frac{u_m}{c}R^2 \simeq -\frac{1}{c^2}\left(\frac{\vec{w}^2}{c^2}+c^2B^2\right)\left(p^k+\frac{u^k}{c}\right)$$
$$\simeq -\frac{1}{c^4}\left(\vec{w}^2+(\vec{w}\vec{p})^2\right)\left(p^k+\frac{u^k}{c}\right)$$

because

$$B^2=\frac{(1-W)^2}{\rho^2}\simeq\frac{W^2}{R^2}=\frac{(\vec{w}\vec{p})^2}{c^4}$$

The term proportional to p^k can be dropped because any term with this factor depends on an odd number of factors p^i and yields zero when integrated over $d\omega$ (see Appendix D, Item 2). Finally

$$\frac{dP^r}{d\tau}=-\left(\frac{q}{4\pi}\right)^2\frac{1}{c^5}\int d\omega\,(\vec{w}^2+(\vec{w}\vec{p})^2)\,u^r$$

Integration is carried out by using formulas (D.5) and (D.6). Namely:

$$\int \vec{w}^2\,d\omega=\vec{w}^2\int d\omega=4\pi\vec{w}^2,\quad \int w^k\,d\omega=4\pi w^k \tag{15.32}$$

$$\int p^k(\vec{w}\vec{p})\,d\omega=w^l\int p^k p_l\,d\omega=-\frac{4\pi}{3}w^l\left(\delta_l^k-\frac{u^k u_l}{c^2}\right)=-\frac{4\pi}{3}w^k$$

because $\vec{w}\vec{u}=0$. Similarly

$$\int(\vec{w}\vec{p})^2\,d\omega=-\frac{4\pi}{3}\vec{w}^2 \tag{15.32$'$}$$

Equations (15.32) will be used again in § 23. Ultimately we obtain

$$\frac{dP^r}{d\tau}=-\frac{1}{4\pi}\frac{2q^2}{3c^5}u^r\vec{w}^2 \tag{15.33}$$

This formula gives the amount of energy and momentum transferred by radiation field per unit proper time of the charge. These energy and momentum of radiation are due to the charge moving along a segment $d\tau$ of its world line between points *1'* and *2'* (see Fig. 10) when these points are brought together infinitely closely.

§ 16. Energy radiated by a moving charge

16.1. Energy flux of the radiation field can be calculated by formulas (14.6) or (14.9). We want to find the angular distribution of radiation and also its total energy integrated over the angles of emission. Obviously, the results will depend on the choice of reference frame in which they are obtained, that is simply on the value of velocity **v** in expressions for field strength.

We start with a simpler case of calculating total radiated energy in the reference frame in which $\mathbf{v}=0$ at a given moment of time.

In this case formulas (14.4) and (14.3) take the form[16]

$$4\pi\mathbf{E}=\frac{q}{c}\,\frac{\mathbf{n}\times(\mathbf{n}\times\dot{\boldsymbol{\beta}})}{R},\quad 4\pi\mathbf{B}=\frac{q}{c}\,\frac{\dot{\boldsymbol{\beta}}\times\mathbf{n}}{R} \tag{16.1}$$

From formula (14.9) we find energy flux as a function of the angle between directions of vectors $\mathbf{n}$ and $\dot{\boldsymbol{\beta}}$ denoted by ϑ:

$$\mathbf{S}=\frac{1}{(4\pi)^2}\,\frac{q^2}{c^3R^2}\,\dot{v}^2\sin^2\vartheta\cdot\mathbf{n} \tag{16.2}$$

This formula shows that the maximum energy flux in the co-moving reference frame of the charge propagates in the plane orthogonal to acceleration ($\vartheta=\pi/2$), and that no radiation is emitted along $\dot{\mathbf{v}}$ ($\vartheta=0,\ \pi$). Radiation intensity is inversely proportional to R^2, which corresponds to a familiar relation for a pointlike light source.

Total energy emitted by the source per unit time in all directions is found by trivial integration over sphere σ of radius R:

$$\oint\mathbf{S}\cdot\mathbf{n}\,d\sigma=\frac{1}{4\pi}\cdot\frac{2}{3}\,\frac{q^2\dot{v}^2}{c^3} \tag{16.3}$$

In accordance with the condition of retardation for observation carried out at time moment t, the center of sphere σ must be placed at the point where the source was at time moment $t-R/c$. Relation (16.3) is known as the *Larmor formula*.

16.2. Let us analyse the general case when the charge moves at a velocity $\mathbf{v}$ with respect to the observer. The condition of retardation will have to be applied to formula (14.6) for the Poynting vector. This condition states that the left-hand side of the formula determines energy flux referred to the time of observation t, while the right-hand side is calculated from (14.4) at time $t'=t-R/c$. But first of all we are interested in finding radiation power during the interval dt' of emission of this radiation and not during interval dt of observation. Indeed, this is the amount of energy lost by the charge generating electromagnetic field. As a resutl, the power emitted by the charge into solid angle $d\Omega$ is

$$-\frac{\partial w(\vartheta,\ \varphi)}{\partial t'}\,d\Omega=|\,\mathbf{S}\,|\,\frac{dt}{dt'}\,R^2\,d\Omega=|\,\mathbf{S}\,|\,\varkappa R^2\,d\Omega \tag{16.4}$$

where, as in (13.16),

$$\frac{dt}{dt'}=1+\frac{1}{c}\,\frac{dR}{dt'}=1-\boldsymbol{\beta}\cdot\mathbf{n}=\varkappa$$

[16] Superscript 1 of field strength is dropped throughout this section, since only radiation field is meant. Neither is the retardation condition mentioned explicitly, though it must be taken into account everywhere.

Squared length of vector **E** is given by

$$(4\pi)^2 \mathbf{E}^2 = \frac{q^2}{c^2\varkappa^6 R^2}\{\dot{\beta}^2\varkappa^2 - (\mathbf{n}\cdot\dot{\boldsymbol{\beta}})^2(1-\beta^2) + 2\varkappa(\mathbf{n}\cdot\dot{\boldsymbol{\beta}})(\boldsymbol{\beta}\cdot\dot{\boldsymbol{\beta}})\} \qquad (16.5)$$

Let us use spherical coordinates shown in Fig. 11. Azimuthal angle ζ is laid off the plane containing vectors **v** and $\dot{\mathbf{v}}$, and polar angle ϑ is laid off vector **v**. As can be seen from Fig. 11,

$$\mathbf{n}\cdot\mathbf{v} = v\cos\vartheta$$

$$\mathbf{n}\cdot\dot{\mathbf{v}} = \dot{v}(\cos\vartheta'\cos\vartheta + \sin\vartheta' \times \sin\vartheta\cos\zeta)$$

$$\mathbf{v}\cdot\dot{\mathbf{v}} = v\dot{v}\cos\vartheta' \qquad (16.6)$$

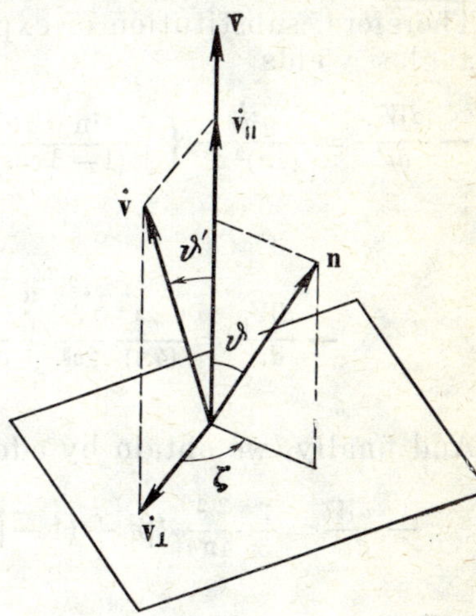

Fig. 11

Substitution of (16.6) into (16.5), (16.5) into (14.6), and (14.6) into (16.4) yields power emitted into solid angle $d\Omega$ in terms of angles ϑ and ζ. Power of emission will also depend on the angle ϑ' between velocity and acceleration. Total radiated power will be obtained by integration over $d\Omega$. We shall begin with this last problem.

For further calculations it will be important to decompose vector of acceleration into the components parallel and orthogonal to velocity:

$$\dot{\mathbf{v}} = \dot{\mathbf{v}}_{\|} + \dot{\mathbf{v}}_{\perp} \qquad (16.7)$$

Let us substitute this decomposition into (16.5) and first of all write out the terms containing products of $\dot{\mathbf{v}}_{\|}$ by $\dot{\mathbf{v}}_{\perp}$. They have the form

$$\frac{2}{c^2\varkappa^6R^2}\{\varkappa(\boldsymbol{\beta}\cdot\dot{\boldsymbol{\beta}}_{\|}) - (1-\beta^2)(\mathbf{n}\cdot\dot{\boldsymbol{\beta}}_{\|})\}(\mathbf{n}\cdot\dot{\boldsymbol{\beta}}_{\perp}) \qquad (16.8)$$

As follows from (16.6),

$$\mathbf{n}\cdot\dot{\mathbf{v}}_{\|} = \dot{v}\cos\vartheta'\cos\vartheta, \quad \mathbf{n}\cdot\dot{\mathbf{v}}_{\perp} = \dot{v}\sin\vartheta'\sin\vartheta\cos\zeta$$

that is expression (16.8) is proportional to $\cos\zeta$ and thus yields zero when integrated over angles ζ. The remaining factors in (16.4) are independent of ζ and therefore the term (16.8) can be ignored in the integration. Two terms are thus left in the integral containing (16.5). One of them depends only on $\dot{\mathbf{v}}_{\|}$, and the other only on $\dot{\mathbf{v}}_{\perp}$. Therefore,

$$\frac{dW}{dt'} = \int \frac{dw(\vartheta,\zeta)}{dt'}\,d\Omega = \frac{dW_{\perp}}{dt'} + \frac{dW_{\|}}{dt'} \qquad (16.9)$$

The integrands are found by using (16.4) and (16.5):

$$-\frac{dw_{\perp}}{dt'}=\frac{1}{(4\pi)^2}\frac{q^2}{c\varkappa^5}\{\varkappa^2\dot{\boldsymbol{\beta}}_{\perp}^2-(1-\beta^2)(\mathbf{n}\cdot\dot{\boldsymbol{\beta}}_{\perp})^2\} \qquad (16.10)$$

$$-\frac{dw_{\parallel}}{dt'}=\frac{1}{(4\pi)^2}\frac{q^2}{c\varkappa^5}\{\dot{\beta}_{\parallel}^2-(\mathbf{n}\cdot\dot{\boldsymbol{\beta}}_{\parallel})^2\} \qquad (16.11)$$

Therefore, substitution of expressions for $\varkappa$, $\mathbf{n}\cdot\dot{\boldsymbol{\beta}}_{\parallel}$, $\mathbf{n}\cdot\dot{\boldsymbol{\beta}}_{\perp}$ in terms of angles yields

$$-\frac{dW_{\perp}}{dt'}=\frac{q^2\dot{v}^2}{(4\pi)^2c^3}\left\{\int\frac{\sin\vartheta\,d\vartheta\,d\zeta}{(1-\beta\cos\vartheta)^3}-(1-\beta^2)\frac{\sin^3\vartheta\cos^2\zeta\,d\vartheta\,d\zeta}{(1-\beta\cos\vartheta)^5}\right\}$$

$$=\frac{1}{4\pi}\cdot\frac{2}{3}\frac{q^2}{c^3}\frac{\dot{v}_{\perp}^2}{(1-\beta^2)^2} \qquad (16.12)$$

$$-\frac{dW_{\parallel}}{dt'}=\frac{q^2}{(4\pi)}\frac{\dot{v}_{\parallel}^2}{2c^3}\int_0^{\pi}\frac{\sin^3\vartheta\,d\vartheta}{(1-\beta\cos\vartheta)^5}=\frac{1}{4\pi}\cdot\frac{2}{3}\frac{q^2}{c^3}\frac{\dot{v}_{\parallel}^2}{(1-\beta^2)^3} \qquad (16.13)$$

And finally, we obtain by adding the two preceding formulas

$$-\frac{dW}{dt'}=\frac{2}{3}\frac{q^2}{4\pi c}\{\dot{\beta}_{\parallel}^2+(1-\beta^2)\dot{\beta}_{\perp}^2\}(1-\beta^2)^{-3}=\frac{2}{3}\frac{q^2}{4\pi c}\frac{\dot{\beta}^2-(\dot{\boldsymbol{\beta}}\times\boldsymbol{\beta})^2}{(1-\beta^2)^3} \qquad (16.14)$$

From the standpoint of relativistic kinematics this result has a very simple meaning (this is somewhat unexpected after the cumbersome derivation). Indeed, recalling formula (5.20), we find that

$$\frac{dW}{dt'}=\frac{1}{4\pi}\frac{2}{3}\frac{q^2\vec{w}^2}{c^3} \qquad (16.15)$$

where $\vec{w}$ is the acceleration 4-vector. A comparison with the Larmor formula (16.3), derived earlier for the co-moving reference frame, shows that it coincides with (16.15) if $\dot{\mathbf{v}}$ is substituted by $\vec{w}$. By the way, the Larmor formula was derived for $\varkappa = 1$, and therefore presumes $dt = dt'$. Equation (16.15) coincides with the time component of equation (15.33) if in this last equation we take into account that $P^0 = W/c$, $u^0 = c\gamma$ and $\gamma\, d\tau = dt'$. The signs in these formulas are opposite because, as was indicated in § 15, formula (15.33) is the energy transferred by radiation, while (16.15) is the energy lost by the emitting source, namely by the accelerated point-like charge.

16.3. Now let us consider separately the case of acceleration directed along velocity, that is $\dot{\mathbf{v}} = \dot{\mathbf{v}}_{\parallel}$, and the case of acceleration orthogonal to velocity, that is $\dot{\mathbf{v}} = \dot{\mathbf{v}}_{\perp}$. In the first of these two cases the angular distribution of radiation power must be found by formula (16.11), and in the second case, by (16.10). Derivation of each of these formulas is, obviously, somewhat simplified if the constraint on the direction of acceleration is introduced directly into the expression for **E**. A comparison of expressions (16.13) and (16.12) for the total radiation power, corresponding to these two cases, shows that for the same magnitude of acceleration the ratio of total emitted radiation for $\dot{\mathbf{v}} \perp \mathbf{v}$ to the total emitted radiation for $\dot{\mathbf{v}} \parallel \mathbf{v}$ is equal to $1 - \beta^2$. Formula (16.11), expressed in terms of angle ϑ (this formula has been already used to calculate integral (16.13)), yields

$$-\frac{dw}{dt'} = \frac{1}{(4\pi)^2} \frac{q^2}{c^3} \frac{\dot{v}^2 \sin^2 \vartheta}{(1-\beta \cos \vartheta)^5} \tag{16.16}$$

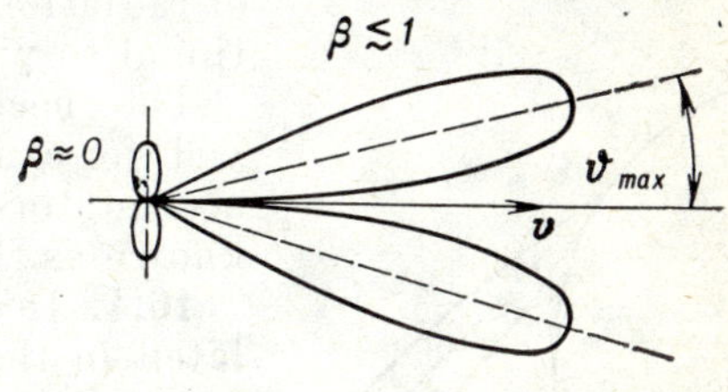

Fig. 12

The dependence on angle ϑ can be studied by usual methods of differential calculus for different values of parameter β. As β increases, distribution varies qualitatively as shown in Fig. 12. Obviously the case of $\beta = 0$ is the one in which the Larmor formula (16.2) is valid. Typically, as β increases, the "lobes" of radiation (or rather, the "cone" of radiation, since the distribution is symmetric with respect to axis **v**) become elongated and are tilted near to the direction of vector **v**. It can be shown[17] that the angle at which the power of emission reaches the maximum is

$$\vartheta_{\max} = \arccos \left(\frac{1}{3\beta} \sqrt{1 + 15\beta^2} - 1 \right)$$

In the limit $\beta = 1$ this angle tends to the value $1/2\gamma$, and the maximum radiation power is proportional to γ^8. An approximate formula is

$$-\frac{dw}{dt'} \simeq \frac{2}{\pi^2} \frac{q^2 \dot{v}^2}{c^3} \gamma^8 \frac{(\gamma\vartheta)^2}{(1+\gamma^2\vartheta^2)^5}, \quad \frac{dW}{dt'} \simeq \frac{1}{6\pi} \frac{q^2}{c^3} \dot{v}^2 \gamma^6 \tag{16.17}$$

The above analysis demonstrates, in particular, that an electron slowed down by an external field is emitting radiation: it generates the *bremsstrahlung* radiation, a phenomenon important in a number of physical problems. The formulas given above are sufficient to study the bremsstrahlung radiation if the electron is slowed down without changing the direction of velocity.

The case $\dot{\boldsymbol{\beta}} \cdot \boldsymbol{\beta} = 0$ represents, for example, the instantaneous emission by a charge moving in a circular orbit. Formula (16.10) valid in this case for $\beta \to 1$ (that is for $\gamma \to \infty$) can be written in an

[17] Cf. J. Jackson, *Classical Electrodynamics* (2nd edition), Wiley, New York, 1975.

approximate form[18]:

$$-\frac{dw}{dt'}\simeq\frac{1}{2\pi^2}\frac{q^2\dot{v}^2}{c^3}\gamma^6\frac{1}{(1+\gamma^2\vartheta^2)^3}\left[1-\frac{4\gamma^2\vartheta^2\cos^2\zeta}{(1+\gamma^2\vartheta^2)^2}\right] \tag{16.18}$$

Here again the emitted radiation is obviously concentrated in the direction of motion (that is for $\vartheta\to 0$). The total power of emission of radiation is given by (16.12) and proportional to γ^4.

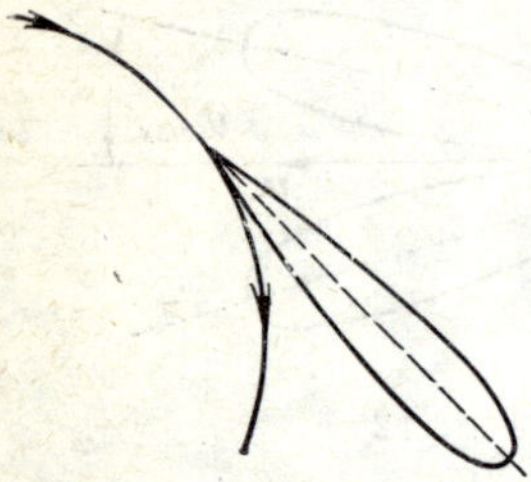

Fig. 13

The "needle" emission of this type is observed, for example, when particles move in cyclic accelerators (Fig. 13). For this reason it is known as the synchrotron radiation[19].

16.4. In addition to energy which we calculated in the preceding sections, the radiation field also possesses a mechanical momentum given by formulas (10.22). In fact, calculation of momentum is completely analogous to that of energy, and the result of calculation agrees with the carlier formula (15.33). Angular momentum $M^{\alpha\beta}$ of the radiation field can be found from the general expression (10.23) which in three-dimensional notation leads to (3.22). Function $\mathbf{r}\times\boldsymbol{\varphi}$ in the integrand of the surface integral in this formula can be interpreted as the flux of angular momentum across boundary σ of the volume under consideration. In § 3 it was shown that vector $\boldsymbol{\varphi}$ is defined by relation $\varphi^\alpha = T^{\alpha\beta}n_\beta$, where $T^{\alpha\beta}$ is the Maxwell tensor of stress, and $\mathbf{n}$ is the normal to surface σ. We readily find by formulas of § 3 for this tensor that

$$\boldsymbol{\varphi}=-\mathbf{n}\frac{E^2+B^2}{2}+\mathbf{E}(\mathbf{n}\cdot\mathbf{E})+\mathbf{B}(\mathbf{n}\cdot\mathbf{B}) \tag{16.19}$$

From this last formula we can calculate density $\mathbf{r}\times\boldsymbol{\varphi}$ of the angular momentum flux in vector form. In particular, let us consider this quantity in the rest frame of the source, that is for $\mathbf{v}=0$, and calculate the total flux of angular momentum through the sphere whose centre coincides with the source; in other words, we want to calculate integral

$$\oint(\mathbf{r}\times\boldsymbol{\varphi})\,d\sigma \tag{16.20}$$

over this sphere, with the normal to this sphere coinciding with vector $\mathbf{n}=\mathbf{r}/r$. However

$$\mathbf{r}\times\boldsymbol{\varphi}=(\mathbf{r}\times\mathbf{E})\,\mathbf{n}\cdot\mathbf{E}+(\mathbf{r}\times\mathbf{B})\,\mathbf{n}\cdot\mathbf{B} \tag{16.21}$$

[18] See the monograph by J. Jackson cited in the preceding footnote.
[19] The theory of radiation is given more fully and in more detail in the monograph by L. D. Landau and E. M. Lifshits *The Classical Theory of Fields* (Course of Theoretical Physics, vol. 2), Pergamon Press, Oxford, 1975.

Fields **E** and **B** being transverse, we have $\mathbf{r} \times \boldsymbol{\varphi} = 0$, which means that the angular momentum of the field in the rest frame is conserved.

16.5. In the preceding section we considered the radiation energy flux into an elementary solid angle $d\Omega$ during interval dt' and then integrated it over the angle. It is often of interest to solve a different problem: to calculate energy flux transferred within a given solid angle during the whole time of emission. Usually it is necessary to find the flux recorded by the observer, that is to find power of radiation as a function of time t, without transformation to time t' carried out in formula (16.4). We shall denote the total radiation energy integrated over time by $\mathscr{E}$. Then

$$\frac{d\mathscr{E}}{d\Omega} = c \int_{-\infty}^{+\infty} R^2 (\mathbf{E}(t))^2 \, dt \tag{16.22}$$

To simplify the expressions, let us introduce the notation $\mathbf{A}(t) \equiv c^{1/2} R \mathbf{E}(t)$. It must be kept in mind that the derived formulas give field **E** with the effect of retardation included, that is for time moment t'. This will be taken into account in further calculations. In addition, using (16.22) we shall assume that the region in which the source is located is seen from the observation point at a small solid angle during all the time of emission.

We shall make use of expansions into the Fourier integral, similar to formulas (E.12) and (E.13):

$$\mathbf{A}(t) = \frac{1}{\sqrt{2\pi}} \int_{-\infty}^{+\infty} \mathbf{A}(\omega) e^{-i\omega t} \, d\omega, \quad \mathbf{A}(\omega) = \frac{1}{\sqrt{2\pi}} \int_{-\infty}^{+\infty} \mathbf{A}(t) e^{i\omega t} \, dt \tag{16.23}$$

For convenience, the normalizing factor in the integrals will be written in a form that used in Appendix E. As $\mathbf{A}(t)$ is a real quantity, that is $\mathbf{A}(t) = \mathbf{A}(t)^*$, then

$$\mathbf{A}(\omega) = \mathbf{A}^*(-\omega) \tag{16.24}$$

Formulas (16.23) represent spectral distribution of the field over possible frequencies ω. Substitution of these formulas into (16.22) yields

$$\frac{d\mathscr{E}}{d\Omega} = \frac{1}{2\pi} \int_{-\infty}^{+\infty} dt \int_{-\infty}^{+\infty} d\omega \int_{-\infty}^{+\infty} d\omega' \, \mathbf{A}(\omega') \cdot \mathbf{A}(\omega) \, e^{i(\omega' + \omega)t}$$

$$= \int_{-\infty}^{+\infty} |\mathbf{A}(\omega)|^2 \, d\omega = 2 \int_{0}^{\infty} |\mathbf{A}(\omega)|^2 \, d\omega \tag{16.25}$$

This expression includes squared modulus of a complex variable, $|\mathbf{A}(\omega)|^2$. Here we have used expansion (C.14) of delta function into the Fourier integral and relation (16.24).

It is also possible to find energy $d\mathscr{E}(\omega)/d\Omega$ emitted into a unit solid angle; first, let us calculate the part of radiation transferred by vibrations with frequencies in the range from ω to $\omega + d\omega$. Formula (16.25) yields

$$\frac{d\mathscr{E}}{d\Omega} = \int\limits_0^\infty \frac{dI(\omega)}{d\Omega}\, d\omega \tag{16.26}$$

where

$$\frac{dI(\omega)}{d\Omega} = 2\,|\mathbf{A}(\omega)|^2 \tag{16.27}$$

is the spectral intensity of emission into a unit solid angle.

Let us modify the general formulas obtained above to a specific case of radiation field determined by term $\mathbf{E}^{(1)}$ in formula (14.4) (we again denote this term simply by $\mathbf{E}$). Equation (16.23) yields

$$\mathbf{A}(\omega) = \frac{1}{4\pi}\frac{q}{\sqrt{2\pi c}} \int\limits_{-\infty}^{+\infty} e^{i\omega t}\, \frac{\mathbf{n}\times[(\mathbf{n}-\boldsymbol{\beta})\times\dot{\boldsymbol{\beta}}]}{\varkappa^3}\Bigg|_{t'=t-R/c} dt$$

$$= \frac{1}{4\pi}\frac{q}{\sqrt{2\pi c}} \int\limits_{-\infty}^{+\infty} e^{i\omega(t'+R/c)}\, \frac{\mathbf{n}\times[(\mathbf{n}-\boldsymbol{\beta})\times\dot{\boldsymbol{\beta}}]}{\varkappa^2}\, dt' \tag{16.28}$$

In order to contract notations we place the origin within a bounded region in which the source moves, and assume that the obversation point is very far from this region, that is $r \gg r'$. Then

$$R(t') \simeq r - \mathbf{n}\cdot\mathbf{r}(t') \tag{16.29}$$

Here $\mathbf{n} = \mathbf{r}/r$, in contrast to preceding cases where this notation was used for unit vector in the direction of $\mathbf{R}$. By substituting (16.29) into (16.28), dropping the constant phase factor $\exp\left(\frac{i}{c}\,\omega r\right)$, and denoting the variable of integration by t, we obtain

$$\mathbf{A}(\omega) = \frac{1}{4\pi}\frac{q}{\sqrt{2\pi c}} \int\limits_{-\infty}^{+\infty} \exp\left\{i\omega\left[t - \frac{\mathbf{n}\cdot\mathbf{r}(t)}{c}\right]\right\} \frac{\mathbf{n}\times[(\mathbf{n}-\boldsymbol{\beta})\times\dot{\boldsymbol{\beta}}]}{\varkappa^2}\, dt \tag{16.30}$$

It can be shown, by a straightforward calculation of the derivative, that

$$\frac{d}{dt}\left(\frac{\mathbf{n}\times(\mathbf{n}\times\boldsymbol{\beta})}{\varkappa}\right) = \frac{\mathbf{n}\times[(\mathbf{n}-\boldsymbol{\beta})\times\dot{\boldsymbol{\beta}}]}{\varkappa^2}$$

(recall that $\varkappa = 1 - \mathbf{n}\cdot\boldsymbol{\beta}$). Therefore the integral in (16.30) can be calculated by parts if we assume that $\boldsymbol{\beta}$ vanishes at the beginning and at the end of the range of integration. The formula for spectral intensity (16.27) then takes the form

$$\frac{dI(\omega)}{d\Omega} = \frac{q^2}{(4\pi)^2\pi c}\,\omega^2\left|\int_{-\infty}^{+\infty} \mathbf{n}\times(\mathbf{n}\times\boldsymbol{\beta})\exp\left\{i\omega\left[t-\frac{\mathbf{n}\cdot\mathbf{r}(t)}{c}\right]\right\}dt\right|^2 \tag{16.31}$$

Note that if q is replaced by $\rho\,dV$, and $\rho\boldsymbol{\beta}$ is assumed equal to $\mathbf{j}/c$, then from (16.31) we can find a formula for radiation emitted by continuously distributed sources:

$$\frac{dI(\omega)}{d\Omega} = \frac{\omega^2}{16\pi^3c^3}\left|\int dt\int dV\,\mathbf{n}\times[\mathbf{n}\times\mathbf{j}(\mathbf{r},t)]\exp\left[i\omega\left(t-\frac{\mathbf{n}\cdot\mathbf{r}}{c}\right)\right]\right|^2 \tag{16.32}$$

Here it is implicitly assumed that all elements of volume of this distribution of sources emit independently of one another.

§ 17. Emission from bounded oscillating sources

17.1. No special assumptions were introduced above to calculate the field generated by a pointlike charge. In principle, formulas (13.11) make it possible to calculate electromagnetic fields generated by arbitrarily distributed sources in infinite space. However, if functions $\rho(\mathbf{r}', t')$ and $\mathbf{j}(\mathbf{r}', t')$ are known, such calculation is usually a very difficult problem, and an exact solution can be found only in very special cases. In what follows we shall consider one of the simplest but at the same time one of the most important cases of approximate calculation of the field by using formulas (13.11). The retardation condition will be taken into account by resorting to delta function.

Both in electronics and in the classical analysis of emission by the atom we have to assume that charges and currents generating radiation are located in a fixed volume. The calculation given below makes it possible to find field in precisely this case, at distances large compared with linear dimensions of the region occupied by sources[20].

Let us expand functions $\mathbf{A}$, φ, $\mathbf{j}$ and ρ into Fourier integrals of type (E.12) and substitute these expansions into formulas (13.11). By using properties of delta function and comparing coefficients of $\exp(-i\omega t)$ in the right- and left-hand sides of these formulas, we easily arrive

[20] This aspect is treated more fully in J. A. Stratton, *Electromagnetic Theory*, McGrow-Hill, New York, 1941. See also the book by J. Jackson.

at relations for amplitudes (they have the form of (E.18)):

$$4\pi\varphi_\omega(\mathbf{r}) = \int dV' \frac{e^{ik|\mathbf{r}-\mathbf{r}'|}}{|\mathbf{r}-\mathbf{r}'|}\rho_\omega(\mathbf{r}')$$

$$4\pi c\mathbf{A}_\omega(\mathbf{r}) = \int dV' \frac{e^{ik|\mathbf{r}-\mathbf{r}'|}}{|\mathbf{r}-\mathbf{r}'|}\mathbf{j}_\omega(\mathbf{r}') \qquad (17.1)$$

The notations here are standard: $k = \omega/c = 2\pi/\lambda$, and λ is the wavelength of harmonic oscillation with cyclic frequency ω propagating at velocity c (we again specify the case of propagation in vacuo). As a rule, in the formulas to follow we drop subscript ω in amplitudes.

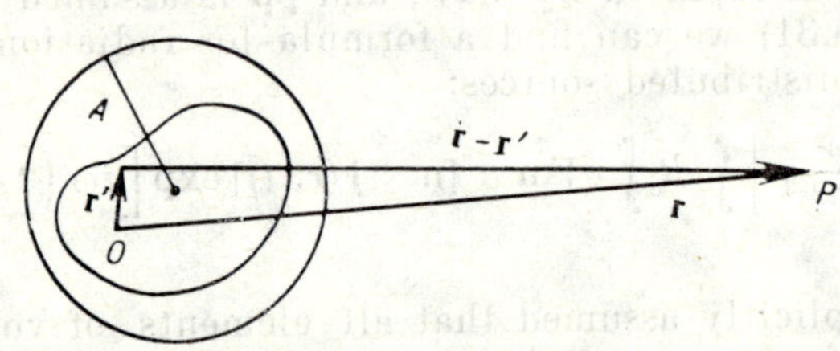

Fig. 14

Assume now that charges and currents generating electromagnetic field do not leave a bounded region in space, at least during an interval of time significant for observation. The size of this region can be characterized by a finite radius A of a sphere inclosing this region. For simplicity, we place the origin of spatial coordinates somewhere inside this region (Fig. 14). We shall calculate the field at the observation point removed very far from the sources. This means that $r \gg A$. And since $r' < A$, $r \gg r'$.

This condition, however, is still insufficient for the application of the approximate technique which we are going to use. We also have to assume that wavelength λ of radiation waves of interest to us is also much larger than the characteristic length A, so that the actual constraints imposed by the approximation can be written in the form

$$A \ll \lambda \text{ and } A \ll r \qquad (17.2)$$

As follows from (17.2), $kr' \ll 1$.

Consider now an expansion of function $\exp(ik|\mathbf{r}-\mathbf{r}'|)/|\mathbf{r}-\mathbf{r}'|$ in the indicated small parameters kr' and r'/r. We shall retain and calculate only the terms containing these parameters to the zero and first power. In this approximation the expression becomes very simple. Indeed, we can therefore neglect ratio $(r'/r)^2$ in expression $|\mathbf{r}-\mathbf{r}'| = \sqrt{r^2 - 2(\mathbf{r}\cdot\mathbf{r}') + r'^2}$ (already under the radical sign) compared to unity, and write

$$|\mathbf{r}-\mathbf{r}'| \simeq \mathbf{r}\sqrt{1-\frac{2}{r}\mathbf{n}\cdot\mathbf{r}'} \simeq r - \mathbf{n}\cdot\mathbf{r}' \qquad (17.3)$$

Here $\mathbf{n}$ denotes a unit vector $\mathbf{r}/r$ and not $\mathbf{R}/R$ as we did above. It must be kept in mind that the addent neglected under the radical sign gives, in a more exact calculation, such additional terms in the integrand as, for example, $(r'/r)^2 \exp(ik\,|\mathbf{r}-\mathbf{r}'|)$. These terms can be ignored only if the required accuracy is not higher than that defined at the beginning of the section.

By using approximation (17.3) in the exponent of the exponential term, expanding this exponent into a series in a small parameter kr', and retaining the terms as agreed above, we obtain

$$\mathbf{A}(\mathbf{r}) = \mathbf{A}^{(1)}(\mathbf{r}) + \mathbf{A}^{(2)}(\mathbf{r}) \tag{17.4}$$

where

$$4\pi\mathbf{A}^{(1)}(\mathbf{r}) = \frac{e^{ikr}}{cr}\int \mathbf{j}(\mathbf{r}')\,dV' \tag{17.5}$$

and

$$4\pi\mathbf{A}^{(2)}(\mathbf{r}) = \frac{e^{ikr}}{cr}\left(\frac{1}{r}-ik\right)\int \mathbf{j}(\mathbf{r}')\,\mathbf{n}\cdot\mathbf{r}'\,dV' \tag{17.6}$$

Of course, similar formulas are obtained for amplitude $\varphi(r)$ if $\mathbf{j}$ in the integrand is replaced by $c\rho$.

It can be seen from formulas relating field strength with potentials that if potentials depend on time as $e^{-i\omega t}$, then field strengths will depend on time in the same way. Substituting expressions $\mathbf{E} = \mathbf{E}_\omega(\mathbf{r})\,e^{-i\omega t}$ and $\mathbf{B} = \mathbf{B}_\omega(\mathbf{r})\,e^{-i\omega t}$ into the Maxwell equations, we obtain

$$\mathbf{E}_\omega = \frac{i}{k}\,\mathrm{curl}\,\mathbf{B}_\omega \tag{17.7}$$

for points in space outside of the region in which sources are located. Magnetic field is calculated by the familiar formula

$$\mathbf{B}_\omega = \mathrm{curl}\,\mathbf{A}_\omega \tag{17.8}$$

As was the case with potentials, in further formulas we drop subscript ω. A factor that can be used in analyzing the results is the relative value of λ and r (we assume, of course, that (17.2) always holds). The range of $r \gg \lambda$ is called the *wave zone*, and the range $r' \ll r \ll \lambda$—the *short-range zone*.

For convenience of calculations, the integrals in (17.5) will be somewhat modified. Namely, we can make use of the continuity equation, written in this case in the form

$$i\omega\rho_\omega = \mathrm{div}\,\mathbf{j}_\omega \tag{17.9}$$

and rewrite the integrand in terms of ρ. By using (B.14$_2$), we can write for each component x'^α of radius vector $\mathbf{r}'$

$$\mathrm{div}'\,(x'^\alpha\mathbf{j}(\mathbf{r}')) = x'^\alpha\,\mathrm{div}'\,\mathbf{j} + (\mathrm{grad}'\,x'^\alpha\cdot\mathbf{j}) = x'^\alpha\,\mathrm{div}'\,\mathbf{j} + j^\alpha \tag{17.10}$$

But the integral over volume in the left-hand side of this equation can be transformed into an integral over a surface, and the surface can be chosen in such manner that the currents on it be equal to zero (for example, we could take the sphere shown in Fig. 14). As a result, we obtain from (17.9) and (17.5)

$$4\pi \mathbf{A}^{(1)}(\mathbf{r}) = -ik\mathbf{p}\,\frac{e^{ikr}}{r} \tag{17.11}$$

where

$$\mathbf{p} = \int \mathbf{r}' \rho(\mathbf{r}')\, dV' \tag{17.12}$$

that is $\mathbf{p}$ is the dipole moment of the distribution of sources, defined precisely as in § 11. We can also write for the considered harmonic component of potential at frequency ω that

$$\mathbf{A}^{(1)}(\mathbf{r}, t) = \mathbf{A}^{(1)}(\mathbf{r})\, e^{-i\omega t} = \dot{\mathbf{p}}(t)\, e^{ikr}/4\pi rc$$

f $\mathbf{p}(t) = \mathbf{p}e^{-i\omega t}$. Obviously, this expression of $\mathbf{A}^{(1)}$ in terms of $\dot{\mathbf{p}}$ remains valid for all those wavelengths which satisfy the conditions defining applicability of the approximation used in the derivation.

17.2. In the first approximation electromagnetic field is found by substituting (17.11) into (17.8), and then applying (17.7). Elementary calculations using ($B.14_3$) yield

$$4\pi \mathbf{B}^{(1)} = k^2 \mathbf{n} \times \mathbf{p}\,\frac{e^{ikr}}{r}\left(1 - \frac{1}{ikr}\right) \tag{17.13}$$

$$4\pi \mathbf{E}^{(1)} = k^2 (\mathbf{n} \times \mathbf{p}) \times \mathbf{n}\,\frac{e^{ikr}}{r} + \{3\mathbf{n}(\mathbf{n}\cdot\mathbf{p}) - \mathbf{p}\}\left(\frac{1}{r^3} - \frac{ik}{r^2}\right) e^{ikr} \tag{17.14}$$

In complete analogy to what was done earlier, and with the same qualifications, we can replace $k^2\mathbf{p}$ by $-\frac{1}{c^2}\ddot{\mathbf{p}}$, and $-ik\mathbf{p}$ by $\frac{1}{c}\dot{\mathbf{p}}$. This also holds for expressions analyzed below.

In the short-range zone formulas (17.13) and (17.14) take the form

$$4\pi \mathbf{B}^{(1)} = ik\mathbf{n} \times \mathbf{p}\,\frac{1}{r^2}, \quad 4\pi \mathbf{E}^{(1)} = [3\mathbf{n}(\mathbf{n}\cdot\mathbf{p}) - \mathbf{p}]\,\frac{1}{r^3} \tag{17.15}$$

The first of them coincides with the Ampère law if the element of current $I\, d\mathbf{s}'$ is replaced by $-i\omega\mathbf{p} = \dot{\mathbf{p}}$. This coincidence takes place at any moment of time, and both sides of the equality depend on time as $\exp(-i\omega t)$. In the second formula the dependence on time is taken into account in the absolutely similar manner. It will be easy to show by using the results of § 11 that amplitude $\mathbf{E}^{(1)}$ in the short-range zone is equal to the strength of static electric field produced by a dipole with dipole moment $\mathbf{p}$.

In the wave zone

$$4\pi \mathbf{B}^{(1)} = k^2 \mathbf{n} \times \mathbf{p}\,\frac{e^{ikr}}{r}, \quad 4\pi \mathbf{E}^{(1)} = k^2 (\mathbf{n} \times \mathbf{p}) \times \mathbf{n}\,\frac{e^{ikr}}{r} \tag{17.16}$$

These expressions show that the wave field has a structure similar to that of the radiation field of a pointlike charge. First, we again arrive at a conclusion that a nonzero energy flux at large distances from sources is characteristic only of the wave field; this conclusion follows from comparing the dependence of fields (17.15) and (17.16) on distance r (cf. p. 112). Second, the wave field is determined by the second derivative $\ddot{\mathbf{p}}(t)$ with respect to time, in analogy to the field of a charge determined by its acceleration. And finally, equations (17.16) directly yield

$$\mathbf{E}^{(1)} = \mathbf{B}^{(1)} \times \mathbf{n}, \quad \mathbf{B}^{(1)} = \mathbf{n} \times \mathbf{E}^{(1)} \tag{17.17}$$

The wave field is therefore transverse, and the mutual arrangement of vectors $\mathbf{n}$, $\mathbf{E}^{(1)}$ and $\mathbf{B}^{(1)}$ at each point of the field is shown in Fig. 7. One important distinction is the fact that the left- and right-hand sides in equations (17.13)-(17.16) refer to the same time moment t, that is to the moment of observation, while in (14.14), for example, retardation must be explicitly taken into account.

17.3. We should not forget, in the calculation of energy flux in the case under discussion, that field strengths are complex-valued. The definition of energy flux given on p. 112 operates with real strengths. Of course, we could separate real parts in formulas (17.16). However, in the case of harmonic dependence on time, it is simpler to apply these formulas directly, assuming at the same time that the observation covers a time interval Δt much longer than the period of the observed oscillations of electromagnetic field. Then the quantity which interests the observer is in fact the mean value of flux $(\mathbf{S}\cdot\mathbf{n})\, r^2$ per unit solid angle over the indicated interval of time. Orthogonality conditions (17.17) give $(\mathbf{E} \times \mathbf{B})\cdot\mathbf{n} = EB$. However, we are interested not in this complex variable but in the product $\operatorname{Re} E \cdot \operatorname{Re} B$, a quantity having physical meaning. As $\operatorname{Re} E = (1/2)\,(E + E^*)$, and a similar expression can be written for $\operatorname{Re} B$, then

$$\operatorname{Re} R \cdot \operatorname{Re} B = \frac{1}{4}(EB + E^*B^* + E^*B + EB^*)$$

In the case of a purely harmonic dependence on frequency ω, the first two terms contain $\exp(-2i\omega t)$ and $\exp 2i\omega t$, respectively and the exponential factors of the remaining two terms cancel out. Denoting by $\langle\ldots\rangle_{\Delta t}$ the operation of averaging over time Δt, we find that

$$\langle \operatorname{Re} E \cdot \operatorname{Re} B\rangle_{\Delta t} = \frac{1}{4}(E^*B + EB^*) = \frac{1}{2}\operatorname{Re}(EB^*) \tag{17.18}$$

and we can assume that E and B in the right-hand side of this formula are complex amplitudes. Consequently, we can rewrite the mean power of emission from an electric dipole $\mathbf{p}$, in the notation of the preceding section, in the form

$$-\frac{dw}{dt}\,d\Omega=\frac{c}{2}\,\mathrm{Re}\,[r^2(\mathbf{E}\times\mathbf{B}^*)\cdot\mathbf{n}]=\frac{c}{2\,(4\pi)^2}\,k^4|\mathbf{n}\times(\mathbf{n}\times\mathbf{p})|^2 \qquad (17.19)$$

in which formulas (17.16) were used.

Denoting the angle between vectors **n** and **p** by ϑ and assuming that components of dipole moment **p** oscillate in phase, we arrive at a formula for angular distribution of power lost by the dipole on emission of radiation:

$$-(4\pi)^2\frac{dw}{dt}=\frac{c}{2}\,k^4|\mathbf{p}|^2\sin^2\vartheta \qquad (17.20)$$

Integration of the above formula over angles yields the total power:

$$-\frac{dW}{dt}=\frac{1}{4\pi}\cdot\frac{1}{3}\,ck^4|\mathbf{p}|^2 \qquad (17.21)$$

17.4. The next approximation of field is found quite similarly, on the basis of the formula for potential (17.6). The integrand must be presented as a sum of symmetric and antisymmetric parts:

$$\mathbf{j}\,(\mathbf{n}\cdot\mathbf{r}')=\frac{1}{2}\,[\mathbf{j}\,(\mathbf{n}\cdot\mathbf{r}')+\mathbf{r}'\,(\mathbf{n}\cdot\mathbf{j})]+\frac{1}{2}\,[\mathbf{j}\,(\mathbf{n}\cdot\mathbf{r}')-\mathbf{r}'\,(\mathbf{n}\cdot\mathbf{j})] \qquad (17.22)$$

and each of the addends must be considered separately.

The integral with the symmetric component of (17.22) can be transformed by using the relation

$$n_\beta\,\mathrm{div}'\,(x'^\alpha x'^\beta\mathbf{j})=n_\beta x'^\alpha x'^\beta\,\mathrm{div}'\,\mathbf{j}+[j^\alpha\,(\mathbf{n}\cdot\mathbf{r}')+x'^\alpha\,(\mathbf{n}\cdot\mathbf{j})] \qquad (17.23)$$

and taking into account, in complete analogy to the earlier transformation of formula (17.10), that integration of the left-hand side yields zero. By using also the continuity equation in the form (17.9), we obtain the corresponding part of vector potential:

$$4\pi\mathbf{A}^{(2)}_{\mathrm{sym}}(\mathbf{r})=-\frac{k^2}{2}\,\frac{e^{ikr}}{r}\left(1-\frac{1}{ikr}\right)\int\mathbf{r}'\,(\mathbf{n}\cdot\mathbf{r}')\,\rho\,(\mathbf{r}')\,dV' \qquad (17.24)$$

A comparison with (11.19) allows to rewrite the integral in (17.24) in the form $n^\alpha d_{\alpha\beta}$. Consequently, the field produced by the "symmetric" component of vector potential can be called the *quadrupole field*.

Quadrupole electric and magnetic fields can be calculated from formulas (17.7) and (17.8) by making use of (17.24). Here we give only selected results.[21] In the wave zone

$$\mathbf{B}^{(2)}=ik\,(\mathbf{n}\times\mathbf{A}^{(2)}),\quad \mathbf{E}^{(2)}=ik\,(\mathbf{n}\times\mathbf{A}^{(2)})\times\mathbf{n} \qquad (17.25)$$

[21] See, for example, Chapter 9 of Jackson's monograph cited on p. 127. As above, we used the Heaviside-Lorentz system of units.

If we use definition (11.19_2) of quadrupole moment $Q_{\alpha\beta}$ and introduce vector $\mathbf{Q}$ by a definition $Q_\alpha = Q_{\alpha\beta}n^\beta$, then

$$4\pi\mathbf{B}^{(2)} = -\frac{ik^3}{6}\frac{e^{ikr}}{r}\mathbf{n}\times\mathbf{Q}(\mathbf{n}) \tag{17.26}$$

Calculation of the mean power of radiation is similar to that given on p. 136 and yields formulas

$$-(4\pi)^2\frac{dw}{dt} = \frac{c}{72}k^6|\mathbf{n}\times\mathbf{Q}(\mathbf{n})|^2, \quad -4\pi\frac{dW}{dt} = \frac{ck^6}{360}\sum_{\alpha,\beta}|Q_{\alpha\beta}|^2, \tag{17.27}$$

Note that in contrast to dipole emission whose power is proportional to k^4, here the mean power is proportional to k^6.

The antisymmetric term in (17.22) is transformed straightforwardly to the form $(1/2)\,\mathbf{n}\times(\mathbf{j}\times\mathbf{r}') = (1/2)\,(\mathbf{r}'\times\mathbf{j})\times\mathbf{n}$. Let us recall now the definition of magnetic moment (12.13) (and set $\alpha = c$) and make use of (12.5). Clearly the expression

$$\mathbf{m} = \frac{1}{2c}\mathbf{r}\times\mathbf{j} \tag{17.28}$$

has the meaning of volume density of magnetic moment. The total magnetic moment obtained by integrating density (17.28) over the whole volume will be denoted by $\mathcal{M}$. Hence,

$$4\pi\mathbf{A}_{ac}^{(2)} = ik\mathbf{n}\times\mathcal{M}\frac{e^{ikr}}{r}\left(1-\frac{1}{ikr}\right) \tag{17.29}$$

With qualifications already indicated with respect to (17.15), we can also write $i\omega\mathcal{M} = -\dot{\mathcal{M}}(t)$. A comparison of the obtained result with formula (17.13) for $\mathbf{B}^{(1)}$ shows that these expressions are formally identical to the accuracy of a constant factor if $\mathbf{p}$ in (17.13) is replaced by $\mathcal{M}$. Therefore, calculation of magnetic field $\mathbf{B}^{(2)}$ by formulas (17.8) and (17.29) gives a result which coincides, again to the accuracy of a constant factor, with a formula for $\mathbf{E}^{(1)}$ derived on the basis of (17.7). Hence,

$$4\pi\mathbf{B}^{(2)} = k^2(\mathbf{n}\times\mathcal{M})\times\mathbf{n}\frac{e^{ikr}}{r} + [3\mathbf{n}(\mathbf{n}\cdot\mathcal{M})-\mathcal{M}]\left(\frac{1}{r^3}-\frac{ik}{r^2}\right)e^{ikr} \tag{17.30}$$

We can now easily show that electric field has the form

$$4\pi\mathbf{E}^2 = -k^2\mathbf{n}\times\mathcal{M}\frac{e^{ikr}}{r}\left(1-\frac{1}{ikr}\right) \tag{17.31}$$

that is can be formally obtained from (17.13) by replacing $\mathbf{p}$ by $\mathcal{M}$ and changing the sign of the left-hand side.

Electromagnetic field found from magnetic moment $\mathcal{M}$ by formulas (17.30) and (17.31) is called the *field of a magnetic dipole*. In the chosen approximation the field of a bounded distribution of sources can be expressed therefore as a sum of electric and magnetic dipoles and an electric quadrupole.

A conclusion that can be drawn from the abovementioned formal similarities between the fields of electric and magnetic dipoles is that in the case of magnetic dipole all the formulas, including that for the power of emission, are obtained from the corresponding expressions for the electric dipole by substituting **B** for **E**, −**E** for **B**, and $\mathcal{M}$ for **p**. However, the fields of electric and magnetic dipoles differ in the direction of the electric vector with respect to vectors **n** and **p** in the first case and **n** and $\mathcal{M}$ in the second case, that is differ in polarization (Fig. 15).

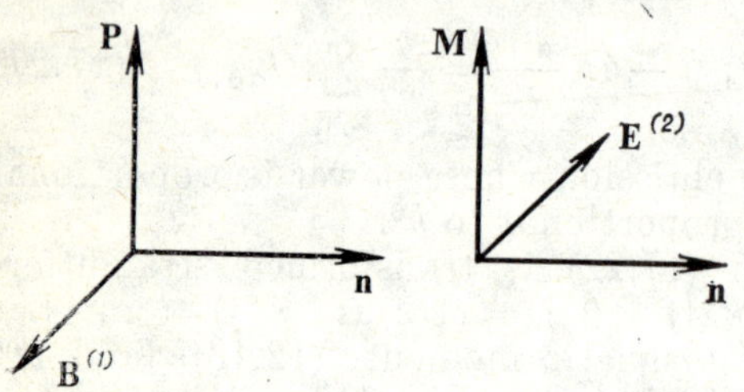

Fig. 15

The method of calculating radiation fields deserves one additional general remark. It is clear from the above presentation that the continuity equation plays an important role in the calculation of the fields. The properties of the distribution of sources producing the field are often such that the continuity equation can be taken into account at the very beginning. Indeed, let us assume that there exists a vector $\mathbf{P}_0$ satisfying relations

$$\mathbf{j}=\frac{\partial \mathbf{P}_0}{\partial t}, \quad \rho=-\operatorname{div} \mathbf{P}_0 \tag{17.32}$$

The continuity equation then becomes an identity, and the Maxwell equations take the form (2.14_1). It was shown in § 2 that they can be solved by analyzing the second-order wave equation (2.16) for the Hertz vector **Π** defined by relations (2.15). We rewrite these relations in SI assuming $\alpha = 1$. The wave equation can be resolved by a standard method; in particular, the following formula holds for harmonic component of **Π** with wave number $k = \omega/v$:

$$4\pi\varepsilon\Pi(\mathbf{r})=\int\limits_V \mathbf{P}_0(\mathbf{r}')\frac{e^{ikR}}{R}\,dV' \tag{17.33}$$

In complete analogy to what has been carried out above for potentials, we can use this equation to work out approximate solutions. With the Hertz vector known, we calculate potentials from formula (2.15) or fields from formulas

$$\mathbf{E}=\operatorname{grad}\operatorname{div}\mathbf{\Pi}-\varepsilon\mu\frac{\partial^2\mathbf{\Pi}}{\partial t^2}, \quad \mathbf{H}=\varepsilon\operatorname{curl}\frac{\partial\mathbf{\Pi}}{\partial t} \tag{17.34}$$

This method is often used, for example, to analyze radiation emitted by antennas.

CHAPTER 4

PROPERTIES OF RADIATION IN ISOTROPIC MEDIA

§ 18. Plane waves. Reflection and refraction. Interference

18.1. We have shown in the preceding sections (see also Appendix E) how to construct a solution of the nonhomogeneous wave equation by means of Green's function for which we choose the solution of the homogeneous wave equation having a singularity of the required type for $R \to 0$. Special attention must be paid, in particular, to formula (E.17) which determines the form of a spherically symmetric Green's function for the Helmholtz equation. The corresponding solution of wave equation depending on time harmonically and describing a field propagating from a pointlike source has the form

$$\frac{A}{R} e^{i(kR-\omega t)} \tag{18.1}$$

and is called the *spherical wave*. If an arbitrary spheric reference frame is chosen with the origin at the point where the source is located, then amplitude A may be a function of coordinate angles ϑ and φ of this frame.

Formulas (17.11) and (17.16) which were derived for potential and field strengths corresponding to electromagnetic field of electric dipole (this field diminishes at the lowest rate in the wave zone) also have the form of spherical wave (18.1); amplitude A is determined by the properties of the source. As a sufficiently good approximation, spherical waves can be considered as plane waves at very large distances from the source. This is quite obvious from the geometric point of view and can also be demonstrated on the basis of equation (18.1). Let us choose a new observation point P with radius vector $\mathbf{R}_1$ (see Fig. 16), so that $\mathbf{R} = \mathbf{R}_1 + \mathbf{r}$ and $\mathbf{R}_1 = \text{const}$. If R_1 is so much larger than r that, to a good approximation, we can set $1/R = 1/R_1$ and assume vector $\mathbf{k}$ to be constant in the sufficiently small region under consideration, then expression (18.1) will differ from function $e^{i(\mathbf{k}\cdot\mathbf{r}-\omega t)}$, defining a plane wave, only in a constant factor $\frac{A}{R_1} e^{i\mathbf{k}\cdot\mathbf{R}_1}$.

The properties of plane waves are very important in the theory of electromagnetic field. This is obvious already from the role played by the Fourier expansion in plane waves (see Appendix E).

The relation between potentials and field strengths takes on an especially simple form for electromagnetic field of this type. Indeed, by setting

$$\mathbf{A}_\omega = \mathbf{A}_{\omega k} e^{i(\mathbf{k}\cdot\mathbf{r}-\omega t)}, \quad \varphi_\omega = \varphi_{\omega k} e^{i(\mathbf{k}\cdot\mathbf{r}-\omega t)} \tag{18.2}$$

where $\mathbf{A}_{\omega k}$ and $\varphi_{\omega k}$ are constant amplitudes and $k = \omega/c$ [1], we can write the Lorentz condition imposed on potentials in order that the wave equations for them were of identical form as

$$\varphi_\omega = \mathbf{n}\cdot\mathbf{A}_\omega \tag{18.3}$$

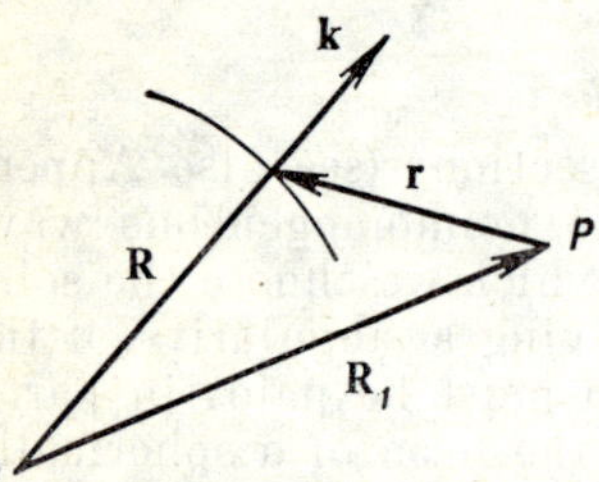

Fig. 16

Here and below, when speaking of plane waves, we denote by **n** a unit vector $\mathbf{k}/k$. It is clear from the above analysis that vector **n** is orthogonal to the plane coinciding with the wave front.

Substitution of expressions (18.2) into the basic formulas (2.1) and (2.2) yields

$$\mathbf{E}_\omega = -i(\mathbf{k}\varphi_\omega - k\mathbf{A}_\omega) = -ik\mathbf{n}\times(\mathbf{n}\times\mathbf{A}_\omega)$$
$$\mathbf{B}_\omega = ik\mathbf{n}\times\mathbf{A}_\omega \tag{18.4}$$

Note that it will be convenient to assume, in analogy to condition (E.6), that complex amplitudes have the following properties:

$$\mathbf{A}_{\omega\mathbf{k}} = \mathbf{A}^*_{-\omega,\,-\mathbf{k}}, \quad \varphi_{\omega\mathbf{k}} = \varphi^*_{-\omega,\,-\mathbf{k}} \tag{18.5}$$

Of course, only real parts of formulas (18.4) have physical meaning. By constructing sums $(1/2)(\mathbf{E}+\mathbf{E}^*)$ and $(1/2)(\mathbf{B}+\mathbf{B}^*)$, we obtain

$$\operatorname{Re}\mathbf{E}_\omega = k\mathbf{n}\times(\mathbf{n}\times\operatorname{Im}\mathbf{A}_\omega),\ \operatorname{Re}\mathbf{B}_\omega = -k\mathbf{n}\times\operatorname{Im}\mathbf{A}_\omega \tag{18.6}$$

so that $\operatorname{Re}\mathbf{E}_\omega = \operatorname{Re}\mathbf{B}_\omega\times\mathbf{n}$ and $\operatorname{Re}\mathbf{B}_\omega = \mathbf{n}\times\operatorname{Re}\mathbf{E}_\omega$. Here $\operatorname{Im}\mathbf{A}_\omega = -\frac{i}{2}(\mathbf{A}_\omega - \mathbf{A}^*_\omega)$. We find that electromagnetic plane wave is transverse: field strength vectors **E** and **B** lie in the plane of its wave front. Moreover, these vectors are mutually orthogonal. In fact, these properties of mutual arrangement of vectors **n**, **E**, **B** are immediately obvious in the complex notation (18.4). Recall that plane waves with real values of **k**, considered here, are solutions of equations of electromagnetic field in homogeneous isotropic dielectrics.

Let relations $\mathbf{D} = \varepsilon\mathbf{E}$ and $\mathbf{B} = \mu\mathbf{H}$ be valid for constant ε and μ. Then representing vectors **B** and **E** as plane waves propagating at velocity v in a medium with constant amplitudes, and taking into account that $k = \omega/v = \sqrt{\mu\varepsilon}\,\omega/\alpha$, we can obtain from the Maxwell

[1] Both k and ω are assumed real-valued.

equations (M.3) and (M.4)

$$\mathbf{E}=-\sqrt{\frac{\mu}{\varepsilon}}\,\mathbf{n}\times\mathbf{H},\quad \mathbf{H}=\sqrt{\frac{\varepsilon}{\mu}}\,\mathbf{n}\times\mathbf{E} \tag{18.7}$$

and from this

$$\sqrt{\mu}H=\sqrt{\varepsilon}\,E \tag{18.8}$$

Obviously, this result could also be obtained from the expressions for potentials given above.

18.2. Equations (18.7) show that vectors **E**, **B**, **n** always form a right-handed trihedral. Following Fresnel, the plane containing vectors **B** and **n** is usually called the plane of polarization, and the direction of vector **B**, the direction of polarization. A wave is referred to as linearly polarized if the direction of polarization in its wave front does not change as the wave propagates. In the general case, the end of vector **B** traces a curve in this plane in the course of propagation. Let us find the form of this curve. First, introduce Cartesian coordinates in the plane of wave front. In these coordinates, at the most, only the initial phase of the components of vector **B** may change. The real parts of equations yield

$$B_1=a_1\cos(\tau+\delta_1),\quad B_2=a_2\cos(\tau+\delta_2) \tag{18.9}$$

where $\tau\equiv\mathbf{k}\cdot\mathbf{r}-\omega t$. Clearly, (18.9) describes a process of interference of two such oscillations which were defined above as linearly polarized (in the directions of the first and second axes). The equation of the curve in question will be obtained by eliminating variable τ from (18.9). It has the form

$$\left(\frac{B_1}{a_1}\right)^2+\left(\frac{B_2}{a_2}\right)^2-2\,\frac{B_1}{a_1}\,\frac{B_2}{a_2}\cos\delta=\sin^2\delta \tag{18.10}$$

where $\delta=\delta_1-\delta_2$. Standard techniques of analytical geometry immediately show that equation (18.10) represents an ellipse inscribed into a rectangle whose sides are parallel to coordinate axes and are equal to $2a_1$ and $2a_2$, respectively (Fig. 17). The end of magnetic vector tracing this ellipse may follow it in any of the two possible directions. It follows from (18.7) that electric vector also traces in this case an ellipse whose parameters are easily found. This general case is referred to as the elliptically polarized plane wave. If $\delta=$ $=\pi/2$ and $a_1=a_2$, we obtain a particular case of circular polarization, and if $\delta=0$, of linear polarization. In the first case the ellipse degenerates into a circle, and in the second case it is turned into two mutually orthogonal straight lines each of which defines one of the possible directions of oscillations. In the general case polarization is called right-handed if the observer facing the direction of propagation finds the end of magnetic vector moving clockwise; in the opposite case it is called the left-handed polarization.[2]

18.3. Let us find now what happens when a plane wave is incident on an interface S of two media differing in values of ε and μ (Fig. 18). Denote by $\boldsymbol{\nu}$ a unit vector of the normal to interface S, directed from medium I into medium II, and for convenience choose the origin somewhere on this plane. The equation of interface plane will then be $\boldsymbol{\nu}\cdot\mathbf{r} = 0$.

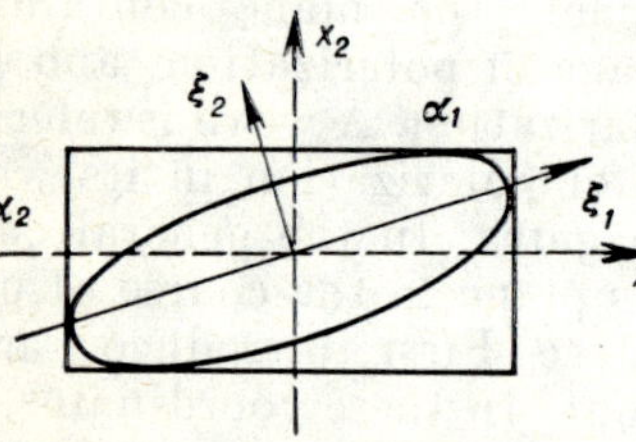

Fig. 17

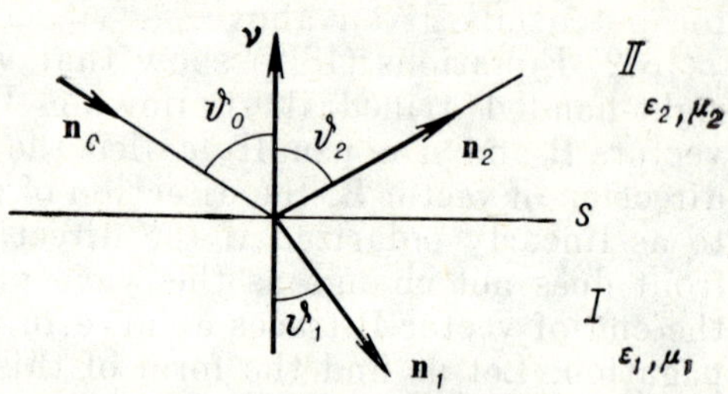

Fig. 18

Assume that the plane wave incident on S propagates from medium II. Field strengths $\mathbf{E}_{\text{in}}$ and $\mathbf{H}_{\text{in}}$ of this wave can be written in the form

$$\mathbf{E}_{\text{in}} = \mathbf{E}_0 e^{ik_2\mathbf{n}_0\cdot\mathbf{r}-i\omega t}, \quad \mathbf{H}_{\text{in}} = \frac{ck_2}{\omega\mu_2}\,\mathbf{n}_0\times\mathbf{E}_{\text{in}} \tag{18.11}$$

The second of these formulas follows from the second formula of (18.7) if we take into account that wave propagation velocity is $v_2 = c\,(\varepsilon_2\mu_2)^{-1/2}$ and $k_2 = \omega/v_2$. By $\mathbf{n}_0$ we denote a unit vector in the direction of propagation of the incident wave. The plane containing vectors $\mathbf{n}_0$ and $\boldsymbol{\nu}$ is called the plane of incidence. At interface S the field of plane wave in medium II must satisfy boundary conditions (4.14). Assume that plane S contains no currents and no charges, that is, in formulas (4.14) $\lambda = 0$ and $\mathbf{i} = 0$. First, we find from boundary conditions that a wave incident on the interface must also produce electromagnetic field in medium I; this field will be denoted by $\mathbf{E}_{\text{refr}}$, $\mathbf{H}_{\text{refr}}$ and called the refracted wave. It can be shown, however, that boundary conditions can be satisfied only if one additional component of field, the so-called reflected wave, appears in medium II. It will be denoted by $\mathbf{E}_{\text{refl}}$, $\mathbf{H}_{\text{refl}}$. Let us try to satisfy boundary conditions assuming that the refracted and reflected waves are also plane waves and have the same frequency

[2] Note that it is often the electric vector and not the magnetic vector which is used to define the plane of polarization.

ω as the incident wave. With obvious notations (see also Fig. 18)

$$\mathbf{E}_{\text{refr}} = \mathbf{E}_1 e^{ik_1\mathbf{n}_1\cdot\mathbf{r}-i\omega t}, \quad \mathbf{H}_{\text{refr}} = \frac{ck_1}{\omega\mu_1}\mathbf{n}_1 \times \mathbf{E}_{\text{refr}}$$

$$\mathbf{E}_{\text{refl}} = \mathbf{E}_2 e^{ik_2\mathbf{n}_2\cdot\mathbf{r}-i\omega t}, \quad \mathbf{H}_{\text{refl}} = \frac{ck_2}{\omega\mu_2}\mathbf{n}_2 \times \mathbf{E}_{\text{refl}} \tag{18.12}$$

Conditions of continuity of tangential components of the field can be written, for example, in the form $\mathbf{v} \times \mathbf{E}_1\, e^{ik_1\mathbf{n}_1\cdot\mathbf{r}} = \mathbf{v} \times \mathbf{E}_0\, e^{ik_2\mathbf{n}_0\cdot\mathbf{r}} + \mathbf{v} \times \mathbf{E}_2\, e^{ik_2\mathbf{n}_2\cdot\mathbf{r}}$ ($\mathbf{r}$ lies in plane S). Vectors $\mathbf{E}_0$, $\mathbf{E}_1$ and $\mathbf{E}_2$ are assumed independent of $\mathbf{r}$. Hence, the required continuity will take place only if the exponents in (18.12) coincide for $\mathbf{r} \in S$, that is when $k_2\mathbf{n}_0\cdot\mathbf{r} = k_1\mathbf{n}_1\cdot\mathbf{r}$ and $\mathbf{n}_0\cdot\mathbf{r} = \mathbf{n}_2\cdot\mathbf{r}$. Let us use now an identity $\mathbf{r} = (\mathbf{r}\cdot\mathbf{v})\,\mathbf{v} - \mathbf{v} \times (\mathbf{v} \times \mathbf{r})$, whence $\mathbf{r} = -\mathbf{v} \times (\mathbf{v} \times \mathbf{r})$ on S, when $\mathbf{v}\cdot\mathbf{r} = 0$. With this we obtain

$$k_2\mathbf{n}_0\cdot[\mathbf{v} \times (\mathbf{v} \times \mathbf{r})] = k_2\mathbf{n}_2\cdot[\mathbf{v} \times (\mathbf{v} \times \mathbf{r})] = k_1 n_1\cdot [\mathbf{v} \times (\mathbf{v} \times \mathbf{r})]$$

From (B.5) we obtain $\mathbf{n}_0\cdot[\mathbf{v} \times (\mathbf{v} \times \mathbf{r})] = (\mathbf{v} \times \mathbf{r})\cdot(\mathbf{n}_0 \times \mathbf{v})$ and similar formulas for the remaining two vector products. Therefore,

$$(\mathbf{n}_0 \times \mathbf{v} - \mathbf{n}_2 \times \mathbf{v})\cdot(\mathbf{v} \times \mathbf{r}) = 0$$

$$(k_2\mathbf{n}_0 \times \mathbf{v} - k_1\mathbf{n}_1 \times \mathbf{v})\cdot(\mathbf{v} \times \mathbf{r}) = 0 \tag{18.13}$$

This means that all vectors $\mathbf{v}$, $\mathbf{n}_0$, $\mathbf{n}_1$, and $\mathbf{n}_2$ lie in the same plane (that is in the plane of incidence). With angles denoted as shown in Fig. 18, we obtain

$$\sin\vartheta_2 = \sin(\pi - \vartheta_0) = \sin\vartheta_0, \quad k_2 \sin\vartheta_0 = k_1 \sin\vartheta_1 \tag{18.14}$$

These relations give *Shell's laws of reflection and refraction.*

By using boundary conditions, we can move further and find relations between amplitudes[3] which make it possible to calculate relative intensities of the incident, reflected, and refracted waves. Then we could analyze changes in polarization of plane waves caused by refraction and reflection. Note that Shell's laws derived above are also valid when media *I* and *II* have nonzero electric conduction. Thus, for example, the incidence and reflected angles are equal both when light is incident on a smooth water surface and on a metal amalgam of a conventional mirror. The abovementioned relations between intensities of waves can be very different in these cases[4].

18.4. Let us consider the process of superposition of fields of two plane monochromatic waves, that is the simplest case of interference. Assume the two waves to have the same frequency ω and repre-

---

[3] These are known as the *Fresnel formulas.*

[4] The derivation and analysis of these relations would take too much space and for this reason, unfortunately, cannot be given here. See, for example, M. Born and E. Wolf, *Principles of Optics*, Pergamon Press, Oxford, 1975. The phenomenon of total internal reflection is also considered in this monograph.

sent them in a complex form: $\mathbf{E}_1 = \mathbf{A}_1 e^{-i\omega t}$, $\mathbf{E}_2 = \mathbf{A}_2 e^{-i\omega t}$. Here $\mathbf{A}_1$ and $\mathbf{A}_2$ are complex amplitudes, and we can assume that they include vector $e^{i\mathbf{k}\cdot\mathbf{r}}$. By virtue of the principle of superposition, which is a corollary of linearity of the Maxwell equations (in this case, of the homogeneous wave equation), the resultant field will be written in the form $\mathbf{E} = \mathbf{E}_1 + \mathbf{E}_2$. Amplitude of field $\mathbf{E}$ will be denoted by $\mathbf{A}$. The quantity of interest is, of course, the energy flux transferred by field $\mathbf{E}$. We shall assume that field $\mathbf{E}$ is observed during a time interval sufficiently long compared to the period of oscillations, so that the observer is interested in the mean energy flux during this interval. The arguments discussed on p. 135 in § 17 and leading to formula (17.18) are then valid. In the general case of homogeneous isotropic medium for which relations (18.18) are valid, we can represent the absolute value of the Poynting vector for the radiation field in the form

$$S = v\langle w\rangle = c\sqrt{\varepsilon/\mu}\,\langle(\mathrm{Re}\,\mathbf{E})^2\rangle$$

As in § 17, angle brackets stand for time averaging. In our case equality (17.18) becomes

$$\langle(\mathrm{Re}\,\mathbf{E})^2\rangle = (1/2)\,|\mathbf{E}|^2 = (1/2)\,|\mathbf{E}_1+\mathbf{E}_2|^2$$
$$= (1/2)|\mathbf{A}_1|^2 + (1/2)|\mathbf{A}_2|^2 + \mathrm{Re}\,(\mathbf{A}_1\cdot\mathbf{A}_2^*) \qquad (18.15)$$

The first two terms are the intensities of each individual wave, and the last term is called the interference term. Note that formula (17.18) is rewritten in terms of vectors instead of their magnitudes, and it is easy to verify that the conditions of derivation are not violated by this. Vectors of the type $e^{i\mathbf{k}\cdot\mathbf{r}}$ mentioned above cancel out in the right-hand side of (18.15), so that in what follows they can also be ignored. Now we shall determine complex variables $\mathbf{A}_1$ and $\mathbf{A}_2$, separating their real and imaginary parts in the form $A_{1\alpha} = a_\alpha e^{ig_\alpha}$, $A_{2\alpha} = b_\alpha e^{ih_\alpha}$, where a_α, b_α, g_α, h_α are real and $\alpha = 1, 2, 3$. With these notations,

$$\mathrm{Re}\,(\mathbf{A}_1\cdot\mathbf{A}_2^*) = \sum_\alpha a_\alpha b_\alpha\,\mathrm{Re}\,[e^{i(g_\alpha-h_\alpha)}] = \sum_\alpha a_\alpha b_\alpha \cos(g_\alpha - h_\alpha) \qquad (18.16)$$

Let us assume now that the conditions of experiment are such that differences $g_\alpha - h_\alpha$ are equal for all α: $g_\alpha - h_\alpha = \delta$ $(\alpha = 1, 2, 3)$[5]. Quantity δ is called the phase difference of waves arriving at the point of observation. The interference term then becomes

$$J_{12} \equiv \mathrm{Re}\,(\mathbf{A}_1\mathbf{A}_2^*) = \mathbf{a}\cdot\mathbf{b}\cos\delta \qquad (18.17)$$

[5] These conditions are provided in classical experiments with Fresnel biprism, Billet split lens, and others (see any university course of general physics) in which two waves emitted by the same source are made to interface.

We have obtained all these results without using the fact that electromagnetic waves are purely transverse[6]. Let us consider a particular case of interference of two waves without assuming in advance that they are transverse; let both waves propagate in direction 3, with electric vector of one of them in plane (1, 3) and that of the second, in plane (2, 3). In our notation, $a_2 = 0$, $b_1 = 0$ and $J_{12} = a_3 b_3 \cos\delta$. Fresnel and Arago have demonstrated experimentally that in reality in these conditions $J_{12} = 0$, regardless of the value of δ, so that interference is absent. We then have to conclude that $a_3 = b_3 = 0$, that is, the waves are indeed transverse in accordance with the general theory.

Assuming now that both waves are linearly polarized and that their electric vectors are parallel to axis 1, let us write in this particular case the formulas for field intensity **E**. Here $a_2 = a_3 = b_2 = b_3 = 0$, so that the total intensity is $I = I_1 + I_2 + 2\sqrt{I_1 I_2} \times \cos\delta$, where $I_1 = a_1^2/2$ and $I_2 = b_1^2/2$. The interference will be observed therefore as an alteration of intensity maxima at $|\delta| = 0, 2\pi, 4\pi, \ldots$ and minima at $|\delta| = \pi, 3\pi, 5\pi, \ldots$. It is convenient to introduce by the relation $\Delta l = \lambda_0 \delta / 2\pi$ the so-called optical path difference Δl of two waves; here λ_0 is the wave-length in vacuo corresponding to frequency ω.

So far we were discussing interference of two monochromatic waves. It is important to determine the conditions in which electromagnetic waves generated by real physical sources can interfere. Assume that radiation observed at a point in space is emitted by N sources. Each of the sources produces at this point field $\mathbf{E}_k$, $\mathbf{H}_k$ ($k = 1, 2, \ldots, N$). As electromagnetic field obeys the principle of superposition, the whole set of sources produces a field

$$\mathbf{E} = \sum_{k=1}^{N} \mathbf{E}_k \quad \text{and} \quad \mathbf{H} = \sum_{k=1}^{N} \mathbf{H}_k$$

The energy flux of the field which, as we saw earlier, characterizes the radiation, is equal to the magnitude of the Poynting vector

$$|\mathbf{S}| = c|\mathbf{E} \times \mathbf{H}| = c \left| \sum_{k,\,l} \mathbf{E}_k \times \mathbf{H}_l \right|$$

We have already mentioned that usually the mean value of the Poynting vector over a time interval is observed (the averaging is denoted by angle brackets). Therefore, intensity I_P of radiation field at point P is given by

$$I_P = |\langle \mathbf{S} \rangle| = c \left| \sum_{k} \langle \mathbf{E}_k \times \mathbf{H}_k \rangle + \sum_{k \neq l} \langle \mathbf{E}_k \times \mathbf{H}_l \rangle \right|$$

If $\sum_{k \neq l} \langle \mathbf{E}_k \times \mathbf{H}_l \rangle = 0$, that is if $I(P) = |\sum_k \langle \mathbf{S}_k \rangle|$, where $\mathbf{S}_k$ is the Poynting vector corresponding to kth source, then the fields **E**, **H**

[6] Indeed, the derivation of these results did not involve the theory of electromagnetic field, so that they are valid for any type of waves.

are called *noncoherent*. In principle, interference is possible if the condition of noncoherence is not satisfied (unfortunately, this "negative" definition will have to suffice). We have shown a little earlier that two monochromatic waves interfere. The conditions of coherence in the general case are too complicated to treat them here; this would require a detailed study of averaging of fields[7]. Generally, a field is not coherent if its components can be considered mutually statistically independent. This condition is satisfied for natural light sources whose radiation field is usually generated by an enormous number of independently emitting atoms. At present, however, there exist and are widely used such optical and electronic setups (lasers, for instance) which generate nearly monochromatic radiation. Functioning of such instruments is based on the quantum theory.

§ 19. Relativistic transformations of plane waves

19.1. Let us consider the description of electromagnetic field of a plane wave in vacuo from the standpoint of different inertial reference frames. It was mentioned in the preceding chapter that such features of radiation field as mutual orthogonality of vectors **B** and **E** and equality of their magnitudes are relativistically invariant. We can note here that the Lorentz condition for plane waves given by formula (18.3) can be rewritten in the form

$$k^i\Phi_i = 0 \tag{19.1}$$

if we denote $k^0 = \omega/c$ and recall the definition of the four-dimensional vector potential (7.3). We have shown in § 7 that the Lorentz condition formulated for the potential of the general type is relativistically invariant. Hence, relation (19.1) for potentials of a plane wave is also invariant. But this can be true only in the case when the four values k^i are components of a space-time (i.e. four-dimensional) vector. This also means that the phase of a plane wave can be written in the form $\mathbf{k}\cdot\mathbf{r} - \omega t = \mathbf{k}\cdot\mathbf{r} - k^0x^0 = -k^ix_i$ (that is, it is relativistically invariant). Four-dimensional vector k^i is then a zero vector, that is, satisfies the relation $k^ik_i = 0$.

Let us form now a scalar product of vector **E**, defined by (18.4), and a unit vector **n**: $\mathbf{E}\cdot\mathbf{n} = -i\,(k\varphi_\omega - \mathbf{k}\cdot\mathbf{A}_\omega) = 0$. Evidently, this scalar product is relativistically invariant. Therefore, the transverse nature of an electric vector (and similarly, of a magnetic vector) with respect to the direction of propagation of a plane wave is also a relativistically invariant property.

Let a plane wave be characterized in an inertial reference frame K by quantities **k** and ω. The same plane wave in reference frame K'

[7] See the monograph by M. Born and E. Wolf cited in footnote 4 on p. 143, Ch. 10.

moving at a velocity $\mathbf{v}$ with respect to K will have wave vector $\mathbf{k}'$ and frequency ω'. Our previous arguments show that $\mathbf{k}'$ and ω' are related to $\mathbf{k}$ and ω by the Lorentz transformations of type (5.12), and that $\mathbf{k}$ is transformed as radius vector $\mathbf{r}$ while ω/c is transformed as time coordinate x^0. These transformation formulas are conveniently written as follows:

$$\mathbf{k}' = (1-\gamma)\,\mathbf{k}_\perp + \gamma\left(\mathbf{k} - \frac{\mathbf{v}}{c^2}\,\omega\right) \tag{19.2$_1$}$$

$$\omega' = \gamma\,(\omega - \mathbf{k}\cdot\mathbf{v}) \tag{19.2$_2$}$$

As usual, symbol $\perp$ marks a component orthogonal to relative velocity $\mathbf{v}$. Here $\mathbf{k} = \frac{\omega}{c}\,\mathbf{n}$ and $\mathbf{k}' = \frac{\omega'}{c}\,\mathbf{n}'$. The first of these formulas can be rewritten in a somewhat different form:

$$\mathbf{k}' = \mathbf{k} + (\gamma - 1)\,\mathbf{k}_{||} - \gamma\,\frac{\mathbf{v}}{c^2}\,\omega$$

whence we see that vector $\mathbf{k}'$ lies in a plane containing vectors $\mathbf{k}$ and $\mathbf{v}$.

Frequency transformation (19.2$_2$) is a relativistic formulation of the *Doppler effect*. Denoting by ϑ the angle between vectors $\mathbf{k}$ and $\mathbf{v}$, we can rewrite (19.2$_2$) in a more detailed form:

$$\omega' = \omega\,\frac{1-\beta\cos\vartheta}{\sqrt{1-\beta^2}} \tag{19.3}$$

Note a property of relativistic formula (19.3) which constitutes a qualitative difference from nonrelativistic description of the Doppler effect. If $\vartheta = \pi/2$, that is if the light wave arrives to the observer's reference frame in a direction orthogonal to its velocity $\mathbf{v}$ with respect to the source, the Doppler effect does not vanish, as would be the case in the classical theory, but is given by the formula $\omega' = \omega/\sqrt{1-\beta^2}$ (*transverse Doppler effect*). This frequency relation is a result of the relativistic *dilation of time* in a moving reference frame. We can imagine, for example, that the "stationary" reference frame K is fixed to a source emitting light waves at a frequency ω. Then the equality $\omega'\,dt' = \omega\,dt$ must hold, because the number of, for example, maxima of waves counted by the two observers during the corresponding time intervals must be identical. But $dt' = dt\,\sqrt{1-\beta^2}$. When the motion has also a radial component, we have to take into account the *classical Doppler effect*[8].

Denote by ϑ' an angle formed by vector $\mathbf{k}'$ and the direction of vector $\mathbf{v}$ (we have seen above that this angle is measured in the same plane as angle ϑ). Projections of transformation (19.2$_1$) onto direc-

[8] The transverse Doppler effect was experimentally confirmed by Ives and Stillwell in 1938.

tions orthogonal and parallel to $\mathbf{v}$ have the form

$$\omega' \sin \vartheta' = \omega \sin \vartheta, \ \omega' \cos \vartheta' = \gamma\omega (\cos \vartheta - v/c) \tag{19.4}$$

By dividing the first of these equalities by the second, we obtain

$$\tan \vartheta' = \frac{\sqrt{1-\beta^2} \sin \vartheta}{\cos \vartheta - \beta} \tag{19.5}$$

Assume that $\mathbf{k} \cdot \mathbf{v} \simeq 0$, that is $\vartheta \simeq \pi/2$, and that velocity of relative motion is nonrelativistic, so that $\beta \ll 1$. At the same time, however, let $\cos \vartheta \ll \beta$ (for instance, we can consider the case when angle ϑ is exactly equal to $\pi/2$). Then $\tan \vartheta' = -\beta^{-1}$. In order to find an interpretation of this formula, we can choose as a light source a star located on the celestial sphere in such manner that the angle between the direction of motion of the Earth in its orbit and the beam of light emitted by the star is, in the reference frame co-moving with the star, equal to ϑ. Then formula (19.5) gives the angle at which this star is seen from the moving Earth. What this nonrelativistic approximation means is quite clear from Fig. 19. This effect is observed in astronomy and is known as the astronomic aberration. Corrections of the order of β^2, predicted by the exact formula (19.5), are too small to be recorded by the existing instruments.

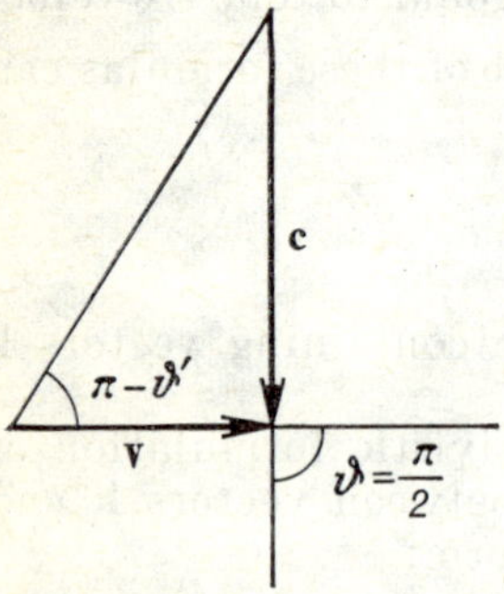

Fig. 19

Relativistic aberration (19.5) must be regarded as an effect produced by addition of light velocity and the velocity of the observer with respect to the reference frame in which the source is at rest. The addition must be relativistic, by formula (5.13) which clearly shows that velocity $\mathbf{u}'$ lies in a plane defined by vectors $\mathbf{u}$ and $\mathbf{v}$. Denoting the angle between these vectors by ϑ and that between $\mathbf{u}'$ and $\mathbf{v}$ by ϑ', it is possible to project both sides of (5.13) onto the direction of $\mathbf{v}$ and onto a plane orthogonal to $\mathbf{v}$, and then divide the second of the obtained expressions by the first. We obtain

$$\tan \vartheta' = \frac{u \sin \vartheta \sqrt{1-\beta^2}}{u \cos \vartheta - v} \tag{19.6}$$

In a particular case of $u = c$ formulas (19.6) and (19.5) coincide.

The formula for addition velocities is obviously relevant for explanation of results of the *Fizeau experiment* in which the velocity of light was measured in a medium moving with respect to the observer at a velocity v. We have already noted that velocity of light with respect to a medium at rest is c/n, where $n \equiv \sqrt{\varepsilon\mu}$ and c is the velocity of light in vacuo. Without loss of generality we can assume that the medium moves along axis x in the observer's reference frame and the light propagates in the same (or in the opposite) direction. Then we can apply the partial Lorentz transformation from the reference frame co-moving with the medium to the observer's reference frame and use the appropriate formula of addition of velocities (5.13′). Denoting the light velocity with respect to the observer by c', we obtain from (5.13′)

$$c' = \frac{c/n \pm v}{1 \pm v/cn} \simeq \left(\frac{c}{n} \pm v\right)\left(1 \mp \frac{v}{cn}\right) \simeq \frac{c}{n} \pm v\left(1 - \frac{1}{n^2}\right) \quad (19.7)$$

for nonrelativistic velocities v of the medium. Therefore, it is not sufficient to add the velocity of light c/n to that of the medium, v, (this would follow from the Galileo transformation) since we have to multiply this velocity first by the so-called Fresnel coefficient $1-1/n^2$. Here this result is a direct corollary of relativistic kinematics and is valid, of course, not only for plane waves but also for light pulses of arbitrary shape propagating in the medium at a group velocity c/n.[9]

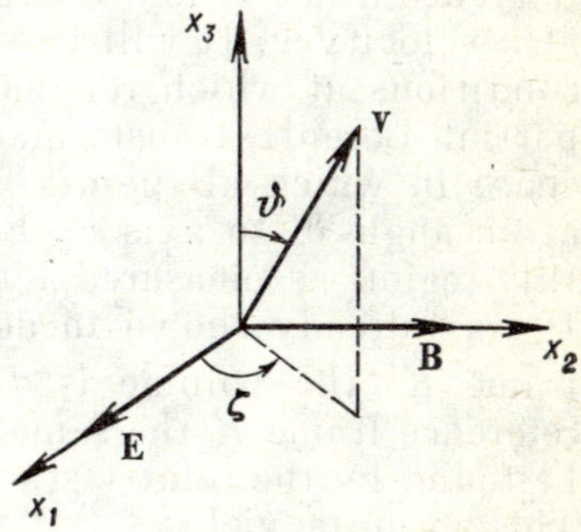

Fig. 20

19.2. Let us find how a transition to a new reference frame transforms the magnitude of the electric field vector in the wave zone. Ignoring for the moment the shape of the wave, we make use only of the transversality condition. Let us choose the origin of spatial coordinates (reference frame K) at a point of space through which light wave propagates; the axes are directed as shown in Fig. 20. The direction of vector $\mathbf{v}$ of the relative velocity of new reference frame K' will be determined in terms of polar angle ϑ and azymuthal angle ζ. First raise the first of formulas (7.14) to the second power. Denoting $\mathbf{v}_1 \equiv \mathbf{v}/v$, we obtain $\mathbf{E}'^2 = -\beta^2\gamma^2(\mathbf{E}\cdot\mathbf{v}_1)^2 + \gamma^2\left(\mathbf{E} + \frac{\mathbf{v}}{c} \times \mathbf{B}\right)^2$. But

$$\mathbf{E}\cdot\mathbf{v}_1 = E\cos\zeta\sin\vartheta$$

$$\left[\frac{\mathbf{v}}{c}\times\mathbf{B}\right]^2 = \beta^2E^2\left(1-\cos^2\widehat{(\mathbf{B},\mathbf{v})}\right) = \beta^2E^2\left(1-\sin^2\zeta\sin^2\vartheta\right)$$

$$2\mathbf{E}\cdot\left(\frac{\mathbf{v}}{c}\times\mathbf{B}\right) = 2\frac{\mathbf{v}}{c}(\mathbf{B}\times\mathbf{E}) = -2\frac{v_z}{c}E^2 = -2\beta E^2\cos\vartheta$$

and in the last two equations we have taken into account that in the wave zone $E = B$. Finally, after collecting like terms, we have

$$E' = \gamma E\,(1 - \beta\cos\vartheta) \quad (19.8)$$

[9] For the concept of group velocity see § 39.

By using Doppler's formula (19.3) the result for plane waves then becomes

$$E'/\omega' = E/\omega \tag{19.9}$$

The corresponding energy flux density is given by the magnitude of the Poynting vector:

$$S' = cE'^2 = S\,(1-\beta\cos\vartheta)^2/(1-\beta^2)$$

whence

$$S'/S = (\omega'/\omega)^2 \tag{19.10}$$

It will be interesting to find what energy is contained in a finite volume moving with the plane wave. For this we shall need some auxiliary arguments. We introduce two inertial reference frames: reference frame K and reference frame K' moving with respect to K at a velocity $\mathbf{v}$. It will be sufficient for our purposes to assume that conditions at which reference frames K and K' are related by the partial Lorentz transformation are satisfied. Consider a region in space in which all points have, with respect to K', a velocity $\mathbf{u}'$, at an angle ϑ' to axis x'. By dV_0 we denote the "proper" volume of this region as measured by an observer at rest with respect to it. It was already shown in deriving equation (5.10) that in reference frame K' the volume is $dV' = dV_0\sqrt{1-u'^2/c^2}$. With respect to reference frame K the same moving region has velocity $\mathbf{u}$ which can be found by the relativistic formula of addition of velocities (5.13'). This formula yields

$$u^2 = \frac{u'^2 + 2u'v\cos\vartheta' + v^2 - (u'^2v^2/c^2)\sin^2\vartheta'}{[1+(u'v/c^2)\cos\vartheta']^2}$$

whence $\sqrt{1-u^2/c^2} = [1+(u'v/c^2)\cos\vartheta']^{-1}\sqrt{1-u'^2/c^2}\sqrt{1-\dfrac{v^2}{c^2}}$. But $dV = dV_0\sqrt{1-u^2/c^2}$ is the volume in reference frame K, so that $dV = dV'\sqrt{1-\beta^2}\left(1+\dfrac{u'}{c}\beta\cos\vartheta'\right)^{-1}$. Note that if $u'\to c$, then simultaneously $u\to c$ and volumes dV and dV' tend to zero. The abovederived formula shows, however, that ratio dV/dV' tends to a definite value. Namely, at $u'\to c$, $dV = dV'\sqrt{1-\beta^2}/(1+\beta\cos\vartheta')$, that is $dV\cdot\omega = dV'\cdot\omega'$.[10]

The energy of the field in volume V is given by $W = (1/2)\int(E^2 + H^2)\,dV = \int E^2\,dV$. But we have found that E^2 is transformed proportionally to ω^2. Using the relation for V obtained above, we find that

$$W/\omega = W'/\omega' \tag{19.11}$$

[10] Here we have used a formula obtained from (19.3) by substituting $-\beta$ for β, the operation exchanging the role of ω and ω'.

The value of this ratio, which is an invariant of the Lorentz transformations, will be denoted by D. The total momentum of the field within volume V is equal, on the basis of formulas in § 3, to

$$\mathbf{P}=\frac{1}{c}\int \mathbf{E}\times\mathbf{H}\,dV=\frac{1}{c}\,\mathbf{n}\int E^2\,dV=\frac{\omega}{c}\,\mathbf{n}D$$

Consequently, quantities W/c and $\mathbf{P}$ form together a 4-vector of zero length ($W^2/c^2 - P^2 = 0$) proportional to vector k^i. Quantum theory states that $D = Nh/2\pi$, where N is an arbitrary integer and h is Planck's constant. When $N = 1$, we arrive at relations holding for an individual quantum of radiation field, a photon. In particular, it follows from the definition of wavelength $\lambda = c/\nu = 2\pi c/\omega$ that

$$G = h/\lambda, \quad \mathscr{E} = h\omega/2\pi \tag{19.12}$$

Here $\mathscr{E}$ is the energy and G is the momentum of a photon.

19.3. To conclude this section, we shall determine the characteristics of a light wave, reflected by a moving mirror, on the basis of relativistic transformation formulas. A mirror moves at a velocity $\mathbf{v}$ in reference frame K; in this reference frame a light wave incident on the mirror is characterized by frequency ω, and its direction of propagation is at an angle ϑ_0 with axis x. Formulas (19.2) give frequency ω' and wave vector $\mathbf{k}'$ in reference frame K' moving together with the mirror. To make these formulas clear, we rewrite them here in a form corresponding to the partial Lorentz transformation from K to K':

$$\omega'=\omega_0\,\frac{1-\beta\cos\vartheta_0}{\sqrt{1-\beta^2}}\,,\quad n'_{x'}=\frac{\cos\vartheta_0-\beta}{1-\beta\cos\vartheta_0}\,,\quad n'_{y'}=\frac{\sin\vartheta_0\sqrt{1-\beta^2}}{1-\beta\cos\vartheta_0}$$

Let the mirror plane be perpendicular to axis x and assume that the standard Snellius law of reflection (18.14), derived above, holds in reference frame K'. Then the reflected wave is obtained in this reference frame simply by replacing $n'_{x'}$ by $-n'_{x'}$, that is for this wave $n'_{x'} = -(\cos\vartheta_0 - \beta)/(1 - \beta\cos\vartheta_0)$. And now for the reflected wave considered already in reference frame K we use transformation formulas inverse with respect to (19.2), and find

$$\omega=\omega'\,\frac{1+\beta n'_{x'}}{\sqrt{1-\beta^2}}=\omega_0\,\frac{1-2\beta\cos\vartheta_0+\beta^2}{1-\beta^2}\simeq\omega_0\,(1-2\beta\cos\vartheta_0)$$

$$\cos\vartheta=n_x=\frac{n'_{x'}+\beta}{1+\beta n'_{x'}}=-\frac{\cos\vartheta_0-2\beta+\beta^2\cos\vartheta_0}{1-2\beta\cos\vartheta_0+\beta^2}\simeq-\cos\vartheta_0+2\beta\sin^2\vartheta_0$$

$$\sin\vartheta=n_y=\frac{n'_{y'}\sqrt{1-\beta^2}}{1+\beta n'_{x'}}=\frac{\sin\vartheta_0\,(1-\beta^2)}{1-2\beta\cos\vartheta_0+\beta^2}\simeq\sin\vartheta_0+2\beta\sin\vartheta_0\cos\vartheta_0 \tag{19.13}$$

At the end of each line we give the nonrelativistic approximation. The first of these formulas plays an important role in the theory of thermal radiation, being fundamental for deriving Wien's displacement law discussed in § 22.

§ 20. Huygens principle. Fundamentals of the theory of diffraction

20.1. Let us analyze now what happens when radiation field propagating in space encounters an obstacle (for instance, a screen with apertures). This problem (which makes the subject of the theory of diffraction) is very important in optics and in the theory of propagation of radiowaves. The reader is familiar with the elementary formulation of the Huygens principle and with the method of investigating diffraction by resorting to the Fresnel zones, as presented in the general physics courses. Here we can use the already familiar solution of the wave equation and investigate the formulation of the problem of diffraction in a more general form.

Let us recall the solution of a nonhomogeneous wave equation as given by formulas (13.8) and (13.9). We have mentioned at the beginning of § 13 that any of the Cartesian components of vectors **E** and **B** can be chosen for function ψ. Let the region in which we want to determine the electromagnetic field strength have no sources of this field and be bounded by a surface σ. Then the abovementioned formulas state that the value of the field at the observation point and its values on the closed boundary surface σ enclosing this point must be related by the following formula:

$$\psi(\mathbf{r}, t) = \\ = \frac{1}{4\pi} \oint_{\sigma} \mathbf{n} \times \left[\frac{\text{grad}' \, \psi(r', t')}{R} - \frac{\mathbf{R}}{R^3} \psi(\mathbf{r}', t') - \frac{\mathbf{R}}{cR^2} \frac{\partial \psi(\mathbf{r}', t')}{\partial t'} \right]_{t'=t-R/c} d\sigma$$

We remind that $\mathbf{R} = \mathbf{r} - \mathbf{r}'$.

By assuming that function ψ is a harmonic function of time, that is $\psi(\mathbf{r}, t) = \psi(\mathbf{r}) \exp(-i\omega t)$, we can immediately rewrite the preceding formula as a relation between amplitudes:

$$\psi(\mathbf{r}) = \frac{1}{4\pi} \int_{\sigma} \frac{e^{ikR}}{R} \mathbf{n} \cdot \left[\text{grad}' \, \psi(\mathbf{r}') + ik \left(1 + \frac{i}{kR} \right) \frac{\mathbf{R}}{R} \psi(\mathbf{r}') \right] d\sigma \tag{20.1}$$

In deriving (20.1), it should be noted that in the region of interest function $\psi(\mathbf{r})$ must be a solution of the homogeneous Helmholtz equation[11]. Now we apply to this region Green's formula (B.28),

[11] See formula (E.14).

assuming that function $\varphi = G$ is Green's function for the Helmholtz equation, that is $(\Delta + k^2) G(\mathbf{r}, \mathbf{r}') = -\delta(\mathbf{r} - \mathbf{r}')$. Then, integrating, by analogy with § 13, over the primed variable, we obtain

$$\psi(\mathbf{r}) = \oint \mathbf{n} \cdot (G \operatorname{grad}' \psi - \psi \operatorname{grad}' G) \, d\sigma \qquad (20.2)$$

If we substitute into (20.2) for Green's function the fundamental solution of the Helmholtz equation—the diverging spherical wave given by formula (E.17) with the plus sign in the exponent—we return to relation (20.1). Formula (20.2) is, however, more general since an arbitrary Green's function G can be obtained as a sum of the fundamental solution used in the derivation of (20.1) and the general solution of the homogeneous equation. It is also easy to trace the analogy between relation (20.1) and the formulas of § 11 which relate electrostatic field at the observation point with the single and double layers present on the boundary surface. It is clear from the preceding argument that the integral in (20.2) must have the form determined by sources which, by assumption, are located outside of σ but produce the field inside this surface.

The uniqueness of solution ψ in the region enclosed by surface σ is assured, as was the case for a static field (see § 11), by fixing on σ either only the values of ψ (Dirichlet's problem) or of $\partial\psi/\partial n$ (Neumann's problem). A proof of this fact is quite analogous to that traced in § 11 if instead of the Poisson equation we take the Helmholtz equation. Therefore the values of ψ and $\partial\psi/\partial n$ cannot be fixed on the surface at the same time in an arbitrary manner, and (20.1) is an integral relation whose consistency must be verified in each specific case.

Formula (20.2) can be interpreted as the rule which allows to construct electromagnetic field at the point of observation as a "sum" of elementary G-type waves emitted by each element of boundary surface σ (functions ψ and $\operatorname{grad}' \psi$ in the integrand must be considered known, fixed as boundary values on σ). In particular, these elementary waves are spherical in (20.1). With this interpretation, relation (20.2) is a general formulation of the *Huygens principle*.

Assume that an infinite surface, to which certain physical properties are assigned, divides the space into two parts. Hereafter we refer to this surface as the screen. Let all sources of electromagnetic field lie to one side of the screen, and the screen have a certain number of holes. When the radiation field produced by the sources is incident on the screen, a fraction of the field is reflected backward but part of the energy penetrates through the screen into the other half-space. It is of interest to compare the transmitted wave with that incident on the screen. If the properties of this radiation, as well as those of the screen, are known, then the finding of the transmitted wave is one of the basic problems in the theory of diffraction.

A closed surface enclosing sources (Fig. 21) can also be considered as a screen. In what follows we mark the region enclosing the sources by numeral *I* and the region in which the diffracted radiation is recorded—by numeral *II*.

We begin with the first variant of the diffraction problem (Fig. 21, *a*). Let us form a closed surface $\sigma_1 + \sigma_2$ whose component σ_2 is a hemisphere contained in *II* and cutting from the screen its part σ_1. Formula (20.1) determines the field at any point inside this

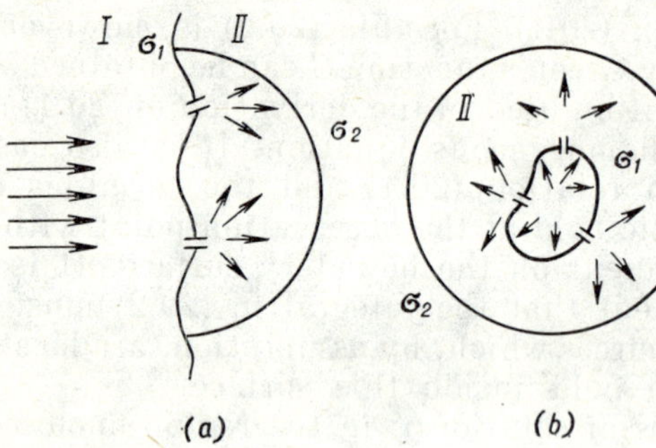

Fig. 21

surface if we know the value of the field on screen σ_1 and on the hemisphere σ_2. Therefore, $\psi(\mathbf{r}) = I[\sigma_1] + I[R]$, where I is the surface integral in the right-hand side of (20.1), and R is the radius of hemisphere σ_2. Let's take now $\lim\limits_{R\to\infty} I(R)$. Obviously, the value of this limit depends on the assumed properties of function ψ in the integrand. These assumptions single out the class of functions in which the solution of the problem is sought. We consider such functions ψ which satisfy the condition of radiation, namely, describe a diverging spherical wave far from the screen, that is for sufficiently large R

$$\psi \simeq f(\vartheta, \varphi)\frac{e^{ikR}}{R}, \quad R\to\infty \tag{20.3}$$

It is easy to calculate that on σ_2

$$\frac{\partial\psi}{\partial n}\,\frac{e^{ikR}}{R} = \frac{\partial\psi}{\partial R}\,\frac{e^{ikR}}{R} \simeq \left(ik-\frac{1}{R}\right)\left(\frac{e^{ikR}}{R}\right)^2 f(\vartheta, \varphi)$$

$$\psi\frac{\partial}{\partial n}\,\frac{e^{ikR}}{R} \simeq \left(ik-\frac{1}{R}\right)\left(\frac{e^{ikR}}{R}\right)^2 f(\vartheta, \varphi)$$

and that terms of the order of $1/R$, $1/R^2$, and $1/R^3$ in the integrand cancell out. The remaining terms decrease more rapidly than $1/R^3$ and therefore $\lim\limits_{R\to\infty} I(R) = 0$ if the radiation condition is satisfied. After the limit transition is realized, integral $I\ [\sigma_1]$ is extended to the whole screen (including, of course, the holes in it).

Similarly, if the radiation condition is satisfied, the integral over sphere σ_2 in region *II* vanishes (see Fig. 21, *b*). Therefore, the preceding arguments are completely valid in the given case if at first the region is bounded by two surfaces σ_1 and σ_2 and in the limit only the integral over surface σ_1 coinciding with the screen does not vanish. It must also be taken into account that the positive direction of normal in the above formulas is defined as the direction from region *II* towards region *I*, in accordance with the customary choice of direction when the Gauss theorem and its corollaries are used. From the physical point of view it is logical to reverse this direction. With this new choice, we obtain

$$\psi(\mathbf{r}) = -\frac{1}{4\pi}\int\limits_{\sigma_1} \frac{e^{ikR}}{R}\,\mathbf{n}\cdot\left[\operatorname{grad}'\psi + ik\left(1+\frac{i}{kR}\right)\frac{\mathbf{R}}{R}\psi\right]d\sigma \qquad (20.4)$$

In principle, therefore, the field in region *II* can be found if the field on surface σ_1 is known. No doubt, calculation of the field on σ_1 can be made only in exceptional cases even if all properties of the sources are known. This field depends on physical properties of the screen, and the interaction between the incident field and the screen should be taken into account when this field is calculated. However, a very large number of problems in classical optics prove to be resolvable with sufficient accuracy if the so-called *Kirchhoff approximation* is used. This approximation involves the following assumptions:

1. Functions ψ and grad ψ vanish on σ_1 everywhere on the side of the surface adjacent to region *II*, with the exception of holes.
2. Within the holes in the screen functions ψ and grad ψ are equal to the corresponding values of the incident wave in the absence of the screen.

Strictly speaking, the conditions as formulated are mathematically incompatible. Indeed, it has been mentioned already that the values of ψ and grad ψ cannot be fixed independently at the points of the boundary surface. It must also be noted that if the screen has holes, then assumptions 1 and 2 in fact state that function ψ has a discontinuity along the boundary of each of the holes and that Green's theorem holds only for functions which are continuous everywhere on the boundary surface σ. Application of the Kirchhoff approximation to optical problems is successful mostly because the ratio of the wavelength to the characteristic size of holes in these problems is small[12]. As a result, diffracted radiation mostly retains

[12] Usually the interaction between the screen material and radiation can be considered strong at distances of about a wavelength.

the direction of the incident wave, and the assumption about vanishing of the field on the shaded side of the screen is nearly correct.

Kirchhoff's formula (20.4) can be replaced, in principle, by a different relation which strictly satisfies the theorem of uniqueness. For this we have to take again equation (20.2) and, instead of choosing the fundamental solution of the Helmholtz equation for function G, to define this function by the following additional conditions:

$$G=0 \text{ on } \sigma_1, \quad R\left(\frac{\partial G}{\partial n}-ikG\right)\to 0 \text{ for } R\to\infty \tag{20.5}$$

By virtue of the second of these conditions,

$$\int\limits_{\sigma_2}\left(G\frac{\partial\psi}{\partial n}-\psi\frac{\partial G}{\partial n}\right)R^2\,d\Omega=\int R^2\left(\frac{\partial\psi}{\partial n}-ik\varphi\right)G\,d\Omega\to 0$$

for $R\to\infty$ if we again assume that function ψ satisfies the condition of radiation. The first of conditions (20.5) expresses the difference between function G and the spherical wave which was used earlier. Owing to this condition, the term in the integrand multiplied by $\partial\psi/\partial n$ vanishes, and the result takes the form

$$\psi(\mathbf{r})=-\int\limits_{\sigma_1}\psi\frac{\partial G}{\partial n}\,d\sigma \tag{20.6}$$

In this case it is sufficient, therefore, to know only boundary conditions for function ψ, and hence Kirchhoff's assumption dealing with this function can be used without any contradictions.

From the physical standpoint, however, application of Kirchhoff's conditions in the described case can be justified only approximately since electromagnetic radiation and the screen always interact, and so the field in the hole always differs from the incident wave. Besides, it cannot vanish completely on the shaded side of the screen. From the standpoint of practical calculations, however, the application of formula (20.6) entails the need of determining a Green's function satisfying conditions (20.5). In fact, the first of them requires that the Green function should be found for each specific shape of the screen. Considerable mathematical difficulties are encountered when such problems are solved, and the closed form of solution is obtainable only in the case of a flat screen. Moreover, the inaccuracy of Kirchhoff's formula is not significant in the approximation in which it is applicable and is compensated by the general character of the formula.

Let us consider now a geometric surface σ_1 which is not a screen (i.e. is devoid of any physical properties) and dividing the space into regions I and II, as shown in Fig. 21. If the radiation condition is satisfied, field $\psi(\mathbf{r})$ in region II will then be given by formula (20.4) where integration is carried out over the whole surface σ_1, and the integrand is the expression for incident wave. Now we replace surface σ_1 with a material screen with holes as before. By $\sigma_1^{(a)}$ we denote the continuous part of the screen σ_1, and by $\sigma_1^{(b)}$ its geometric part formed by apertures, so that we can write $\sigma_1 = \sigma_1^{(a)} + \sigma_1^{(b)}$. If Kirchhoff's approximation is valid, field ψ_b (r) in region *II* is determined only by the integral over $\sigma_1^{(b)}$. Now replace the continuous part of the screen $\sigma_1^{(a)}$ by apertures, and the apertures $\sigma_1^{(b)}$ by a continuous surface. In this case field ψ_a (r) in region *II* is determined by the integral taken only over $\sigma_1^{(a)}$. It then follows immediately that $\psi_a(\mathbf{r}) + \psi_b(\mathbf{r}) = \psi(\mathbf{r})$. It is logical to refer to the screens for which diffracted fields ψ_a and ψ_b are calculated as to complementary screens. The derived above equation is known as the *Babinet principle* for diffraction on such screens.

It is important to mention that all the formulas derived above are scalar formulas, that is all components of vectors **E** and **B** are considered independently of one another. Such scalar theory of diffraction does not make it possible to analyze effects caused by changes in the polarization of the diffracted wave. Often such effects are significant, as for example in the study of diffraction of electromagnetic waves in the range of radio frequencies. There exists the vector generalization of Kirchhoff's formula making such calculations possible.

The simplest method of realizing this generalization is to write formulas (20.2) for each component of vector **E** and then to find their vector sum; this gives

$$\mathbf{E}(\mathbf{r}) = \int_{\sigma} [G(\mathbf{n}\cdot\text{grad}')\,\mathbf{E} - \mathbf{E}(\mathbf{n}\cdot\text{grad}'\,G)]\,d\sigma \tag{20.7}$$

A similar expression can be written for **B**. However, formula (20.7) is inconvenient for calculations. Different formulas are usually applied in practice[13].

20.2. We cannot go here into the details of vector formulas but shall attempt to achieve some progress in the physical interpretation of relation (20.6) for the simplest case. Namely, we assume that the screen is flat, and its plane is represented by the equation $z = 0$ and that the field must be found at a point P with coordinates x, y, z (Fig. 22). Let point S be a mirror image of point P with respect to the screen, that is S has coordinates x, y, $-z$. For an arbitrary point Q with coordinates ξ, η, ζ (shown in the figure in the plane of the screen, that is for $\zeta = 0$) we compose a function

$$G(\xi, \eta, \zeta) = \frac{e^{ikr_1}}{4\pi r_1} - \frac{e^{ikr_2}}{4\pi r_2} \tag{20.8}$$

---

[13] See, for example H. Hönl, A. W. Maue, and K. Westpfahl, *Theorie der Beugung*.

It is easy to find that this function satisfies conditions determining Green's function, including conditions (20.5). The derivation to follow will clearly show that the singularity of function G at point S (i.e. for $r_2 = 0$) is of no consequence. As follows from (20.8)

$$4\pi \frac{\partial G}{\partial \zeta} = \frac{d}{dr_1}\left(\frac{e^{ikr_1}}{r_1}\right)\frac{\partial r_1}{\partial \zeta} - \frac{d}{\partial r_2}\left(\frac{e^{ikr_2}}{r_2}\right)\frac{\partial r_2}{\partial \zeta}$$

If point Q lies on the screen, then (see Fig. 22) $r_1 = r_2 = r$, $\frac{\partial r_2}{\partial \zeta} = -\frac{\partial r_1}{\partial \zeta} = \cos(\widehat{\mathbf{n}, \mathbf{r}})$, that is $\frac{\partial G}{\partial n} = -\frac{\partial G}{\partial \zeta} = 2\frac{d}{dr}\left(\frac{e^{ikr}}{4\pi r}\right)\cos(\widehat{\mathbf{n}, \mathbf{r}})$. But $\frac{d}{dr}\left(\frac{e^{ikr}}{r}\right) = ik\frac{e^{ikr}}{r}\left(1 - \frac{1}{ikr}\right) \simeq \frac{2\pi i}{\lambda}\frac{e^{ikr}}{r}$, if we assume that point P is at a large distance from the screen, namely that $kr = \frac{2\pi r}{\lambda} \gg 1$. Now the substitution of Green's function (20.8) into (20.6) transforms it to

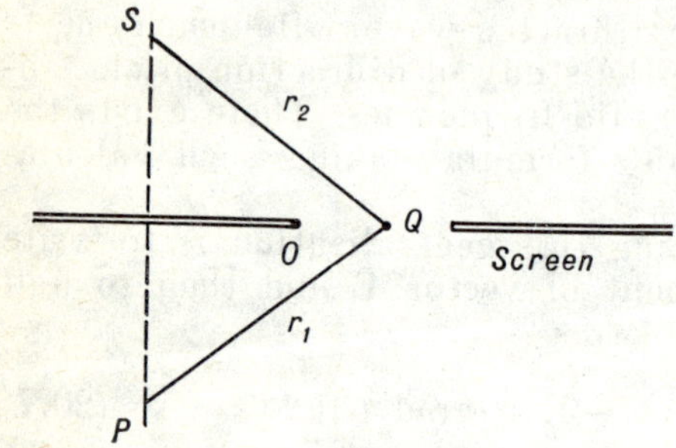

Fig. 22

$$i\lambda\psi(P) = \int_\sigma \frac{e^{ikr}}{r}\cos(\widehat{\mathbf{n}, \mathbf{r}})\,\psi\, d\sigma \tag{20.9}$$

where integration, in accordance with Kirchhoff's approximation, is carried out only over apertures in the flat screen. Hence, formula (20.9) states that each element $d\sigma$ of an aperture emits a spherical wave whose amplitude and phase are determined by the incident wave ψ. The field at the point of observation is found as a superposition of these spherical waves. We have arrived again to Huygens' principle in its clearest formulation.

For function ψ in the integrand we substitute the values corresponding to the incident spherical wave with amplitude A; this wave is emitted by a pointlike source located at distance r' from the aperture. Then

$$i\lambda\psi(P) = \int e^{ik(r+r')}\frac{\cos(\widehat{\mathbf{n}, \mathbf{r}'})}{rr'}\,d\sigma \tag{20.10}$$

If characteristic size of the aperture, that is, of the range of integration in the right-hand side of (20.10), are small compared with r and r', then factor $\cos(\widehat{\mathbf{n}, \mathbf{r}})/rr'$ is nearly constant and can be factured out of the integral. Kirchhoff's approximation will be valid if we additionally assume that k is large and so function $e^{ik(r+r')}$ oscillates very rapidly. Expand r in powers of ξ and η with $\zeta = 0$. If R is a value of r corresponding to origin O (as in the preceding case, we place it in the plane of the screen), and α and β are direction cosines of the beam, connecting point O with the observation point, with respect to axes ξ and η, then the expansion takes the form

$$r=[(x-\xi)^2+(y-\eta)^2+z^2]^{1/2}\simeq R-\frac{x}{R}\xi-\frac{y}{R}\eta+\frac{1}{2R}(\xi^2+\eta^2)$$
$$-\frac{1}{2R^3}(x\xi+y\eta)^2\simeq R-\alpha\xi-\beta\eta+\frac{1}{2R}[\xi^2+\eta^2-(\alpha\xi+\beta y)^2] \qquad (20.11)$$

Denote the direction cosines of the beam connecting the pointlike source with the origin O by α_0, β_0. Let the distance of this source from point O be R'. An approximate equation for r' is similar to (20.11), in which R is replaced by R' and α, β by $-\alpha_0$, $-\beta_0$. Whence $\exp[ik(r+r')] = \exp[ik(R+R')]\exp(-ik\Phi)$, where Φ is a quadratic form in ξ and η, which is easily obtained in explicit form by making use of the above expansions of the type of (20.11). Assuming that the slowly varying factor in the integrand is equal to its value at point O, we obtain

$$i\lambda\psi(P)=\frac{A}{RR'}\cos\widehat{(\mathbf{n},\mathbf{R})}\,e^{ik(R+R')}\int e^{-ik\Phi}\,d\xi\,d\eta \qquad (20.12)$$

As a result, the calculation of the diffracted wave under all the enumerated assumptions is reduced to calculating the integral in (20.12) over the area of aperture. If it can be additionally assumed that $R\to\infty$ and $R'\to\infty$, then the expression for Φ is reduceable to a linear function: $\Phi \simeq (\alpha-\alpha_0)\,\xi + (\beta-\beta_0)\,\eta$, and the calculation is substantially simplified. This case is known as the *Fraunhofer diffraction*. The general equation (20.12) describes the *Fresnel diffraction*.

As in the preceding sections, our study of the diffraction will be restricted to the formulation of the problem and to a brief discussion of its physical meaning as given above. A large number of methods whose detailed description can be found in special treatises[14] have been developed for specific problems of mathematical physics dealing with diffraction.

§ 21. Geometrical optics approximation

21.1. In the limit of very short wavelengths ($\lambda\to 0$) the description of radiation field can be reduced to a number of geometric relationships.

---

[14] See, for example, the monograph cited in footnote 13 on p. 157.

Assume the medium to be isotropic and electrically nonconductive. A case of special interest for us will be that of inhomogeneous medium. As we have seen in the preceding chapters, the Maxwell equations for monochromatic radiation with no sources take the form

$$\operatorname{curl} \mathbf{H} + ik_0\varepsilon\mathbf{E} = 0, \quad \operatorname{curl} \mathbf{E} - ik_0\mu\mathbf{H} = 0$$
$$\operatorname{div} \varepsilon\mathbf{E} = 0, \qquad \operatorname{div} \mu\mathbf{H} = 0 \tag{21.1}$$

Here we can assume that **H** and **E** are amplitudes depending on **r** only. In addition, $k_0 = 2\pi/\lambda_0 = \omega/c$, where λ_0 is wavelength in vacuo, and ε and μ may be functions of **r**.

Now let us try to find an electromagnetic field at very large distances from the sources in the complex form:

$$\mathbf{E} = \mathbf{e}(\mathbf{r})\, e^{ik_0 C(\mathbf{r})}, \quad \mathbf{H} = \mathbf{h}(\mathbf{r})\, e^{ik_0 C(\mathbf{r})} \tag{21.2}$$

where **e**, **h**, C are real functions[15]. Substitute (21.2) into (21.1). By using formulas of vector analysis (see Appendix B) we can rephrase the obtained relations to the form

$$\operatorname{grad} C \times \mathbf{h} + \varepsilon\mathbf{e} = -\frac{1}{ik_0} \operatorname{curl} \mathbf{h}$$

$$\operatorname{grad} C \times \mathbf{e} - \mu\mathbf{h} = -\frac{1}{ik_0} \operatorname{curl} \mathbf{e}$$

$$\mathbf{e} \cdot \operatorname{grad} C = -\frac{1}{ik_0} (\mathbf{e} \cdot \operatorname{grad} (\ln \varepsilon) + \operatorname{div} \mathbf{e})$$

$$\mathbf{h} \cdot \operatorname{grad} C = -\frac{1}{ik_0} (\mathbf{h} \cdot \operatorname{grad} (\ln \mu) + \operatorname{div} \mathbf{h}) \tag{21.3}$$

The limit of interest here is for $k_0 \to \infty$. If the right-hand sides of the above equations can be neglected in this limit, we arrive at relations

$$\operatorname{grad} C \times \mathbf{h} + \varepsilon\mathbf{e} = 0, \quad \operatorname{grad} C \times \mathbf{e} - \mu\mathbf{h} = 0$$
$$\mathbf{e} \cdot \operatorname{grad} C = 0, \qquad \mathbf{h} \cdot \operatorname{grad} C = 0 \tag{21.4}$$

Obviously, only the first two of these equations are independent. By eliminating function **h**, we obtain

$$(\mathbf{e} \cdot \operatorname{grad} C) \operatorname{grad} C - \mathbf{e} (\operatorname{grad} C)^2 + \varepsilon\mu\mathbf{e} = 0$$

We then obtain from the third of equations (21.4)

$$(\operatorname{grad} C)^2 = n^2 \tag{21.5}$$

where we denote $n = \sqrt{\varepsilon\mu}$.

[15] It can be shown (see the monograph by M. Born and E. Wolf given in the footnote 4 on p. 143) that by assuming **e** and **h** real-valued, we restrict the analysis to linearly polarized waves; in the general case **e** and **h** must be complex-valued.

Assumptions involved in the derivation of equations (21.4) are justified if ε, μ and $|\operatorname{grad} C|$ have values of the order of 1, and the variations of functions $\mathbf{e}$ and $\mathbf{h}$ over distances comparable to wavelength are small compared with the values of the functions themselves. If radiation field tends to zero on the boundary of a certain region (for example, where the shadow begins) sufficiently rapidly, then this condition is violated. Neither does it hold in the neighbourhood of foci of optical instruments, where intensity of radiation sharply increases.

Function C introduced by definition (21.2) is called the optical path or *eikonal*, and equation (21.5) is known as the eikonal equation. Surfaces described by condition

$$C(\mathbf{r}) = \text{const} \tag{21.6}$$

are called the geometric *wavefronts*, and curves orthogonal to these surfaces—geometric *light beams*. Let $\mathbf{r} = \mathbf{r}(s)$ be a parametric equation of light beam, with the length of the arc along the beam measured with respect to an arbitrary origin chosen as parameter s. Then $d\mathbf{r}/ds \equiv \mathbf{s}$ is a unit vector tangent to the beam, and the condition of beam orthogonality to the wavefront is given by equation

$$n\mathbf{s} = n\frac{d\mathbf{r}}{ds} = \operatorname{grad} C \tag{21.7}$$

Let us clarify the physical meaning of vector $\mathbf{s}$, namely, let us show that it defines the direction of propagation of the radiation energy flux. In the limit of very high frequencies which we consider here this flux must be calculated by using the averaged formula of the type of (17.18):

$$\langle \mathbf{S} \rangle = \frac{c}{2}\,\mathbf{e} \times \mathbf{h} \tag{21.8}$$

Substitute here $\mathbf{h}$ from the second equation of (21.4). Making use of the third equation in (21.4), as well as (21.7), equation (21.8) can be transformed to

$$\langle \mathbf{S} \rangle = v\,\langle w \rangle\,\mathbf{s} \tag{21.9}$$

Average value of energy (21.9) is equal, on the basis of (17.18), to

$$\langle w \rangle = 2\,\langle w_e \rangle = \frac{\varepsilon}{2}\,\langle \mathbf{e}^2 \rangle$$

and v is the velocity of wave propagation in the medium $v = c/\sqrt{\mu\varepsilon}$. Hence, we have proved our statement. It will be instructive to compare formula (21.9) with the exact formula (14.9) for radiation emitted by a pointlike charge and with formulas obtained on p. 137 by similar time-averaging for a bounded distribution of sources.

If radiation contained in the region of propagation performs no mechanical work, it is readily found that the energy conservation law (3.3) takes the form div **S** = 0. Hence, the time-averaged value is also zero, div ⟨**S**⟩ = 0. The intensity of radiation is found as the absolute value of the averaged Poynting vector: $I = |\langle \mathbf{S} \rangle| = v \langle w \rangle$. Hence, div $(I\mathbf{s}) = 0$. Consider a pencil formed by light rays and bounded by transverse surfaces $d\sigma_1$ and $d\sigma_2$ (Fig. 23). Application of Gauss' theorem to this pencil immediately yields

$$I_1\, d\sigma_1 = I_2\, d\sigma_2 \tag{21.10}$$

if we take into account that a normal to the side surface of the pencil is everywhere orthogonal to vector **s**. Formula (21.10) is called the *law of intensity* in geometrical optics. In particular, if light beams originate at one point and are straight (we shall see soon that the rays are straight in homogeneous media), then, by choosing for $d\sigma_1$ and $d\sigma_2$ the elements of spherical surfaces with common center in the pointlike source, we arrive at $I(R) = \text{const}/R^2$, that is the law of inverse quadratic dependence of radiation intensity on distance.

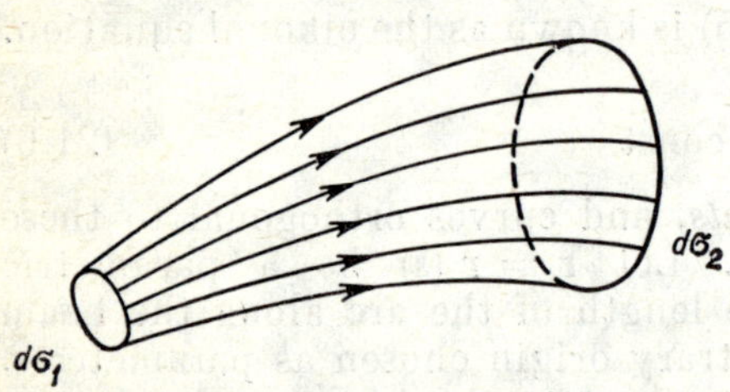

Fig. 23

Equation (21.7) can also be given in the form

$$\frac{dC}{ds} = n \tag{21.11}$$

where the left-hand side is the derivative of C in the direction of the light beam. The integral of n along the light beam between two points P_1 and P_2 on the beam, $\int_{P_1}^{P_2} n\, ds = C(P_2) - C(P_1)$, is called the *optical length* of this segment of the beam. We have seen above that $n\, ds = c\, ds/v = c\, dt$, where dt is the time interval during which energy propagates along the beam through a distance ds. Hence, optical length of a segment of light beam divided by light velocity in vacuo is equal to the time required for the radiation to cover the length of this segment.

21.2. Let us further investigate the properties of equation (21.7). First of all, function C can be eliminated by using (21.5) and (B.17). Namely,

$$\frac{d}{ds}\left(n\frac{dr}{ds}\right) = \frac{d}{ds}(\operatorname{grad} C) = (\mathbf{s}\cdot\operatorname{grad})\operatorname{grad} C$$

$$= \frac{1}{n}(\operatorname{grad} C\cdot\operatorname{grad})\operatorname{grad} C = \frac{1}{2n}\operatorname{grad}(\operatorname{grad} C)^2 = \frac{1}{2n}\operatorname{grad} n^2$$

$$\frac{d}{ds}\left(n\frac{d\mathbf{r}}{ds}\right) = \operatorname{grad} n \tag{21.12}$$

In particular, it follows that $d^2\mathbf{r}/ds^2 = 0$ for $n = \text{const}$, that is light beams in a homogeneous medium are straight.

It also follows directly from equation (21.7) that

$$\operatorname{curl}(n\mathbf{s}) = 0 \tag{21.13}$$

As, according to formula (B.14$_3$), $\operatorname{curl}(n\mathbf{s}) = \operatorname{grad} n \times \mathbf{s} + n \operatorname{curl} \mathbf{s}$, it also follows from (21.13) that $\mathbf{s} \cdot \operatorname{curl} \mathbf{s} = 0$. In a homogeneous medium $n = \text{const}$, so that $\operatorname{curl} \mathbf{s} = 0$. Consider now an arbitrary open surface σ with contour l. Integrate over this surface the component of curl ($n\mathbf{s}$) normal to this surface and apply Stokes' theorem. By virtue of formula (21.13),

$$\oint n\mathbf{s} \cdot d\mathbf{l} = 0 \tag{21.14}$$

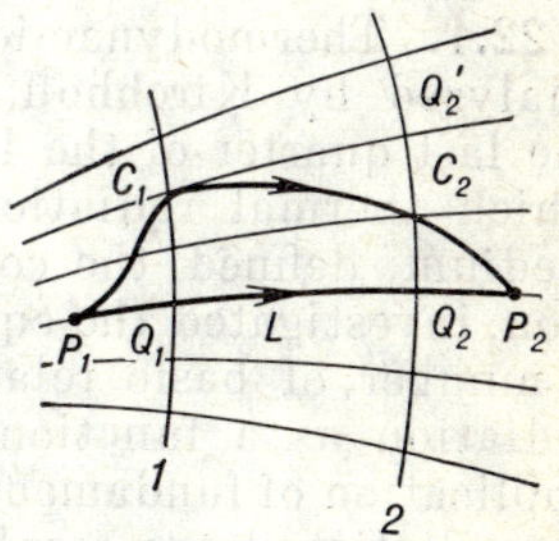

Fig. 24

The integral in the left-hand side of (21.14) is called the *Lagrangian integral invariant*. The origin of this term is clear: as follows from (21.13), the value of an integral of the type $\int_{P_1}^{P_2} n\mathbf{s} \cdot d\mathbf{l}$ is independent of the shape of the curve connecting points P_1 and P_2.

Actual light rays satisfy the *Fermat variational principle* which is a corollary of Lagrange's theorem (21.14). An assumption necessary for this derivation is that the considered region in space is regular, that is, it contains no points at which light rays could cross. Let us calculate optical lengths of segment P_1P_2 of one of light rays and of curve C connecting the same two points (Fig. 24). Points Q_1, Q_2, and C_1, C_2 are taken at the intersections of the light ray and of curve C, respectively, with wavefronts *1* and *2*, as shown in the figure. Point Q_2' is at the intersection of wavefront *2* with light beam propagating from point C_1. Apply now equation (21.14) to closed contour $C_1C_2Q_2'C_1$. Namely,

$$n\mathbf{s} \cdot d\mathbf{l}\,|_{C_1C_2} + n\mathbf{s} \cdot d\mathbf{l}\,|_{C_2Q_2'} - n\,ds_{C_1Q_2'} = 0 \tag{21.15}$$

On wavefront *2* orthogonality relation $\mathbf{s} \cdot d\mathbf{l}\,|_{Q_2Q_2'} = 0$ is valid since $d\mathbf{l}$ lies on this wavefront, and $\mathbf{s}$ is the direction of light beam. Then, it is found from (21.11) that $n\,ds\,|_{C_1Q_2'} = n\,ds\,|_{Q_1Q_2}$. And finally, the definition of scalar product yields $n\mathbf{s} \cdot d\mathbf{l}\,|_{C_1C_2} \leqslant n\,ds\,|_{C_1C_2}$. Therefore, (21.15) becomes $n\,ds\,|_{Q_1Q_2} \leqslant n\,ds\,|_{C_1C_2}$, so that

$$\int_{\Delta} n\,ds < \int_{C} n\,ds \tag{21.16}$$

Equality in (21.16) would take place only if $\mathbf{s}\cdot d\mathbf{l} = ds$ everywhere on curve C, that is, if this curve coincides with the actual light beam. But this case is excluded by virtue of the above-introduced assumption that only one beam passes through a point within the considered region. Formula (21.16) constitutes the mathematical expression of Fermat's principle.

Geometric investigation of light beams constitutes the foundation on which the theory of optical images rests; we cannot deal with this theory within the scope of this book.

§ 22. Fundamentals of radiation thermodynamics

22.1. Thermodynamic properties of radiation fields were first analyzed by Kirchhoff, Stefan, Boltzmann, Wien and Planck in the last quarter of the 19th century. They derived conditions under which thermal radiation is in equilibrium with the surrounding medium, defined the concept of temperature of equilibrium radiation, investigated the equation of state of this radiation, and derived a number of basic relations describing spectral energy density of radiation as a function of wavelength. The reader is aware that application of fundamental concepts of classical statistical mechanics to radiation later resulted in contradiction resolved only by the quantum theory of matter. In the present section we are going to study briefly the basic aspects of radiation thermodynamics directly related to classical electrodynamics.

It is of paramount importance for the study of thermodynamic properties of radiation to find the pressure exerted by radiation on the matter. Let us first calculate this pressure in a simple particular case. Let matter fill the halfspace to the right of plane $x^1 = 0$, and let a plane wave propagating along axis x^1 be incident on this plane from the left. Consider a cylinder in the right-hand halfspace with base $d\sigma$ on plane $x^1 = 0$ and height equal to h. We shall assume that all radiation penetrating the matter is completely absorbed at a distance h from the interface, so that at this distance fields $\mathbf{E}$ and $\mathbf{H}$ can be considered equal to zero.

Recall now equations (3.9)-(3.16) which determine the forces applied by the electromagnetic field to the medium which we assume to be linear so that $\mathbf{D} = \varepsilon\mathbf{E}$ and $\mathbf{B} = \mu\mathbf{H}$. Namely,

$$f_\alpha = \frac{\partial T_{\alpha\beta}}{\partial x^\beta} - \frac{\mu\varepsilon}{c}\frac{\partial}{\partial t}(\mathbf{E}\times\mathbf{H})_\alpha \qquad (22.1)$$

where

$$T_{\alpha\beta} = (E_\alpha D_\beta + H_\alpha B_\beta) - (1/2)\,\delta_{\alpha\beta}(\mathbf{E}\cdot\mathbf{D} + \mathbf{H}\cdot\mathbf{B}) \qquad (22.2)$$

In our case the transversality of electromagnetic field in a plane wave is given by equalities $E_1 = 0$, $H_1 = 0$. Taking this into account, we obtain the bulk force f_1 applied orthogonally to interface $x^1 = 0$, from formulas (22.1) and (22.2):

$$f_1 = -\frac{1}{2}\frac{\partial}{\partial x^1}(\mathbf{E}\cdot\mathbf{D}+\mathbf{H}\cdot\mathbf{B}) - \frac{\mu\varepsilon}{c}\frac{\partial}{\partial t}(\mathbf{E}\times\mathbf{H})_1 \qquad (22.3)$$

As follows from relation (3.19), surface force φ_1 is found from the relation $\int f_1\, dV = \int \varphi_1\, d\sigma$. For the abovementioned cylinder we then obtain $\varphi_1 = \int_0^h f_1\, dx$. However, the observed quantity is not the instantaneous value of this force but its time-averaged value which can be defined by the following formula:

$$\langle\varphi_1\rangle = \lim_{T\to\infty}\frac{1}{2T}\int_{-T}^{T}\varphi_1\, dt$$

Consequently, we want to find $\langle\varphi_1\rangle = \int_0^h \langle f_1\rangle\, dx$. In calculating the mean value $\langle f_1\rangle$ from (22.3) we shall take into account that the Poynting vector of a plane wave is a periodic function of time. This means that quantity

$$\int_{-T}^{T}\frac{\partial}{\partial t}(\mathbf{E}\times\mathbf{H})_1\, dt = (\mathbf{E}\times\mathbf{H})_1\Big|_{-T}^{T}$$

is bounded, so that the limit calculated according to the definition of the mean value is zero. Integrating the first term in (22.3) over dx and taking into account that field vanishes at depth h, we obtain for the pressure of radiation on the boundary of the matter:

$$p = \langle\varphi_1\rangle = (1/2)\,\langle\mathbf{E}\cdot\mathbf{D}+\mathbf{H}\cdot\mathbf{B}\rangle = w \qquad (22.4)$$

where w is the energy density of electromagnetic field at the interface.

Consider now a different case. Namely, take a closed cavity surrounded by material walls and filled with radiation. The properties of the matter inside the cavity will be considered identical to those of vacuum, so that in the Gaussian system of units we are using here $\mathbf{D} = \mathbf{E}$ and $\mathbf{B} = \mathbf{H}$. We can assume that radiation is either emitted by the walls or by emitters within the cavity, occupying a negligible fraction of its volume. Further, we assume the radiation field to be isotropic, that is its properties are assumed identical in all directions on the average (during a sufficiently long interval of time). Components of the field in any two mutually perpendicular directions are assumed statistically independent. Strictly speaking, this means by definition that equalities of the type $\langle E_1E_2\rangle = 0$ hold. In addition, we assume without proving it that mean values of derivatives are equal to derivatives of mean values. Much more profound analysis of the statistical properties of electromagnetic field than could be undertaken here would be required to construct a foundation to these assumptions. But once they are chosen, calculation of pressure becomes trivial.

As in the derivation of formula (22.4), it can also be shown that $\left\langle \frac{\partial}{\partial t} (\mathbf{E} \times \mathbf{H}) \right\rangle = 0$ since components of **E** and **H** are finite. Moreover, the assumption of isotropy means that $\langle E_\alpha^2 \rangle = (1/3) \langle \mathbf{E}^2 \rangle$. Consider an element of wall surface and direct axis x^1 inward, along the normal to this element. As follows from symmetry arguments (as for an ideal gas in a vessel), only component f_1 of the force can differ from zero. With all these arguments, equation (22.3) gives

$$\langle f_1 \rangle = \frac{\partial}{\partial x^1} \left\langle E_1^2 + H_1^2 - \frac{1}{2} (\mathbf{E}^2 + \mathbf{H}^2) \right\rangle = -\frac{1}{6} \frac{\partial}{\partial x^1} \langle \mathbf{E}^2 + \mathbf{H}^2 \rangle$$

whence pressure p is

$$p = \int \langle f_1 \rangle \, dx = \frac{1}{6} \langle \mathbf{E}^2 + \mathbf{H}^2 \rangle = \frac{1}{3} w \qquad (22.5)$$

An absolutely identical result will be obtained if we assume that the walls reflect the incident radiation completely. We shall soon show that relation (22.5) between pressure and energy density makes it possible to derive fundamental laws of radiation thermodynamics.[16] In this analysis it plays the role of the equation of state of the "photon gas". In what follows we shall consider only such radiation in the cavity for which this equation was derived.

22.2. In order to use thermodynamic arguments we have to define the concept of *temperature* of radiation. It is logical to assume that radiation is in thermal equilibrium with walls having a temperature θ if the walls absorb per unit time the same amount of energy as they emit. Radiation is then called equilibrium radiation, and its state is characterized by the same temperature θ.

Radiation enclosed in a cavity is made up of waves which, in the general case, may have any frequency. If $w_\nu \, d\nu$ denotes the amount of radiation energy per unit volume, carried by waves with frequencies from ν to $\nu + d\nu$, then

$$w = \int_0^\infty w_\nu \, d\nu \qquad (22.6)$$

It is function w_ν (*spectral energy density*) introduced here that will be of principal interest in further analysis.

[16] Detailed discussion of radiation thermodynamics can be found in the following monographs: M. Planck, *The Theory of Heat Radiation*, Blakiston, Philadelphia, 1914, and R. Becker, *Theorie der Electrizität*. Band II. Electronentheorie; Leipzig und Berlin, Taubner, 1933.

At the first glance, it would seem natural to assume that energy distribution of equilibrium radiation over frequency must depend on the properties of the matter with which this radiation is in equilibrium. However, Kirchhoff had demonstrated that this assumption is in contradiction with the second law of thermodynamics: were it true, one could construct a perpetuum motion machine of the second kind. Therefore *Kirchhoff's law* is valid: function w_ν depends only on temperature θ and is independent of any characteristics of specific properties of the wall material.[17]

Before investigating the properties of function $w_\nu(\theta)$ any further, we have to introduce additional quantities closely related to this function. Inside the cavity filled with radiation consider an infinitesimal element of surface $d\sigma$ (in a particular case it may coincide with an element of wall surface, but generally it is an infinitesimal part of an arbitrary geometric surface). As radiation passes through element $d\sigma$ in all directions, we can consider the properties of radiation emitted from this element of surface. Let solid angle $d\Omega$ radiate from element $d\sigma$ at an angle ϑ to the normal. Energy flux propagating within this solid angle will be proportional, in particular, to $\cos\vartheta$ (since radiation is assumed isotropic, and therefore energy flux is proportional to the cross-sectional area of the pencil into which enters the energy propagating from $d\sigma$ at an angle ϑ). Energy entering into solid angle $d\Omega$ during time dt must therefore be considered equal to $K\cos\vartheta\, d\Omega\, d\sigma\, dt$. Coefficient K is called the *radiance* and, as usual, $d\Omega = \sin\vartheta\, d\vartheta\, d\zeta$.

Let us find a relation between K and w. Let $d\sigma$ be an element of the wall surface, and V—an arbitrary but very small volume within the cavity at a distance r from element $d\sigma$. Let us take a solid angle $d\Omega$ such that the cone corresponding to it intersects volume V; as volume V is very small, we obtain a nearly cylindrical solid. Denote the cross-sectional area by df and the height of this cylinder by h. By the definition of solid angle, $d\Omega = df/r^2$. If element $d\sigma$ of the wall emits some energy per unit time in the direction of $d\Omega$ then a fraction h/c of this energy is within the cylinder, that is, according to the definition in the preceding paragraph, it contains the amount of energy equal to $\frac{h}{c} K\cos\vartheta\, d\sigma \frac{df}{r^2}$. Let us sum now over all bundles of beams the radiation emitted from $d\sigma$ and intersecting volume V. With sufficient accuracy $\sum h\, df = V$, so that we can calculate energy contained in the whole of volume V; it is equal to $VK\, d\sigma \frac{\cos\vartheta}{cr^2}$. Now we can integrate over the whole boundary

---

[17] The derivation of Kirchhoff's law from the second law of thermodynamics can be found in the monographs cited in footnote 16 on p. 166.

of the volume. The result can be written in the form

$$wV = \frac{VK}{c}\int \frac{\cos\vartheta \, d\sigma}{r^2}$$

The integral in the right-hand side is the sum of solid angles at which the wall is seen from volume V; it is equal to 4π. We thus obtain a relation between K and w in the form of the following simple equality: $w = \frac{4\pi}{c} K$. Obviously, it can be rewritten as a relation between spectral densities:

$$w_\nu \, d\nu = \frac{4\pi}{c} K_\nu \, d\nu \tag{22.7}$$

A concept which is often used is that of unilateral emission. This term means radiation emitted by an element of the wall into a hemisphere (i.e. into the solid angle 2π). The corresponding energy L is

$$K \int_0^{\pi/2} d\vartheta \cos\vartheta \sin\vartheta \int_0^{2\pi} d\varphi = \pi K \tag{22.8}$$

From this and from (22.7) we obtain

$$L = (1/4)\, cw \tag{22.9}$$

22.3. With the thermodynamic state of equilibrium radiation defined, we can apply to it all the methods of thermodynamics. Thus, for example, we can consider equilibrium radiation as a "working substance" of a hypothetical heat engine and investigate the corresponding Carnot cycle. This will give us entropy of thermal radiation. If the state of radiation is determined by the volume of the cavity and by radiation temperature, then the basic thermodynamic relation can be written in the conventional form

$$\theta \, d\mathscr{S} = d\mathscr{W} + p\, dV \tag{22.10}$$

Here $\mathscr{S}$ is the total entropy within volume V, and $\mathscr{W}$ is energy. Obviously we have to assume that in our conditions radiation energy is distributed uniformly over the volume it occupies, so that $\mathscr{W} = wV$.

Making use of the equation of state of radiation (22.5) and equation (22.10), we can derive the *Boltzmann law of radiation* (Stefan-Boltzmann law): energy density of radiation is proportional to the fourth power of absolute temperature. As w is a function only of θ (Kirchhoff's law), then

$$d\mathscr{W} = w\, dV + \frac{dw}{d\theta}\, d\theta \tag{22.11}$$

Substitution of (22.11) and (22.5) into (22.10) yields

$$d\mathscr{S} = \frac{V}{\theta}\frac{dw}{d\theta}\, d\theta + \frac{4}{3}\frac{w}{\theta}\, dV \tag{22.12}$$

so that $\left(\frac{\partial \mathscr{S}}{\partial \theta}\right)_V = \frac{V}{\theta}\frac{dw}{d\theta}$ and $\left(\frac{\partial \mathscr{S}}{\partial V}\right)_\theta = \frac{4}{3}\frac{w}{\theta}$. Since $\mathscr{S}$ is a function of state, equation $\frac{\partial^2 \mathscr{S}}{\partial V\, \partial \theta} = \frac{\partial^2 \mathscr{S}}{\partial \theta\, \partial V}$ must hold. Having calculated these second derivatives, we obtain $dw/d\theta = 4w/\theta$ and hence

$$w = \sigma\theta^4 \tag{22.13}$$

Here σ is the universal Stefan-Boltzmann constant. For unilateral radiation, formulas (22.13), (22.7), and (22.9) yield $L = \frac{\sigma c}{4}\theta^4$. It is this quantity equal to energy flux through a small aperture in the wall of a cavity that is measured in experiments; the measured value of constant $\sigma c/4$ is

$$5.670 \times 10^{-5}\ \text{erg}/(\text{s}\cdot\text{cm}^3\cdot\text{deg}^4) = 5.670 \times 10^{-8}\ \text{W}/(\text{m}^2\cdot\text{deg}^4)$$

For pressure we obtain $p = (1/3)\ \sigma\theta^4$ and for entropy $\mathscr{S} = (4/3)\ \sigma\theta^3 V$.

We can pass now to *Wien's law* which reduces the determination of spectral density $w_\nu(\theta)$ to finding another function which depends only on ν/θ. In this proof, radiation is considered in a cavity whose walls are assumed totally reflecting (ideal mirrors). The following remark is in order here. As follows from the derivation of Kirchhoff's law and from the law itself, equilibrium radiation is formed if the emitter is able to absorb all those frequencies which it emits, and intensity of absorption of each frequency must be equal to the intensity of its emission. This emitter is known as the *blackbody*, and the equilibrium radiation as the blackbody radiation. In deriving Wien's law for a cavity with mirror walls it is assumed that a blackbody emitter is introduced into this cavity and that its dimensions are very small ("a speck of carbon black"). Radiation will be equilibrium precisely owing to the interaction with this emitter. It is also assumed that if the volume of the cavity undergoes infinitely slow adiabatic variation, radiation will remain equilibrium in the course of this variation.[18] The cavity can be modelled by a cylinder with a plunger.

Let us analyze what should happen to function w_ν when the plunger advances. The surface of the plunger facing radiation is a slowly moving mirror. In § 19 we have obtained some results which show that frequency of radiation reflected from a moving mirror changes. This Doppler effect is given by the first of formulas (19.13) (here we need the nonrelativistic approximation of this formula). As the unilateral radiation of the wall is given by (22.8), and as frequency ν of the reflected radiation changes, energy $w_\nu\, d\nu \cdot V$ corresponding to frequency ν diminishes during time dt by $\pi K_\nu A\, d\nu\, dt$, where A is the total area of the moving mirror. However, the reflection from the moving mirror will, during the same interval of time,

[18] See Planck's monograph cited in footnote 16 on p. 166.

shift other frequencies into the range of interest, from ν to $\nu + d\nu$. Let us calculate an increment in energy corresponding to this range of frequencies and produced owing to this effect. We denote by ν' the frequency of radiation incident on the plunger. If radiation is emitted within a solid angle $d\Omega$ at an angle ϑ to the normal to the plunger surface, then the definition of $K_{\nu'}$ shows that the radiation energy in the range from ν' to $\nu' + d\nu'$ over time dt is equal to $AK_{\nu'} \cos\vartheta \, d\Omega \, dt \, d\nu'$. After reflection this radiation has frequency ν if ν and ν' are related by (19.13). But we have also shown in § 19 that energies and frequencies of radiation measured in different reference frames are related by the equation $W/\nu = W'/\nu'$. Consequently, for calculation of energy in a reference frame fixed with respect to the walls the above expression must be additionally multiplied by a ratio

$$\frac{\nu}{\nu'} = 1 + \frac{2v}{c}\cos\vartheta \tag{22.14}$$

To summarize, the above arguments show that energy available in the frequency range $(\nu, \nu + d\nu)$ is equal to

$$A \int_{\Omega} K_{\nu'} \cos\vartheta \frac{\nu}{\nu'} \, d\Omega \, d\nu' \, dt \tag{22.15}$$

If the difference $\nu' - \nu$ is very small, we can make use of relation (22.14) and obtain $K_{\nu'} = K_\nu + \frac{\partial K}{\partial \nu}(\nu' - \nu) = K_\nu - \frac{\partial K_\nu}{\partial \nu}\frac{2v\nu}{c} \times$ $\times \cos\vartheta$. Substitution of this value of $K_{\nu'}$ into (22.15) and integration of (22.15) over the hemisphere yields

$$A\left(\pi K_\nu - \frac{\partial K_\nu}{\partial \nu}\frac{2v\nu}{c}\frac{2\pi}{3}\right) d\nu \, dt$$

Now we have to subtract the mentioned earlier loss of energy $w_\nu \, d\nu$ because some frequencies leave the interval under consideration. Therefore

$$d(Vw_\nu) = -\frac{4\pi}{3} A \frac{v\,dt}{c} \nu \frac{\partial K_\nu}{\partial \nu}$$

But it is clear that $Av\,dt = -dV$ and therefore, if we take account of (22.7), $d\ (Vw_\nu) = \frac{\nu}{3}\frac{\partial w_\nu}{\partial \nu} dV$. Since the dependence of w_ν on V represents, in fact, the dependence of w_ν on time in the course of displacement of the plunger, the last relation leads to the equation

$$V\frac{\partial w_\nu}{\partial V} = \frac{\nu}{3}\frac{\partial w_\nu}{\partial \nu} - w_\nu \tag{22.16}$$

In order to find function w_ν, it will be convenient to introduce new variables: $x = V$, $y = \nu^3 V$. If w_ν is considered now to be a function of these variables, equation (22.16) can be transformed to $\frac{\partial}{\partial x}(xw_\nu) = 0$, whence $xw_\nu = \psi(y)$. Function ψ cannot be specified on the basis of the indicated assumptions only. Using former variables, we obtain

$$w_\nu(V) = \frac{1}{V}\psi(\nu^3 V) \equiv \nu^3\varphi(\nu^3 V) \qquad (22.17)$$

where we denoted $\nu^3\varphi = \psi/V$.

So far we were considering w_ν as a function of V. But actually we are interested in w_ν as a function of θ. In order to derive this relation, let us make use of the first law of thermodynamics for an adiabatic process associated with the motion of the plunger:

$$d(Vw) + p\,dV = 0 \qquad (22.18)$$

By using the Stefan-Boltzmann law (22.13) and the expression for p associated with this law, we can transform (22.18) to the form

$$d(V\theta^4) + (1/3)\,\theta^4\,dV = 0$$

Hence, $V\theta^3 = \text{const}$. Therefore (22.17) can be rephrased in a different form:

$$w_\nu(\theta) = \nu^3 f(\nu/\theta) \qquad (22.19)$$

This equation represents *Wien's law*. From (22.19) we can derive another law, the so-called *Wien displacement law*. To do this, let us make a transition from distribution of energy over frequency to a distribution over wavelength: $w_\lambda d\lambda$. As $|d\nu| = \frac{c}{\lambda^2}|d\lambda|$, we immediately find from (22.19) that $w_\lambda(\theta) = \lambda^{-5}g(\lambda\theta)$. Let us denote by λ_m the wavelength for which function $w_\lambda(\theta)$ reaches a maximum at a given temperature θ. Condition $\partial w_\lambda/\partial\lambda = 0$ becomes: $5g(\lambda_m\theta) = \lambda_m\theta g'(\lambda_m\theta)$. In other words, product $\lambda_m\theta$ must be equal to a universal constant found as a root of equation $5g(\xi) = \xi g'(\xi)$. Equation $\lambda_m\theta = \text{const}$ is precisely the displacement law.

22.4. The arguments provided by classical electrodynamics and thermodynamics proved to be insufficient for a complete determination of function $w_\nu(T)$. Furthermore, additional efforts led to the Rayleigh-Jeans formula of this function which proved to be in utter contradiction with the physics of the phenomenon. Only the quantum theory of radiation developed by M. Planck and A. Einstein has shown a way out of these principal difficulties. The energy distribution obtained by Planck allows to calculate, in particular, the constant which enters the displacement law. The result of this calculation can be written in the form $\lambda_m T \approx \frac{h}{k}\frac{c}{4.965}$. Here h is Planck's constant, and k is the Boltzmann constant.

The abovementioned Rayleigh-Jeans formula, which determines function $w_\nu(\theta)$ completely (but incorrectly), was derived by a more sophisticated analysis of the properties of radiation filling a cavity with mirror walls, the one used to obtain Wien's law. Radiation field was represented as an infinite sum of standing waves. A similar method is also used in modern electrodynamics, for example, in quantization of an electromagnetic field. In the last case the method differs from that of Jeans in that travelling waves have to be considered in addition to standing waves. However, the results are essentially the same if we consider the radiation produced in vacuo by very remote sources. It is assumed in the analysis that a volume (for example, cubic) can be singled out in the field, such that the properties of the field remain unaltered under a sufficiently large number of translations by the edge length of this cube (in the direction of any of the edges). This is called the *periodicity condition.*

Since the method of radiation field expansion in plane waves is very important for the theory, let us discuss it in more detail. Formulas (18.3) and (18.4) show that in the case of transverse electromagnetic waves we can set $\varphi_\omega = 0$ because with this condition

$$\mathbf{n}\cdot\mathbf{A}_\omega = 0, \quad \mathbf{E}_\omega = ik\mathbf{A}_\omega \quad \text{and} \quad \mathbf{B}_\omega = ik\mathbf{n}\times\mathbf{A}_\omega$$

Therefore, condition $\varphi = 0$ or relation $\operatorname{div}\mathbf{A} = 0$, which follows from the Lorentz gauge condition, express gauge conditions allowed for the field of transverse electromagnetic waves. In this case

$$\mathbf{E} = -\frac{1}{c}\frac{\partial\mathbf{A}}{\partial t}, \quad \mathbf{H} = \operatorname{curl}\mathbf{A} \tag{22.20}$$

If the unit of periodicity in the field is a cube with edge L, then

$$\mathbf{A}(x+L,\ y,\ z) = \mathbf{A}(x,\ y+L,\ z) = \mathbf{A}(x,\ y,\ z+L) = \mathbf{A}(x,\ y,\ z) \tag{22.21}$$

In our specific case

$$\left(\Delta - \frac{1}{c^2}\frac{\partial^2}{\partial t^2}\right)\mathbf{A} = 0 \tag{22.22}$$

We shall seek a solution of this equation in the form

$$\mathbf{A} = \sum_\lambda [q_\lambda(t)\,\mathbf{A}_\lambda + q_\lambda^*(t)\,\mathbf{A}_\lambda^*] \tag{22.23}$$

where $\mathbf{A}_\lambda$ depend only on $\mathbf{r}$. Potential $\mathbf{A}(\mathbf{r},\ t)$ here is definitely real, and complex functions q_λ and $\mathbf{A}_\lambda$ are denoted more conveniently by subscripts λ and not ω as we did in § 18. Substitution of (22.23) into (22.22) separates the variables, and the following equations must hold:

$$(\Delta + \nu_\lambda^2/c^2)\,\mathbf{A}_\lambda = 0, \quad \ddot{q}_\lambda + \nu_\lambda^2 q_\lambda = 0 \tag{22.24}$$

for any λ. Here $-\nu_\lambda^2/c^2$ is the constant defining separation of the variables. The solutions of these equations are

$$\mathbf{A}_\lambda = c\boldsymbol{\varepsilon}_\lambda e^{i(\boldsymbol{\varkappa}_\lambda\cdot\mathbf{r})}, \quad q_\lambda = |q_\lambda| e^{-i\nu_\lambda t} \tag{22.25}$$

Here $|\boldsymbol{\varkappa}_\lambda| = \nu_\lambda/c$ and $\boldsymbol{\varepsilon}_\lambda$ is a unit vector representing the direction of polarization; by virtue of transversality, $\boldsymbol{\varepsilon}_\lambda\cdot\boldsymbol{\varkappa}_\lambda = 0$. Therefore, $q_\lambda \mathbf{A}_\lambda$ is a wave propagating in direction $+\boldsymbol{\varkappa}_\lambda$. Hereafter we denote the wave whose propagation direction is that of vector $-\boldsymbol{\varkappa}_\lambda$ by $\mathbf{A}_{-\lambda}$, with $\boldsymbol{\varepsilon}_{-\lambda} = \boldsymbol{\varepsilon}_\lambda$.

If we impose the requirement of periodicity (22.21), then components of vectors $\boldsymbol{\varkappa}_\lambda$ are given by formulas

$$\varkappa_{\lambda\alpha} = \frac{2\pi}{L} n_{\lambda\alpha} \quad (\alpha = 1, 2, 3) \tag{22.26}$$

where $n_{\lambda\alpha}$ are arbitrary integers (positive or negative).

As follows from (22.25) and from the periodicity condition, the orthogonality and normalization conditions hold:

$$\int \mathbf{A}_\lambda\cdot\mathbf{A}_\mu^* \, dV = \int \mathbf{A}_\lambda\cdot\mathbf{A}_{-\mu} \, dV = c^2\delta_{\lambda\mu} \tag{22.27}$$

Here and below we assume for simplicity that $L = 1$.

Introduce now real variables

$$Q_\lambda = q_\lambda + q_\lambda^*, \quad P_\lambda = -i\nu_\lambda (q_\lambda - q_\lambda^*) = \dot{Q}_\lambda \tag{22.28}$$

Function

$$\mathscr{H}_\lambda = 2\nu_\lambda^2 q_\lambda q_\lambda^* = (1/2)(P_\lambda^2 + \nu_\lambda^2 Q_\lambda^2) \tag{22.29}$$

has the form of the Hamiltonian of an oscillator vibrating at frequency ν_λ.

Let us show that energy of the field can be represented as a sum of energies of such oscillators with all possible frequencies, that is

$$\mathscr{W} = \frac{1}{2}\int (\mathbf{E}^2 + \mathbf{H}^2)\, dV = \sum_\lambda \mathscr{H}_\lambda \tag{22.30}$$

We see from (22.20) and (22.23) that $\mathbf{E} = -\frac{1}{c}\sum_\lambda (\dot{q}_\lambda \mathbf{A}_\lambda + \dot{q}_\lambda^* \mathbf{A}_\lambda^*)$. By using the last formula, equation (22.27), and the first of definitions in (22.28), the calculation of integral $\int \mathbf{E}^2\, dV$ yields $\sum_\lambda \dot{Q}_\lambda^2 = \sum_\lambda P_\lambda^2$.[19]

In order to calculate $\int \mathbf{H}^2\, dV$, let us first consider an integral of the type $\int (\operatorname{curl} \mathbf{A}_\lambda \cdot \operatorname{curl} \mathbf{A}_\mu)\, dV$. The integrand can be transformed

---

[19] Note that both function $\mathbf{A}_\lambda$ and function $\mathbf{A}_{-\lambda}$ in expansion (22.23) are multiplied, as can be seen from the derivation, by the same coefficient q_λ.

by using the formula

$$\operatorname{curl} \mathbf{A}_\lambda \cdot \operatorname{curl} \mathbf{A}_\mu = \operatorname{div} (\mathbf{A}_\mu \times \operatorname{curl} \mathbf{A}_\lambda) + (\mathbf{A}_\mu \cdot \operatorname{curl} \operatorname{curl} \mathbf{A}_\lambda)$$

As a result of periodicity, $\oint \mathbf{n} [\mathbf{A}_\mu \times \operatorname{rot} \mathbf{A}_\lambda] \, d\sigma = 0$ (integration over the boundary surface of the cube). Therefore,

$$\int \operatorname{curl} \mathbf{A}_\lambda \cdot \operatorname{curl} \mathbf{A}_\mu \, dV = \frac{\nu_\lambda^2}{c^2} \int \mathbf{A}_\lambda \cdot \mathbf{A}_\mu \, dV$$

Substitution of $\mathbf{H} = \operatorname{curl} \mathbf{A}$ into relations (22.27) yields $\int \mathbf{H}^2 \, dV = \sum_\lambda \nu_\lambda^2 Q_\lambda^2$. After gathering all these results we see that equation (22.30) indeed holds.

Let us calculate now the number of oscillators of the field corresponding to a specific direction of polarization $\boldsymbol{\varepsilon}_\lambda$, to vector $\boldsymbol{\varkappa}_\lambda$ within an element of solid angle $d\Omega$, and to frequencies from ν to $\nu + d\nu$. As follows from (22.26),

$$\nu_\lambda^2 = (2\pi c/L)^2 (n_{\lambda_1}^2 + n_{\lambda_2}^2 + n_{\lambda_3}^2) \tag{22.31}$$

Each frequency lower than a certain frequency ν_λ corresponds to a point with integral coordinates n_1, n_2, n_3 within a sphere in the three-dimensional space. A rough estimate of the number of such points within a spherical layer from ν_λ to $\nu_\lambda + d\nu_\lambda$ can be obtained if we assume that points in this layer are distributed nearly continuously. Then $n^2 \, dn \, d\Omega = \nu^2 \, d\nu \, d\Omega L^3/(2\pi c)^3$, so that the number of oscillators of the field per unit volume is proportional to ν^2. If $\overline{\mathscr{E}}_\nu$ is the mean energy per each field oscillator, then $w_\nu \sim \nu^2 \overline{\mathscr{E}}_\nu$. We have mentioned that thermodynamic properties of radiation were studied in a cavity with mirror walls; in this section we formulated the problem somewhat differently. In § 38, however, we shall see that the number of standing waves in such a cavity is given by an absolutely identical relation. The *Rayleigh-Jeans formula*, which states that the total energy density given by (22.6) is infinite at any temperature, results precisely from the theorem of classical statistics on uniform distribution of energy over degrees of freedom $\overline{\mathscr{E}}_\nu = \overline{\mathscr{E}} = k\theta$. Correct results will be obtained if Planck's formula for the mean energy is used:

$$\overline{\mathscr{E}}_\nu = \frac{h\nu}{\exp(h\nu/k\theta) - 1}$$

CHAPTER 5

THE LORENTZ-DIRAC EQUATION. SCATTERING AND ABSORPTION OF ELECTROMAGNETIC FIELD

§ 23*. The Lorentz-Dirac equation. Radiative reaction

23.1. Fields generated by a moving electric charge must interact with this charge and thus affect its motion. In the most elementary way this effect can be estimated by applying the law of energy conservation to the phenomenon of emission analyzed in § 16. We have found there that in the nonrelativistic approximation the energy emitted in all directions per unit time by a system consisting of a charge and a field generated by this charge is given by the Larmor formula (16.3)[1]. Let us assume that energy is conserved owing to the deceleration of motion of the emitting charge; this deceleration is a result of an additional force $\mathbf{F}_{\text{rad}}$, analogous to the force of friction, applied to the charge by the radiation field. This force is known as the radiative friction or *radiative reaction*. Let us impose the requirement that the work done by this force compensates the loss of energy caused by radiation. Then we can write on the basis of (16.3)

$$\int_{t_1}^{t_2} \mathbf{F}_{\text{rad}} \cdot \mathbf{v}\, dt = -\frac{1}{4\pi}\frac{2}{3}\frac{q^2}{c^3}\int_{t_1}^{t_2} \dot{\mathbf{v}} \cdot \dot{\mathbf{v}}\, dt$$

$$= \frac{1}{4\pi}\frac{2}{3}\frac{q^2}{c^3}\left(-\mathbf{v}\cdot\dot{\mathbf{v}}\Big|_{t_1}^{t_2} + \int_{t_1}^{t_2} \ddot{\mathbf{v}}\cdot\mathbf{v}\, dt\right)$$

Here we have applied integration by parts. Assume now that for some reasons the nonintegral term in the above formula equals zero. If we consider, for example, oscillatory motion of a charge when $\mathbf{v}$ and $\dot{\mathbf{v}}$ are approximated sufficiently well by periodic functions, we can compare the left- and right-hand sides of the above formula after averaging it over a large number of periods. Then the mean value of the nonintegral term can indeed be assumed equal to zero, and friction can be expressed as

$$\mathbf{F}_{\text{rad}} = m\tau_0\ddot{\mathbf{v}} \tag{23.1}$$

---

[1] The Larmor formula remains valid if the velocity of the emitter differs from zero but remains sufficiently small compared to that of light in vacuo. Thus, a simple estimate shows that it works for radiation emitted by a charge vibrating at optical frequency at an amplitude of several angstroms.

where

$$\tau_0 \equiv \frac{1}{4\pi}\frac{2}{3}\frac{q^2}{c^3 m} \tag{23.2}$$

The dimension of τ_0 is that of time. Taking radiative friction into account, we must rewrite the equation of motion of the charge in the form

$$m(\dot{\mathbf{v}} - \tau_0 \ddot{\mathbf{v}}) = \mathbf{F} \tag{23.3}$$

where $\mathbf{F}$ stands for all external forces applied to the charge (for instance, the Lorentz force), with an exception of the radiative reaction. The value of parameter τ introduced by equation (23.2) is given completely by the charge and the mass of the particle generating radiation. Thus, if q and m are the charge and the mass of the electron, then τ_0 is of the order of 10^{-24} s. Distance covered by light during this interval is of the order of 10^{-13} cm.

23.2. Obviously, the substantiation given above to equation (23.3) is not satisfactory. In addition to, in fact, arbitrary elimination of the nonintegral term, this derivation is relativistically noninvariant, and as a result the final expression (23.3) is equally noninvariant. Nevertheless, the application of this equation to investigate emission by a charge driven by a quasi-elastic force (i.e. oscillator) gives correct results with respect to the spectral composition of this radiation (see below, § 25).

Relativistic analysis of radiative reaction was given by Dirac in 1938[2]. This analysis, however, made use not only of retarded but also of advanced potentials for calculation of electromagnetic field. Moreover, the equation of motion was derived in an asymptotic condition that the action of forces on a charge vanishes both in the remote past and in the remote future.

Here we shall derive the relativistic equation of motion of electrons taking into account radiative reaction (the Lorentz-Dirac equation) by using the methods already applied in § 15 to a different problem. Here we shall not need Dirac asymptotic condition. Let us recall the energy-momentum conservation of electromagnetic field given by relations (15.26) and (15.27). In § 15 we were interested in properties of the field at very large distances from an emitting charge, but here, as should be clear from the above arguments, we must analyze them in the immediate vicinity of the charge, that is for $\varepsilon \to 0$ (in the notation of § 15).

We shall begin with calculating $Q_r(\varepsilon)$ in this limit. Some of the formulas necessary for this calculation have already been obtained in § 15, and among them (15.28). Calculation of the element of timelike hypersurface $\rho = \varepsilon$ is based on formula (D.4) in which derivative

[2] P. A. M. Dirac, *Proc. Roy. Soc.*, **A167**, 148 (1938).

$\partial\rho/\partial x^r$ must be given in the form of (15.16). Assuming then that the direction of the normal to hypersurface $\rho = \varepsilon$ is given sufficiently well by estimate (15.28) (this assumption is equivalent to neglecting infinitesimal terms of higher orders), we can write

$$Q_r(\varepsilon) = \frac{1}{c}\int_{s=s_1}^{s=s_2} ds \int T^{km} p_m \varepsilon^2 (1-W)\, d\omega \qquad (23.4)$$

The integrand in (23.4) is represented by (15.22). It will be useful to express all the terms in the integrand via (15.20) and (15.15) in terms of which W, according to its definition (15.14), is proportional to ρ, that is to ε. Additionally, R^k must be replaced by its expansion (15.6). The terms in the integrand containing a product of an odd number of factors p_i can be dropped, since they will give zero in the subsequent integration over $d\omega$ (see Appendix D, Item 2). Note that W is proportional to p_i. Therefore, the integrand can be transformed to

$$-\frac{1}{c^2\varepsilon}\left[w^k + \frac{3}{2} p^k (\vec{w}\vec{p})\right] + \frac{u^k}{c^5}[\vec{w}^2 + (\vec{w}\vec{p})^2] + O(\varepsilon) \qquad (23.5)$$

Integrals over $d\omega$ are found from (15.32) and (15.32′).

The final result is

$$\int T^{rm} p_m \varepsilon^2 (1-W)\, d\omega = \frac{1}{4\pi}\frac{q^2}{c^2}\left(-\frac{w^r}{2\varepsilon} + \frac{2}{3c^3} u^r \vec{w}^2\right) + O(\varepsilon) \qquad (23.6)$$

A significant fact is that the first term in the right-hand side diverges when $\varepsilon \to 0$. We shall see presently that this is caused by the "proper energy" of the emitting charge.

As follows from (15.26), (15.31), and (23.4),

$$P^r(\tau_2) - P^r(\tau_1) = \int_{\tau_1}^{\tau_2} d\tau \int d\omega\, T^{rm} p_m \varepsilon^2 (1-W) \qquad (23.7)$$

where the inner integral over $d\omega$ has the form of (23.6).

Let us assume that τ_1 and τ_2 have nearly the same value, namely $\tau_2 \equiv \tau_1$, $\tau_1 = \tau - \Delta\tau$, where $\Delta\tau > 0$ and $\Delta\tau = O(\varepsilon)$. In order to calculate the integral over $d\tau$ of expression (23.6), we make use of the mean value theorem. This yields

$$\int_{\tau_1}^{\tau_2} \ldots \simeq \frac{1}{4\pi}\frac{q^2}{c^2}\left[-\frac{w^r(\tau - k\,\Delta\tau)}{2\varepsilon} + \frac{2}{3c^3} u^r(\tau - k\,\Delta\tau)\, \vec{w}^2(\tau - k\,\Delta\tau)\right]\Delta\tau + O(\varepsilon\,\Delta\tau) \qquad (23.8)$$

Here $0 < k < 1$. Expanding the left-hand side of (23.7) and formula (23.8) into the Taylor series and dividing by $\Delta\tau$ we obtain

$$\frac{dP^r}{d\tau} = \frac{q^2}{4\pi c^2}\left[-\frac{w^r(\tau)}{2\varepsilon} + \frac{1}{2}\frac{k\,\Delta\tau}{\varepsilon}\,b^r(\tau) + \frac{2}{3c^3}\,u^r(\tau)\,\vec{w}^2(\tau)\right] + O(\varepsilon)$$

Here $b^r \equiv dw^r/d\tau$.[3] Term b^r is retained because it is proportional to ε. Let us set now $k\,\Delta\tau = 4\varepsilon/3c$. We shall see below that this condition is necessary for the orthogonality of the four-dimensional force to the four-dimensional velocity and therefore, for the constancy of the rest mass of the emitting charge along its world line. This gives

$$\frac{dP^r}{d\tau} = \frac{q^2}{4\pi c^2}\left[-\frac{w^r(\tau)}{2\varepsilon} + \frac{2}{3c}\left(b^r + \frac{u^r\vec{w}^2}{c^2}\right)\right] + O(\varepsilon)$$

The total 4-momentum of the field and particle must be conserved in the transition to the limit $\varepsilon \to 0$. In other words, the equality

$$\frac{d}{d\tau}(\mu u^r) = \frac{dP^r}{d\tau} + F^r_{\text{ext}} \tag{23.9}$$

must be valid. Here $\vec{F}_{\text{ext}}$ denotes external forces applied to the particle and independent whether the particle emits or not. We have mentioned already that derivative $dP^r/d\tau$ contains a diverging term proportional to ε^{-1}. However, this term can be eliminated by means of *renormalization*. Namely, we shall assume that the rest mass μ (the so-called "bare" mass of the particle) is nonobservable. Let us construct a sum

$$m_0 = \mu + \frac{q^2}{8\pi c^2\varepsilon} \tag{23.10}$$

Our theory predicts that for a pointlike particle $m_0 \to \infty$. We shall write this off as a shortcoming of the theory which does not explain the origin of the mass and charge, and formally assume m_0 equal to the observed (finite and constant) value of the rest mass.[4] Equation (23.9) then becomes

$$m_0 w^r = \frac{1}{4\pi}\frac{2q^2}{3c^3}\left(b^r + \frac{u^r\vec{w}^2}{c^2}\right) + F^r_{\text{ext}} \tag{23.11}$$

We have derived the relativistic *Lorentz-Dirac equation* which describes the motion of an emitting charge taking into account the properties of radiation field emitted by this charge. Note that as follows from relation $\vec{w}\vec{u} = 0$,

$$\vec{b}\vec{u} = -\vec{w}^2 \tag{23.12}$$

[3] Pay attention to the "causal" character of the above limit operation (i.e. from earlier to latter values of τ).

[4] See additional remarks at the beginning of the next section.

If, for example, $\vec{F}_{ext}$ is the Lorentz force, it is orthogonal to velocity $\vec{u}$ (see Subsection 8.1). Therefore, as it has been stated above, the projection of the right-hand side of the Lorentz-Dirac equation on $\vec{u}$ equals zero.

The term proportional to $\vec{b}$ is called the *Schott vector*, and the term in the parantheses is called the *Abraham vector*. In the nonrelativistic limit equation (23.11) coincides with (23.3).

§ 24*. Renormalization of mass. Hyperbolic motion of charge

24.1. We have renormalized the mass by (23.10) in the exact Lorentz-Dirac equation for a charge moving with acceleration. In order to clarify the physical meaning of renormalization, let us consider a charge moving at a constant velocity, that is we set $\vec{w}=0$. Let us assume that in its co-moving reference frame this charge is a sphere with radius ε. We want to calculate the energy-momentum vector of the field contained in a spacelike plane orthogonal to the world line of the charge (Fig. 25). Density of energy-momentum is found from (15.24). In the chosen case expressions (15.14), (15.15), and (15.18) are reduced to $W = 0$, $B = 1/\rho$, $\vec{V} = \vec{u}/\rho$, so that $\vec{V}^2 = c^2/\rho^2$, and formula (15.24) takes a simple form:

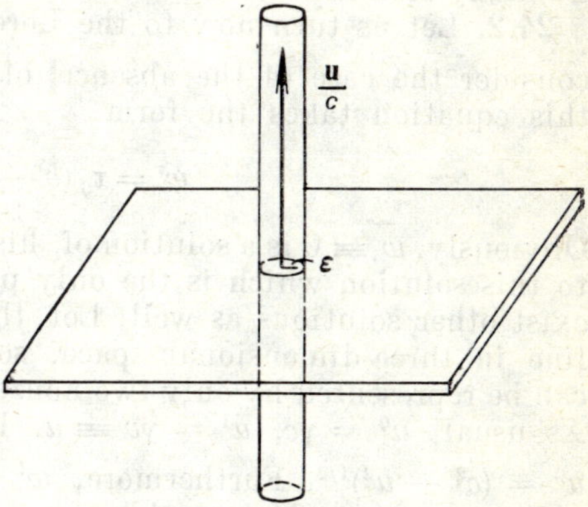

Fig. 25

$$\left(\frac{4\pi}{q}\right)^2 T^{km}\frac{u_m}{c}=\frac{1}{2c\rho^4}\,u^k$$

Therefore,

$$P^k=\left(\frac{q}{4\pi}\right)^2\frac{u^k}{2c^2}\int\frac{d\sigma}{\rho^4}$$

As the factor of u^k must be invariant, integration over $d\sigma$ yields the same result in any reference frame. In particular, we can make use of the co-moving reference frame of the charge in which, as was shown in Appendix D, we can consider ρ as the radius vector in spherical coordinates of the three-dimensional Euclidean space, so that $d\sigma = \rho^2 \, d\rho \, d\omega$ and the integral is equal to $4\pi/\varepsilon$. Finally, $P^k = \frac{q^2}{8\pi c^2 \varepsilon} u^k$. Consequently, the energy-momentum vector of the field around a uniformly moving charge is written in the form $P^k = m_{\rm el} u^k$, where mass $m_{\rm el}$ (the so-called *electromagnetic mass* of the particle) is equal precisely to the additional (renormalizing) term in formula (23.10). For pointlike charges it diverges. The physical meaning of $m_{\rm el}$ stems from the fact that $q^2/8\pi\varepsilon$ is the energy of the electrostatic field of the charge in its rest frame, and factor c^{-2} corresponds to the transition from energy to rest mass in Einstein's formula. Therefore, the total inertia of a particle includes the inertia of the field it entrains. It must be kept in mind that component cP^0 of the derived expression has the meaning of energy only in the co-moving reference frame. Energy $cP^{0\prime}$ of the particle in any other reference frame is related to its energy in the co-moving reference frame by standard transformation formulas: $cP^0 = \gamma \, (cP^{0\prime} + \mathbf{v} \cdot \mathbf{P}')$.

24.2. Let us turn now to the Lorentz-Dirac equation (23.11) and consider the case of the absence of forces, that is $\vec{F}_{\rm ext} = 0$. Then this equation takes the form

$$w^r = \tau_0 \, (b^r + u^r \vec{w}^2/c^2) \tag{24.1}$$

Obviously, $\vec{w} \equiv 0$ is a solution of his equation. However, in addition to this solution which is the only physically meaningful one, there exist other solutions as well. Let the motion be along a straight line in three-dimensional space, so that four-dimensional vectors can be represented by only two nonzero projections (on axes 1 and 0). As usual, $u^0 = \gamma c$, $u^1 = \gamma v \equiv u$. Hence, $v = uc \, (c^2 + u^2)^{-1/2}$ and $u^0 = (c^2 + u^2)^{1/2}$. Furthermore, $w^1 = \dot{u}$, $w^0 = \dot{u} u \, (c^2 + u^2)^{-1/2}$. Finally, equation (24.1) yields

$$\dot{u} = \tau_0 \left(\ddot{u} - \frac{\dot{u}^2 u}{c^2 + u^2} \right)$$

Let us introduce a substitution $u = c \sinh \lambda$. The above equation then reduces to $\dot{\lambda} - \tau_0 \ddot{\lambda} = 0$, whence $\lambda = A e^{\tau/\tau_0} + B$, that is $u = c \sinh \, (A e^{\tau/\tau_0} + B)$. Consequently, as $\tau \to \infty$, velocity u rises infinitely provided $A \neq 0$. If $A = 0$, we obtain the preceding case of uniform motion. Self-accelerated motions of a free particle obtained for $A \neq 0$ are physically meaningless. This shows that in the general case an investigation of the Lorentz-Dirac equation is impossible without additional conditions required to single out physically meaningful solutions of this equation.[5] For example,

---

[5] Despite the fact that these conditions were not used in the derivation of this equation in § 23.

we may require that the velocity of the particle remain finite for $\tau \to \infty$; this requirement is an asymptotic constraint on the charge motion in the infinitely remote future.

Let us consider the case in which the motion of an emitter can be regarded as *uniformly accelerated*. Equation (23.11) was derived under the assumption that the rest mass of the particle is constant. As a result, we have shown that both the right-hand side of this equation, as well as the two components of its terms independently, are orthogonal to vector $\vec{u}$ of four-dimensional velocity. Condition $\dot{\vec{w}} = \vec{b} = 0$ is in contradiction with this assumption. This is readily shown by calculating a scalar product of both sides of equation (23.11) by vector $\vec{u}$ (provided $\vec{w} \neq 0$). In relativistic mechanics, however, the concept of uniformly accelerated motion must be introduced in a different manner. Namely, the motion is referred to as uniformly accelerated if the value of the four-dimensional vector of acceleration, calculated at any moment of proper time in a transition to the rest frame of the moving particle, remains constant. As the components of four-dimensional acceleration are given by formulas (5.17_2), this means that condition $\vec{w}(\tau_2) = \vec{w}(\tau_1)$ reduces to a condition $\mathbf{h} \equiv d\mathbf{a}/dt = 0$ in the rest frame. By differentiating (5.17_2) it is easy to find the components of 4-vector $\vec{b}$ in the general case. These are

$$b^0 = \frac{1}{c}\gamma^5(\mathbf{a}^2 + \mathbf{v}\cdot\mathbf{h}) + \frac{4\gamma^7}{c^3}(\mathbf{v}\cdot\mathbf{a})^2$$

$$\mathbf{b} = \gamma^3\mathbf{h} + \frac{3\gamma^5}{c^2}(\mathbf{v}\cdot\mathbf{a})\,\mathbf{a} + \frac{b^0}{c}\mathbf{v}$$

In the rest frame ($\mathbf{v} = 0$ and $\gamma = 1$) they are $b^0 = \mathbf{a}^2/c$, $\mathbf{b} = \mathbf{h}$. Let us find the component $\vec{b}_\perp$ of 4-vector $\vec{b}$, orthogonal to four-dimensional velocity $\vec{u}$, via relation $\vec{b}_\perp\vec{u} = 0$. Of course, this relation is relativistically invariant. We can conclude from (23.12) that

$$\vec{b}_\perp = \vec{b} - \frac{\vec{b}\vec{u}}{c^2}\vec{u} = \vec{b} + \frac{\vec{w}^2}{c^2}\vec{u}$$

In the rest frame we obtain $\mathbf{b}_\perp = \mathbf{b} = \mathbf{h}$, $b^0_\perp = 0$; as a result, the condition under which the motion can be considered as uniformly accelerated takes the form $\vec{b}_\perp = 0$. But in this form it is invariant and must hold in an arbitrary reference frame. Hence, the uniformly accelerated motion is defined by the relation

$$\vec{b} + \frac{\vec{w}^2}{c^2}\vec{u} = 0 \tag{24.2}$$

that is, in this motion the Abraham vector vanishes. In the three-dimensional notation (in an arbitrary reference frame) the last equation can be rewritten, by using the given above expressions for components of vector $\vec{b}$, in the form

$$\mathbf{h}+\frac{3\gamma^2}{c^2}(\mathbf{v}\cdot\mathbf{a})\,\mathbf{a}=0 \tag{24.3}$$

From (24.2) we obtain $\vec{b}\vec{w}=\vec{w}\dot{\vec{w}}=\frac{1}{2}\frac{d}{d\tau}(\vec{w})^2=0$. Hence, emission power which, according to (15.33) or (16.15), is equal to $\mathcal{R}= =-\frac{1}{4\pi}\frac{2q^2}{3c^3}\vec{w}^2$, is constant, and the Lorentz-Dirac equation is written in the form $m_0\vec{w}=\vec{F}_{\text{ext}}$. The energy-momentum conservation holds, nevertheless, since the obtained relations are derived as direct corollaries of precisely this law in the particular case under discussion. However, the right-hand side $\vec{F}_{\text{ext}}$ of the Lorentz-Dirac equation cannot be arbitrary here but must satisfy condition (24.2) from which we obtain $\dot{\vec{F}}_{\text{ext}}=-(\vec{F}_{\text{ext}}\vec{w})\frac{\vec{u}}{c^2}$. This variation of the applied external force in time compensates dissipation of energy caused by the emission of radiation. The four-dimensional force varies in the Minkowski space only in its direction since

$$\vec{F}\dot{\vec{F}}=\frac{1}{2}\frac{d}{d\tau}\vec{F}^2=0$$

Kinematically, the relativistic uniformly accelerated motion is completely defined. Indeed, equation (24.2) can be rewritten in the form

$$\frac{d^2\vec{u}}{d\tau^2}=\Lambda^2\vec{u}$$

where $\Lambda^2\equiv\mathcal{R}/(\tau_0 m_0 c^2)$ is a constant, $\Lambda>0$. From this

$$\vec{u}=\vec{\alpha}e^{\Lambda\tau}+\vec{\beta}e^{-\Lambda\tau} \tag{24.4}$$

where $\vec{\alpha}$ and $\vec{\beta}$ are independent of τ. Condition $\vec{u}^2=c^2$ shows that these vectors must satisfy relations $\vec{\alpha}^2=\vec{\beta}^2=0$ and $2\vec{\alpha}\vec{\beta}=c^2$. Furthermore, we find from (24.4)

$$\vec{r}=\vec{\gamma}+\Lambda^{-1}(\vec{\alpha}e^{\Lambda\tau}-\vec{\beta}e^{-\Lambda\tau}) \tag{24.5}$$

where $\vec{\gamma}$ is a constant vector.

Consider a simple particular case of unidimensional motion along axis x in an arbitrary reference frame, with $\vec{\gamma}=0$. We choose $\alpha^1= =-\beta^1=c/2$, and $\alpha^0=\beta^0=c/2$. The conditions determining the properties of vectors $\vec{\alpha}$ and $\vec{\beta}$ are assumed satisfied. Then $ct = A \sinh(\Lambda\tau)$ and $x = A \cosh(\Lambda\tau)$ where $A = c\Lambda^{-1}$, whence $c^2t^2 - x^2 = -A^2$. This means that the world line of the particle is at the same time a timelike circle in the Minkowski space and is mapped by a hyperbola on the Minkowski diagram. Owing to this particular case, the relativistic uniformly accelerated motion is sometimes called the *hyperbolic motion*.[6]

§ 25. Spectrum composition of radiation emitted by an oscillator. Scattering and absorption of radiation

25.1. Let us consider again nonrelativistic equation (23.3) taking into account radiative reaction in the description of motion of a charge. First we consider this equation in the case of unidimensional motion and quasielastic external force. Equation (23.3) then transforms to

$$\ddot{x} - \tau_0 \dddot{x} + \omega_0^2 x = 0 \tag{25.1}$$

Dots in (25.1) denote differentiation with respect to time t. If constant τ_0 were zero, this equation would describe harmonic oscillations with frequency ω_0; by properly choosing the origin on axis x and the initial velocity, we can write the solution in the form $x(t) = x_0 \exp(-i\omega_0 t)$, where x_0 is a real constant. If $\tau_0 \neq 0$, we shall find a solution of equation (25.1) by a substitution $x(t) = x_0 \exp(-\alpha t)$, assuming α to be complex. Parameter α will then be found from the equation

$$\tau_0\alpha^3 + \alpha^2 + \omega_0^2 = 0$$

The roots of this cubic equation for α can be found explicitly. One of the roots is real and negative. Clearly, it has to be ignored since it corresponds to the "self-accelerated" solution. It will be convenient to find approximate expressions for complex roots, assuming $\xi \equiv \omega_0\tau_0 \ll 1$. We can also assume, on the basis of the estimate given at the beginning of § 23 for τ_0, that this condition may hold for a wide class of emitters. Introduce a dimensionless quantity $\alpha' = \alpha/\omega_0$. The equation becomes $\xi\alpha'^3 + \alpha'^2 + 1 = 0$. Substitute now the expansion $\alpha' \simeq \beta + \gamma\xi + \delta\xi^2$, retain only the terms of the order not exceeding ξ^2 and set coefficients of ξ^0, ξ^1, ξ^2 equal to zero; this yields $\beta^2 + 1 = 0$, $\beta^2 + 2\gamma = 0$ and $3\gamma\beta^2 + \gamma^2 + 2\beta\delta = 0$, whence we successively find $\beta = \pm i$, $\gamma = 1/2$ and $\delta = \mp (5/8)\, i$. Here the upper and lower indices for β and δ must be chosen simultaneously. Substitution of the obtained coefficients into the expan-

[6] We recommend that the reader read up on the problem of hyperbolic motion in: V. L. Ginzburg, *Theoretical Physics and Astrophysics*, Pergamon Press, Oxford, 1979.

sion for α' yields for α in the approximation considered

$$\alpha = (1/2)\,\Gamma \pm i\,(\omega_0 + \Delta\omega)$$

where

$$\Gamma = \omega_0^2\tau_0, \quad \Delta\omega = (-5/8)\,\omega_0^3\tau_0^2$$

Quantity Γ is called the *natural linewidth*, and $\Delta\omega$—the *shift of the spectral line*. We shall clarify the meaning of this terminology later in this section when analyzing the properties of radiation emitted by an oscillation charge. The equation of motion of such a charge takes the form

$$x(t) = x_0 e^{-\frac{\Gamma}{2}t} e^{\mp i\omega' t}$$

where $\omega' \equiv \omega_0 + \Delta\omega$. As could be expected, we find that radiative reaction results in damped oscillations. Correction $\Delta\omega$ is very small and as a rule is ignored. Accordingly, hereafter we assume $\omega' \approx \omega_0$.

Consider now a spherical oscillator whose equations of motion are given by (25.1) for each of the three spatial coordinates. In the nonrelativistic limit we are considering the emitted energy of radiation as given by the Larmor formula (16.3), that is, it is proportional to $\ddot{\mathbf{r}}^2$. If the oscillator vibrates harmonically at frequency ω_0, its electromagnetic radiation consists of waves with the same frequency. If radiative friction is taken into account, the motion of the oscillator is, as we have shown, exponentially damped. We can assume that the initial deviation of the particle from its equilibrium position at moment $t = 0$ is caused by an external force which instantaneously drops to zero. The damped oscillations of the charge will generate radiation which will be equally nonharmonic, namely, will tend to zero with time. The spectral composition of such radiation is investigated by expansion into the Fourier integral.

Later we shall see that damping of oscillations is caused not only by radiative reaction force but also by the interaction of the oscillator with the surrounding medium; this interaction is described phenomenologically by including into the equation a friction force proportional to $\dot{\mathbf{r}}$. Furthermore, this force which is taken into account in equation (25.11) usually exerts a predominant effect on the motion of the oscillator and therefore effects the emitted electromagnetic field much stronger than radiative reaction. It will, however, be shown in solving equation (25.11) that this effect can also be described in terms of the spectral linewidth.

The results obtained above show that the general solution of the problem of the motion of the oscillator can be written in the form

$$\mathbf{r} = \mathbf{A}e^{-(\Gamma/2 + i\omega')t} + \mathbf{B}e^{-(\Gamma/2 - i\omega')t}$$

As $\mathbf{r}$ is real, that is $\mathbf{r} = \mathbf{r}^*$, it follows that $\mathbf{A} = \mathbf{B}^*$. Let us carry out a spectral analysis of the first term; we demand that $\mathbf{r}(t) = 0$ at $t < 0$ in accord with the assumption on the character of motion. Then, by using a formula of the type of (E.12), we can write

$$\mathbf{A}e^{-\frac{\Gamma}{2}t - i\omega' t} = \int_{-\infty}^{+\infty} \boldsymbol{\alpha}(\omega)\, e^{-i\omega t}\, d\omega \quad \text{for} \quad t > 0$$

$$\int_{-\infty}^{+\infty} \boldsymbol{\alpha}(\omega)\, e^{-i\omega t}\, d\omega = 0 \qquad \text{for} \quad t < 0$$

If both sides of this equation are integrated over time from $-\infty$ to $+\infty$, then the integral in the left-hand side must actually be integrated from 0 to $+\infty$. Prior to integration, we multiply the right- and left-hand sides of the equation by $e^{i\bar{\omega} t}$. Then

$$\mathbf{A}\int_0^{\infty} e^{\left(-\frac{\Gamma}{2} + i(\bar{\omega} - \omega')\right)t}\, dt = \int_{-\infty}^{+\infty} \boldsymbol{\alpha}(\omega)\, d\omega \int_{-\infty}^{+\infty} e^{i(\bar{\omega} - \omega)t}\, dt = 2\pi \boldsymbol{\alpha}(\bar{\omega})$$

where we have made use of formula (C.14) for delta function. Elementary integration in the left-hand side yields (after replacing $\bar{\omega}$ by ω)

$$\boldsymbol{\alpha}(\omega) = \frac{1}{2\pi}\, \frac{\mathbf{A}}{\Gamma/2 - i(\omega - \omega')} \tag{25.2}$$

Clearly, expansion of the second term in the formula for $\mathbf{r}$ into the Fourier integral will differ only in $-\omega'$ replacing ω'.[7] Differentiation with respect to t in the integrand, necessary to calculate $\ddot{\mathbf{r}}$, results in an additional factor $-\omega^2$. Hence,

$$\ddot{\mathbf{r}}(t) = \int_{-\infty}^{+\infty} \mathbf{f}(\omega)\, e^{-i\omega t}\, d\omega \tag{25.3}$$

where

$$\mathbf{f}(\omega) = -\frac{\omega^2}{2\pi}\left[\frac{\mathbf{A}}{\Gamma/2 - i(\omega - \omega')} + \frac{\mathbf{A}^*}{\Gamma/2 - i(\omega + \omega')}\right] \tag{25.4}$$

Besides, $\mathbf{f}^*(-\omega) = \mathbf{f}(\omega)$. Formula (25.3) is therefore rewritten in the form

$$\ddot{\mathbf{r}}(t) = \int_{-\infty}^{+\infty} \mathbf{f}^*(\bar{\omega})\, e^{i\bar{\omega} t}\, d\bar{\omega} \tag{25.5}$$

[7] Formula (25.2) can be interpreted as a result of Fourier transform of function $\theta(t)\,\mathbf{f}(t)$, where $\mathbf{f}(t)$ is the first term in the expression for $\mathbf{r}(t)$ and $\theta(t)$ is a discontinuous Heaviside function defined as follows: $\theta(t) = 0$ for $t \leqslant 0$ and $\theta(t) = 1$ for $t > 0$. This interpretation is to be borne in mind in relation to the subsequent derivation of formula (25.7), when integration over time can be carried out from $t = -\infty$.

The total energy of radiation emitted by the oscillator during a period from $t = 0$ to $t = +\infty$ is equal, according to (16.3), to

$$I = \frac{1}{4\pi}\frac{2q^2}{3c^3}\int_0^\infty \ddot{\mathbf{r}}(t)^2\,dt = \int_0^{+\infty} I(\omega)\,d\omega \tag{25.6}$$

where the right-hand side represents the definition of quantity $I(\omega)$, the so-called *spectral energy density*. As follows from formulas (25.3), (25.5), and (C.14),

$$\int_0^\infty \ddot{\mathbf{r}}^2\,dt = \int_{-\infty}^{\infty}\int_{-\infty}^{+\infty}\int_{-\infty}^{+\infty} \mathbf{f}(\omega)\,\mathbf{f}^*(\overline{\omega})\,e^{i(\overline{\omega}-\omega)t}\,dt\,d\omega\,d\overline{\omega}$$

$$= 2\pi\int_{-\infty}^{+\infty} |\mathbf{f}(\omega)|^2\,d\omega = 4\pi\int_0^\infty |\mathbf{f}(\omega)^2\,d\omega \tag{25.7}$$

because in our case $|\mathbf{f}(-\omega)|^2 = |\mathbf{f}(\omega)|^2$. The integrand must be found by using (25.4). In many cases it can be assumed that $\Gamma \ll \omega_0$, so that we can limit the calculation to a narrow frequency band ω defined by the condition $|\omega - \omega_0| \ll |\omega_0|$. Then the magnitude of the first term in (25.4) is much greater than that of the second term which therefore can hereafter be ignored. Consequently,

$$I(\omega) = \frac{2q^2}{3c^3}|f(\omega)|^2$$

Let us calculate the total radiated energy according to (25.6), integrating it over ω. In accordance with our assumption, factor ω^4 can be replaced by a constant quantity ω'^4. Further we have

$$\int_0^\infty \frac{d\omega}{\frac{\Gamma^2}{4}+(\omega-\omega')^2} = \frac{2}{\Gamma}\int_{-2\omega'/\Gamma}^{\infty}\frac{dx}{1+x^2}$$

However, the lower limit can be replaced by $-\infty$ if we assume that $\Gamma \ll \omega_0$. This yields

$$I \simeq \frac{2q^2}{3c^3}\cdot\frac{\omega'^4}{(2\pi)^2}\cdot\frac{2\pi}{\Gamma}|\mathbf{A}|^2$$

and the formula for spectral energy density takes the form

$$I(\omega) = \frac{I}{\Gamma^2/4+(\omega-\omega')^2}\cdot\frac{\Gamma}{2\pi} \tag{25.8}$$

Energy $I(\omega)$ emitted as electromagnetic waves with frequencies within an interval from ω to $\omega + d\omega$ is plotted in Fig. 26 as a function of frequency ω in accordance with formula (25.8). Obviously, the emitted energy reaches maximum at $\omega' \approx \omega_0$, and the oscillator emits waves with all possible frequencies with intensity $I(\omega)$, while the harmonic oscillator emits only waves with frequency ω_0.

The curve in Fig. 26 shows that parameter Γ is equal to the width of energy distribution at half height, that is for the ordinate equal to $I_{max}/2$. Consequently, Γ is often referred to as half width (although this term is obviously incorrect).

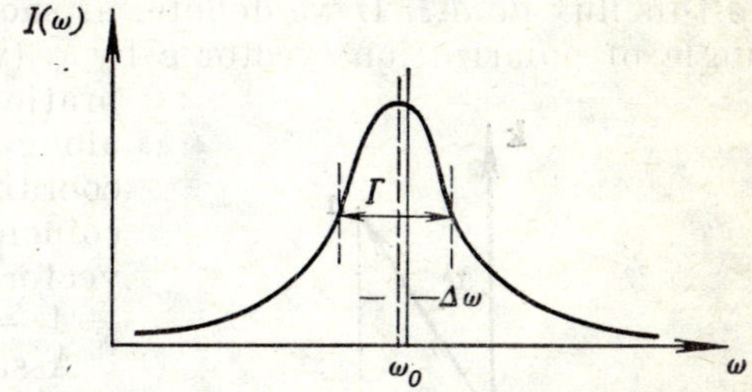

Fig. 26

25.2. Until now we were analyzing the "free" motion of an oscillating emitter subjected to two forces: quasielastic force which tends to restore equilibrium, and forces exerted by the emitted field. Let us turn now to "forced" oscillations caused by external forces applied to the particle during a given time interval. Such external forces may be caused by the interaction between the particle and electromagnetic wave. This interaction results in scattering and absorption of radiation.

First we assume that quasielastic forces are absent and that a plane monochromatic electromagnetic wave interacts with a charged particle which is free in all other respects. The field accelerates the charge and makes it radiate. As throughout this section, we consider the nonrelativistic case. As $v \ll c$, we can ignore the effect of the magnetic field of the incident wave and write the eqution of motion in the form

$$m\dot{\mathbf{v}} = q\mathbf{E} = q\boldsymbol{\varepsilon}Ee^{i(\mathbf{k}\cdot\mathbf{r}-\omega t)}$$

Here $\boldsymbol{\varepsilon}$ is a unit vector of polarization of the wave, and E is its complex amplitude. The waves emitted in these conditions by the accelerated charged particle are called scattered waves. The energy flux of this scattered radiation must be calculated from (16.2) and must take into account (17.18), since acceleration is written in complex form. The last formula can be considered applicable if the emitting charge subjected to the incident wave field performs a sufficiently large number of oscillations during the time of observation of the scattering. Hence,

$$\langle S\rangle = c\left(\frac{q}{4\pi c^2 R}\right)^2 \sin^2\theta \cdot \frac{1}{2}|\dot{\mathbf{v}}|^2$$

and

$$\frac{dI}{d\Omega} = \frac{c}{2}|E|^2\left(\frac{q^2}{4\pi mc^2}\right)^2 \sin^2\theta$$

Here θ is the angle between the acceleration vector (i.e. the polarization $\boldsymbol{\varepsilon}$ of the incident wave) and the direction $\mathbf{n}$ of emission into the solid angle $d\Omega$. As $(1/2)\, c \mid E \mid^2$ is the time-averaged energy flux of the incident wave across a unit surface area (cf. § 17), it is convenient to introduce the *differential cross section of scattering* defined as the ratio of the calculated intensity of scattered waves to this flux $d\sigma/d\Omega$. If we denote, as shown in Fig. 27, the azymuthal angle of polarization vector $\boldsymbol{\varepsilon}$ by ψ (we have seen that forced vibrations of the emitter are directed along this vector) in the spherical coordinate system, then angle θ coincides with the angle between vectors $\mathbf{n}$ and $\boldsymbol{\varepsilon}$, whence $\sin^2\theta = 1 - \sin^2\vartheta \cos^2(\varphi - \psi)$.

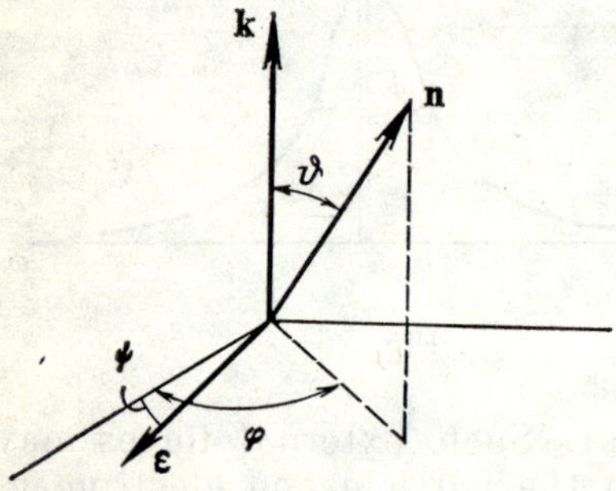

Fig. 27

Assume now that the incident radiation is not polarized and let us average the cross section over angle ψ. Then

$$\frac{d\sigma}{d\Omega} = \left(\frac{q^2}{4\pi mc^2}\right)^2 \cdot \frac{1}{2}(1 + \cos^2\vartheta) \tag{25.9}$$

The obtained expression is called the *Thompson formula*. The total Thompson scattering cross section is calculated by integrating over $d\Omega$:

$$\sigma = \frac{8\pi}{3}\left(\frac{q^2}{4\pi mc^2}\right)^2 \tag{25.10}$$

The quantity in the parantheses is called the *classical radius of a particle* (of the electron or proton, for example, if the appropriate values of charge and mass are substituted into the formula).

Let us take up the case of electromagnetic radiation interacting with a particle bound to the equilibrium position by a quasielastic force. As a first step, we ignore the radiative reaction but assume that the particle undergoes some friction forces applied by the surrounding medium and proportional to velocity. As in the preceding example, we neglect magnetic forces. The equation of motion then takes the form

$$\ddot{\mathbf{r}} + \Gamma'\dot{\mathbf{r}} + \omega_0^2\mathbf{r} = \frac{q}{m}\mathbf{E}(t) \tag{25.11}$$

where Γ' is the friction coefficient. The solution of this nonhomogeneous equation can be obtained by using the Fourier expansion of functions $\mathbf{r}(t)$ and $\mathbf{E}(t)$ in the form (E.12). As $\mathbf{r}$ and $\mathbf{E}$ are real, the corresponding Fourier coefficients satisfy (similarly to cases considered above) the relations $\mathbf{r}(\omega) = \mathbf{r}^*(-\omega)$ and $\mathbf{E}(\omega) =$

$= \mathbf{E}^*(-\omega)$. As follows from (25.11),

$$\mathbf{r}(\omega) = \frac{q}{m} \frac{\mathbf{E}(\omega)}{\omega_0^2 - i\omega\Gamma' - \omega^2} \tag{25.12}$$

The work done by radiation in the process of forced oscillations of the charge is

$$W = q \int_{-\infty}^{+\infty} \dot{\mathbf{r}} \cdot \mathbf{E}\, dt \tag{25.13}$$

Field $\mathbf{E}(t)$ can be that in the equilibrium position of the oscillator if its vibration amplitude is sufficiently small. By differentiating the expansion for $\mathbf{r}(t)$, substituting this expansion and the expansion for $\mathbf{E}(t)$ into (25.13), and using formula (C.14) for delta function we obtain

$$\frac{1}{q} W = -2\pi i \int_{-\infty}^{+\infty} \omega \mathbf{r}(\omega) \cdot \mathbf{E}^*(\omega)\, d\omega = 4\pi \int_0^{\infty} \omega \operatorname{Im}(\mathbf{r}(\omega) \cdot \mathbf{E}^*(\omega))\, d\omega$$

The last equality is a corollary of the fact that $\mathbf{r}(t)$ and $\mathbf{E}(t)$ are real. It is transformed, by using (25.12), to

$$W = \frac{4\pi q^2}{m} \int_0^{\infty} \frac{|\mathbf{E}(\omega)|^2 \omega^2 \Gamma'}{(\omega_0^2 - \omega^2)^2 + \omega^2 \Gamma'^2}\, d\omega \tag{25.14}$$

The integrand has a sharp peak at $\omega = \omega_0$. Consequently, $|\mathbf{E}(\omega)|^2 \omega^2$ in the numerator of the integral can be replaced by $|\mathbf{E}(\omega_0)|^2 \omega_0^2$ and term $\omega^2\Gamma'^2$ in the denominator by $\omega_0^2\Gamma'^2$. Besides, $\omega_0^2 - \omega^2 \simeq$ $\simeq (\omega_0 - \omega)\cdot 2\omega_0$. Then a change of variable $\omega - \omega_0 = \frac{\Gamma'}{2} x$ transforms the integral to $\int_{-2\omega_0/\Gamma'}^{\infty} \frac{dx}{1+x^2}$. Integration is carried out as in the derivation of (25.8). The total work of the radiation incident on the oscillator is equal, therefore, to

$$W = \frac{2\pi^2 q^2}{m} |\mathbf{E}(\omega_0)|^2 \tag{25.15}$$

This expression can be rewritten in a somewhat different form. In complete anology to relation (25.7), we can obtain

$$\int_{-\infty}^{+\infty} A^2(t)\, dt = 4\pi \int_0^{\infty} |A(\omega)|^2\, d\omega \tag{25.16}$$

This formula relates a real function $A(t)$ with its Fourier transform. In particular, we obtain for the radiation forcing the charged particle to vibrate the energy flux per unit surface area:

$$S = c \int_{-\infty}^{+\infty} \mathbf{E}^2\, dt = 4\pi c \int_0^{\infty} |\,\mathbf{E}(\omega)\,|^2\, d\omega \tag{25.17}$$

Hence $dS/d\omega \equiv s(\omega) = 4\pi c \mid \mathbf{E}(\omega) \mid^2$ and

$$W = \frac{\pi}{2} \frac{q^2}{mc} s(\omega_0) = 2\pi^2 r_0 c s(\omega_0) \tag{25.18}$$

Here r_0 is the classical radius of the particle given in equation (25.10). Formula (25.18) gives the energy dissipated by the radiation to excite vibrations of the charge; hence, it represents the absorption of radiation by the oscillator. On the other hand, if we define the differential absorption cross section as

$$\sigma(\omega) \equiv \frac{dW}{d\omega} \cdot \frac{1}{s(\omega_0)}$$

then formulas (25.14) and (25.18) can be rewritten in the form

$$\int_0^{+\infty} \sigma(\omega)\, d\omega = 2\pi^2 r_0 c \tag{25.19}$$

Sometimes this last equation for the total absorption cross section is referred to as the sum rule.

Let us consider now *scattering* and *absorption of radiation by the oscillator* with radiative reaction taken into account, still assuming that the oscillator also experiences friction forces proportional to the velocity of the charged particle. We shall limit the analysis to the case of monochromatic wave for which $\mathbf{E}(t) = \varepsilon E \exp(-i\omega t)$ (we assume that the wave field differs negligibly from its value in the equilibrium position of the oscillator; this point will be chosen as the origin of spatial coordinates). Instead of (25.11), we have to solve equation

$$\ddot{\mathbf{r}} + \Gamma' \dot{\mathbf{r}} - \tau_0 \dddot{\mathbf{r}} + \omega_0^2 \mathbf{r} = \frac{q}{m} \varepsilon E e^{-i\omega t} \tag{25.20}$$

Substitution $\mathbf{r} = \mathbf{r}_0 \exp(-i\omega t)$ yields

$$\mathbf{r}(t) = \frac{q}{m} \varepsilon \frac{E e^{-i\omega t}}{\omega_0^2 - \omega^2 - i\omega\tilde{\Gamma}} \tag{25.21}$$

where

$$\tilde{\Gamma} \equiv \Gamma' + (\omega/\omega_0)^2\, \Gamma \tag{25.22}$$

25.3. Consider first of all the problem of *scattering*. This means that we have to calculate, by analogy with the case of scattering on a free charge, the energy emitted by the oscillator whose motion is described by (25.21).

Electric field of radiation emitted by the oscillator is given by formula (16.1), that is

$$\mathbf{E}_{\text{rad}} = \frac{q}{4\pi c^2 R}\, \mathbf{n} \times (\mathbf{n} \times \ddot{\mathbf{r}})\big|_{\text{ret}}$$

Here **R** is the distance from the equilibrium position of the oscillator to the point at which the radiation is observed. If the distance to the oscillator is sufficiently large, the oscillator can be considered pointlike. The conditions necessary for this were analyzed in detail in § 17. Because of retardation, $\ddot{\mathbf{r}}$ must be represented in the form

$$-\ddot{\mathbf{r}} = \frac{q}{m}\, \omega^2 \frac{E e^{-i\omega t'}}{(\omega_0^2 - \omega^2) - i\omega\tilde{\Gamma}}\, \boldsymbol{\varepsilon}$$

where $t' = t - R/c$ and t is the time of observation.

Assume that we want to calculate that part of the radiation energy which is transferred by a wave polarized in direction $\boldsymbol{\varepsilon}'$. Polarization vector $\boldsymbol{\varepsilon}'$ satisfies the condition of transversality with respect to the propagation direction $\mathbf{n}$ of the wave, that is $\boldsymbol{\varepsilon}' \cdot \mathbf{n} = 0$. Hence, as $-\mathbf{n} \times (\mathbf{n} \times \boldsymbol{\varepsilon}) = \boldsymbol{\varepsilon} - \mathbf{n}\,(\mathbf{n} \cdot \boldsymbol{\varepsilon})$, we find from the above formulas

$$\boldsymbol{\varepsilon}' \cdot \mathbf{E}_{\text{rad}} = \frac{q^2}{4\pi m c^2}\, \frac{\omega^2 E e^{-i\omega t} e^{ikR}}{\omega_0^2 - \omega^2 - i\omega\tilde{\Gamma}}\, \frac{\boldsymbol{\varepsilon} \cdot \boldsymbol{\varepsilon}'}{R}$$

Energy arriving into solid angle $d\Omega$ from the emitter at a distance R is equal to $(1/2)\, c\, |\boldsymbol{\varepsilon}' \cdot \mathbf{E}_{\text{rad}}|^2 R^2\, d\Omega$ (we have taken into account the time averaging). At the same time the energy causing vibrations of the charge, which result in the radiation emitted in all directions, is equal to $(1/2)\, c\, |\mathbf{E}|^2$ (per unit area). The ratio of the first of these quantities to the second gives, as in the preceding problem of free charge, the differential scattering cross section of the radiation polarized in direction $\boldsymbol{\varepsilon}'$:

$$\frac{d\sigma(\omega, \boldsymbol{\varepsilon}')}{d\Omega} = \left| \frac{R}{E}\, \boldsymbol{\varepsilon}' \cdot \mathbf{E}_{\text{rad}} \right|^2 = \left(\frac{q^2}{4\pi m c^2} \right)^2 (\boldsymbol{\varepsilon} \cdot \boldsymbol{\varepsilon}')^2 \frac{\omega^4}{(\omega_0^2 - \omega^2)^2 + \omega^2 \tilde{\Gamma}^2} \quad (25.23)$$

In the case $\omega \ll \omega_0$, when $\tilde{\Gamma} \simeq \Gamma'$, we obtain an approximate formula

$$\frac{d\sigma(\omega, \boldsymbol{\varepsilon}')}{d\Omega} \simeq \left(\frac{q^2}{4\pi m c^2} \right)^2 (\boldsymbol{\varepsilon} \cdot \boldsymbol{\varepsilon}')^2 \left(\frac{\omega}{\omega_0} \right)^4 \quad (25.24)$$

This is the case known as the *Rayleigh scattering*.

Let $\omega \simeq \omega_0$, which means resonance of the incident wave frequency with the natural frequency of the oscillator. Then the denominator of formula (25.23) can be written in the form $(\omega_0 - \omega)(\omega_0 + \omega) \simeq$ $\simeq 2\omega_0 (\omega_0 - \omega)$, and ω^4 in the numerator can be replaced by ω_0^4. Also, as follows from (25.22), $\tilde{\Gamma} \simeq \Gamma' + \Gamma$. Hence, the scattering cross section, to within a coefficient independent of ω, is given by $[(\omega_0 - \omega)^2 + (\tilde{\Gamma}/2)^2]^{-1}$. This scattering is known as the *resonance fluorescence*. Integration over angle $d\Omega$ and frequencies yields the total scattering cross section proportional to $(\Gamma/\tilde{\Gamma})^2$.

If $\omega \gg \omega_0$, formula (25.23) transforms, after averaging over polarization of the incident wave, to the Thompson formula (25.9) (with the definition of angles in these two expressions taken into account).

The next case is *absorption*. We see from the above analysis that absorption is determined by a formula of the type of (25.14), but in which Γ' must be replaced by $\tilde{\Gamma}$. Therefore

$$\frac{dW}{d\omega} = \frac{4\pi q^2}{m} |E(\omega)|^2 \frac{\omega^2 \tilde{\Gamma}}{(\omega_0^2 - \omega^2)^2 + \omega^2 \tilde{\Gamma}^2} \cdot \qquad (25.25)$$

Dividing this by the energy of the incident wave per unit area in a given frequency interval (as can be found from the discussion of formula (25.17), it is equal to $4\pi c\, |E(\omega)|^2$), we obtain a characteristic called the *absorption cross section*:

$$\sigma_{\text{abs}}(\omega) = \frac{q^2}{mc} \frac{\omega^2 \tilde{\Gamma}}{(\omega_0^2 - \omega^2)^2 + \omega^2 \tilde{\Gamma}^2}$$

In accordance with the three cases mentioned above for scattering, we have:

$$\frac{mc}{q^2} \sigma_{\text{abs}}(\omega) = \begin{cases} \omega^2 \tilde{\Gamma}/\omega_0^4 & \text{for } \omega \ll \omega_0 \\ \dfrac{1}{4} \dfrac{\tilde{\Gamma}}{(\omega_0 - \omega)^2 + (\tilde{\Gamma}/2)^2} & \text{for } \omega \simeq \omega_0 \\ \tilde{\Gamma}/\omega^2 & \text{for } \omega \gg \omega_0 \end{cases} \qquad (25.26)$$

Remark: absorption cross section can be called (and is often called) the *total cross section*. Indeed, the energy of scattered waves appears at the expense of the energy of incident radiation transformed into the energy of motion of the emitter. But the formulas for absorption cross section take into account, among other factors, the work done against nonradiative friction forces characterized by Γ'. The difference between the total cross section and scattering cross section is called the *reaction cross section*.

When using the formulas derived in the present section, one must keep in mind that they were obtained on the basis of classical theory. Quantum effects become significant at higher frequencies compared to $mc^2/\hbar$, where mc^2 is the proper energy of the scattering particle, and $\hbar$ is the Planck constant divided by 2π. These effects reveal corpuscular properties of electromagnetic radiation made up of photons, and change essentially the dependence of absorption and scattering of frequency.

The broadening of spectral line characterized by $\tilde{\Gamma}$ has, in fact, a clearly understandable nature. Namely, it occurs because the emitter (an atom, for instance) emits light with frequencies sufficiently close to the mean frequency during a finite interval of time. In addition to the radiative reaction analyzed at the beginning of this section, the process of light emission may be terminated, for example, by collisions between emitting atoms. The term $\Gamma'\dot{\mathbf{r}}$ in the equation of motion, which we interpreted earlier as accounting for friction forces exerted by the surrounding medium, takes into account precisely these processes. The Doppler effect represents quite a different cause of broadening in the observed spectral lines emitted by the atoms. Even if friction is neglected and we assume that the oscillator emits in its co-moving reference frame the light with frequency exactly equal to ω_0, this frequency will be shifted owing to the motion of the oscillator with respect to the observer. If the velocities of the observed emitters are spread (for instance, this spread is described by Maxwell's distribution), the total intensity of the observed radiation will be frequency-dependent. In the case of Maxwell's distribution of velocities we can derive

$$I = I_0 \exp\left[-(\omega - \omega_0)/D\right]^2$$

where $D = \frac{\omega_0}{c}(2R\theta/m)^{1/2}$ is the Doppler shift of the spectral line, R is the gas constant, and θ is temperature.

CHAPTER 6

MOTION OF CHARGED PARTICLES IN ELECTROMAGNETIC FIELDS. SYSTEMS OF INTERACTING CHARGES

§ 26. Integration of the equations of motion

If the electromagnetic field applied to a charged particle can be considered as a function of coordinates and time, the problem of determining the path of a pointlike particle is reducible to integration of equation (8.3) with a known right-hand side. The effect of the field generated by the particle itself on its motion (radiative reaction) is ignored[1]. Recall that the right-hand side of (8.3) has the form of (3.13) (in this section we use the Gaussian system of units, and so set $\alpha = c$).

Integration of the equations of motion can be carried out exactly only in a number of particular cases. In the present section we discuss the most important of them. First, we assume that the conditions of nonrelativistic approximation are valid. Then it follows from the general equation of particle motion (see § 6) that the parameter of proper time τ can be replaced by time t. Hence, in this limit the equation becomes

$$m_0 \frac{d\mathbf{v}}{dt} = q\mathbf{E} + \frac{q}{c}\mathbf{v}\times\mathbf{B} \tag{26.1}$$

The relativistic case will be discussed later, in Subsection 26.4.

26.1. Consider a static uniform magnetic field **B**, and assume $\mathbf{E} = 0$.

Let axis z be in the direction of vector **B**. Equation (26.1) projected on the axes of coordinates gives

$$\frac{dv_x}{dt} = \omega_L v_y, \quad \frac{dv_y}{dt} = -\omega_L v_x, \quad \frac{dv_z}{dt} = 0 \tag{26.2}$$

where we have introduced a parameter

$$\omega_L \equiv \frac{qB}{m_0 c} \tag{26.3}$$

called the *cyclotron*, or *Larmor*, *frequency*[2] which will be important

[1] This means that we ignore the energy lost by the charge on the emission of radiation, i.e. (see § 16) acceleration must be assumed "sufficiently small"; consequently, one must not forget the approximate character of the results of the present section when they are used.

[2] Sometimes the quantity $\omega_L' = qB/(2m_0c)$ is called the Larmor frequency. See pp. 498 and 222. Depending on the sign of charge q, ω_L can be either positive or negative.

in the subsequent analysis. The initial velocity will be denoted by $\mathbf{v}_0$. The third equation in (26.2) describes uniform motion along axis z and thus is of no interest to us here. Integration of the remaining two equations will be simplified if the second of them is multiplied by i, square root of -1, and added to the first equation, term by term. This yields

$$\frac{dw}{dt}+i\omega_{\mathrm{L}}w=0 \tag{26.4}$$

where $w \equiv v_x + iv_y$. Denote $w_0 \equiv v_{0x} + iv_{0y}$. Then $w = w_0 \exp(-i\omega_{\mathrm{L}}t)$. The separation of the real and imaginary parts gives

$$v_x = v_{0x}\cos\omega_{\mathrm{L}}t + v_{0y}\sin\omega_{\mathrm{L}}t$$

$$v_y = v_{0y}\cos\omega_{\mathrm{L}}t - v_{0x}\sin\omega_{\mathrm{L}}t$$

so that $v_x^2 + v_y^2 = v_{0x}^2 + v_{0y}^2 \equiv v_{0\perp}^2$, where symbol $\perp$ indicates orthogonality to vector $\mathbf{B}$. The above formulas can be written in the vector form

$$\mathbf{v}_\perp = \mathbf{v}_{0\perp}\cos\omega_{\mathrm{L}}t - \mathbf{B}_1 \times \mathbf{v}_0 \sin\omega_{\mathrm{L}}t \tag{26.5}$$

Here $\mathbf{B}_1 \equiv \mathbf{B}/B$ is a unit vector along axis z. Denoting $\mathbf{v}_\perp \equiv d\mathbf{r}_\perp/dt$, we immediately obtain

$$\omega_{\mathrm{L}}\mathbf{r}_\perp = \mathbf{B}_1 \times \mathbf{v}_0\cos\omega_{\mathrm{L}}t + \mathbf{v}_{0\perp}\sin\omega_{\mathrm{L}}t \tag{26.6}$$

if the integration constant is set equal to $\mathbf{B}_1 \times \mathbf{v}_0$. This merely indicates a certain choice of origin in plane x, y. Let $v_{0z} = 0$, so that $v_0 = v_{0\perp}$. From (26.6), $r_\perp = v_0/|\omega_{\mathrm{L}}| = cm_0v_0/|q|B$. The particle thus circumscribes in the plane orthogonal to $\mathbf{B}$ a circle with radius $r_\perp$ (the smaller the larger the field strength is), with frequency ω_{L}. The origin is chosen at the center of this circle. Hereafter we use instead of $r_\perp$ a symbol r_{L} (the Larmor radius). Now we find directly from (26.6) and (26.5)

$$\mathbf{v} = \omega_{\mathrm{L}}\mathbf{r}_{\mathrm{L}} \times \mathbf{B}_1$$

A particle rotating in a circle in a magnetic field can be treated as a linear current with intensity $I = qv_0/2\pi r_{\mathrm{L}}$ (since $v = v_0$). Recall now definition (12.13) of such current. In our case it can be written in the form $\boldsymbol{\mu} = \frac{1}{c}\mathbf{n}IS$, where S is the area of a circle with radius r_{L}. The positive direction of normal $\mathbf{n}$ coincides with the direction of vector $\mathbf{r}_{\mathrm{L}} \times \mathbf{j}$, that is $q\mathbf{r}_{\mathrm{L}} \times \mathbf{v} = -q\omega_{\mathrm{L}}r_{\mathrm{L}}^2\mathbf{B}_1$. Consequently,

$$\boldsymbol{\mu} = -\frac{1}{2c}\mathbf{B}_1|q|v_0r_{\mathrm{L}} = -\mathbf{B}\frac{m_0}{2}\left(\frac{v_0}{B}\right)^2 \tag{26.7}$$

As we have seen above, this equation can also be written in the form $\boldsymbol{\mu} = \frac{2q}{c}\mathbf{r}_{\mathrm{L}} \times \mathbf{v}$. Let us compare this formula with the formula for the orbital momentum of a charge orbiting the rotation center. This momentum is equal to $\mu_{mech} = \mathbf{r}_L \times m_0\mathbf{v}$. This gives an important relation between the magnitude of the mechanical angular momentum and the considered above "orbital" magnetic moment:

$$\frac{\mu}{\mu_{mech}} = \frac{q}{2m_0c} \tag{26.7'}$$

For example, the value of this ratio for the electron is determined by universal constants, namely its charge and mass.

It is important to remark that the directions of rotation for particles with opposite signs of charge in a given magnetic field are opposite, while the magnetic moment of the current produced by these particles is, according to the above formulas, directed against the field vector **B**, regardless of their signs of charge.

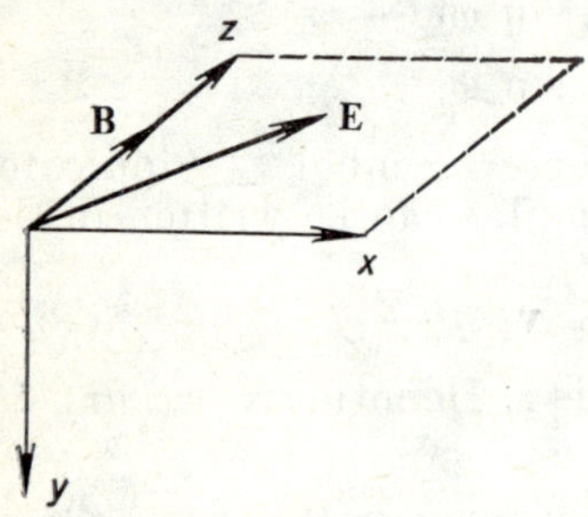

Fig. 28

26.2. Equation (26.1) can also be integrated in a more general case in which magnetic field **B** is again static and uniform but electric field is nonzero; the direction of the electric field is constant but its magnitude may be a function of time. Axis z of the coordinate system will be chosen again along vector **B**, and axes x and y can always be directed in such a manner that plane (x, z) contained vector **E** (Fig. 28). By projecting (26.1) on coordinate axes and introducing a complex variable w, we obtain, as we have done in Subsection 26.1, equations

$$\frac{dw}{dt} = -i\omega_L w + \frac{c\omega_L}{B} E_x(t), \quad \frac{dv_z}{dt} = \frac{q}{m_0} E_z(t) \tag{26.8}$$

As before, parameter ω_L is given by (26.3). Here equation for v_z can, in principle, be integrated if $E_z(t)$ is a known function of time; again, of maximum interest is function w. The solution of the equation satisfied by w is equal to the sum of the solution of homogeneous equation (26.4) and a particular solution w_1 corresponding to the right-hand side given in (26.8). It is readily verified by substitution that such a particular solution can be chosen in the form

$$w_1(t) = e^{-i\omega\ t} \int_0^t E_x(t')\, e^{i\omega_L t'}\, dt' \left(\frac{c\omega_L}{B}\right) \tag{26.9}$$

Consequently,

$$w(t) = w_0 e^{-i\omega_L t} + w_1(t) \tag{26.10}$$

where w_0 has the same meaning as in Subsection 26.1.

Now let $\mathbf{E}$ = const. Then it follows from (26.9) that

$$w_1(t) = ic\frac{E_x}{B}(e^{-i\omega_{\mathrm{L}}t} - 1) \qquad (26.11)$$

By substituting (26.11) into (26.10) and equating the real and imaginary parts, we obtain

$$v_x = v_{0x}\cos\omega_{\mathrm{L}}t + (v_{0y} + cE_x/B)\sin\omega_{\mathrm{L}}t$$

$$v_y + cE_x/B = (v_{0y} + cE_x/B)\cos\omega_{\mathrm{L}}t - v_{0x}\sin\omega_{\mathrm{L}}t \qquad (26.12)$$

This result could be derived from (26.5) by replacing v_y with $v'_y = v_y + cE_x/B$ and v_{0y} with $v'_{0y} = v_{0y} + cE_x/B$. Therefore, vector $\mathbf{v}'_\perp$ with components $v'_x = v_x$ and v'_y corresponds to the rotation in plane x, y with frequency ω_{L} discussed in Subsection 26.1. With our choice of coordinate system, $E_y = 0$ and it is easy to show that

$$\mathbf{v}_\perp = \mathbf{v}'_\perp + \frac{c}{B^2}\mathbf{E}\times\mathbf{B} \qquad (26.13)$$

Consequently, a particle moving in a circular orbit is also displaced in a plane orthogonal to vector $\mathbf{B}$, with a constant velocity equal to the second term in (26.13). This displacement is known as the *electric drift*, and the instantaneous center of rotation, as the *leading center*.

Note that the obtained result follows directly from equation (26.2) if we set $\mathbf{v} = \mathbf{v}_\perp$, $\mathbf{E} = \mathbf{E}_\perp$ and then change for a new reference frame in which the particle moves with respect to the frame at a velocity

$$\mathbf{v}' = \mathbf{v} - \frac{c}{B^2}\mathbf{E}\times\mathbf{B}$$

In another important particular case of formulas (26.9) and (26.10) field $\mathbf{E}(t)$ is a periodic function of time, so that $E_x(t) = E_0\cos\omega t$. Integration in (26.9) is carried out easily. It will be more convenient to represent $\cos\omega t$ in complex form and, after calculating the integral, rewrite $\exp(-i\omega_{\mathrm{L}}t)$ in the form $\exp[-(1/2)i(\omega_{\mathrm{L}} - \omega)t] \times \exp[-(1/2)i(\omega_{\mathrm{L}} + \omega)t]$ and $\exp(i\omega t)$ in the form $\exp[(1/2)i(\omega_{\mathrm{L}} + \omega)t]\cdot\exp[-(1/2)i(\omega_{\mathrm{L}} - \omega)t]$. By assuming $w_1 = v_{1x} - iv_{1y}$ and separating the real and imaginary parts in integral (26.9), we obtain

$$v_{1x} = \frac{c\omega_{\mathrm{L}}E_0}{B}\left[\frac{\cos(1/2)(\omega_{\mathrm{L}} - \omega)t\sin(1/2)(\omega_{\mathrm{L}} + \omega)t}{\omega_{\mathrm{L}} + \omega} + \frac{\sin(1/2)(\omega_{\mathrm{L}} - \omega)t\cos(1/2)(\omega_{\mathrm{L}} + \omega)t}{\omega_{\mathrm{L}} - \omega}\right]$$

$$v_{1y} = \frac{2c\omega_{\mathrm{L}}^2E_0}{B(\omega_{\mathrm{L}}^2 - \omega^2)}\sin\frac{(\omega_{\mathrm{L}} - \omega)}{2}t\cdot\sin\frac{(\omega_{\mathrm{L}} + \omega)}{2}t$$

Assume now that frequency ω of the external field is nearly equal to the Larmor frequency ω_L. This yields $(\omega_L - \omega)^{-1} \sin (1/2) (\omega_L - \omega) t \simeq t/2$. Therefore, in this case, called the *cyclotron resonance*, we have

$$v_{1x} \simeq \frac{cE_0}{2B} \sin \omega_L t + \frac{c\omega_L E_0}{2B} t \cos \omega_L t$$

$$v_{1y} \simeq \frac{cE_0\omega_L}{2B} t \sin \omega_L t$$

If we ignore the contribution of the first term to v_{1x}, which is a periodic function of t (at any rate, this assumption is valid with respect to averaging over time), the kinetic energy K of the transverse motion will increase infinitely:

$$K = \frac{m_0}{2} (v_{1x}^2 + v_{1y}^2) \simeq \frac{c^2\omega_L^2 E_0^2 t^2 m_0}{8B^2} = \frac{(qE_0 t)^2}{8m_0}$$

This effect is used for acceleration of charged particles. Of course, various dissipative forces existing in actual situations restrict kinetic energy of the motion.

26.3. It is of interest to analyze the case of a charge subjected, in addition to a static magnetic field, to a quasielastic force. Let $\mathbf{E} = 0$. The equation of motion then becomes

$$\ddot{\mathbf{r}} + \omega_0^2 \mathbf{r} = \frac{q}{m_0 c} \dot{\mathbf{r}} \times \mathbf{B} \tag{26.14}$$

Denote $\zeta \equiv x + iy$. By calculations absolutely similar to those in (26.1), we obtain the equation $\ddot{\zeta} + \omega_0^2 \zeta + 2i\omega_L' \dot{\zeta} = 0$, where $\omega_L' \equiv \omega_L/2$. The solution of this equation is

$$\zeta = e^{-i\omega_L' t} (Ae^{-i\omega_z t} + Be^{i\omega_z t}) \tag{26.15}$$

where $\omega_z \equiv \sqrt{\omega_L'^2 + \omega_0^2}$ and A, B are arbitrary complex amplitudes. If $\omega_0 \gg \omega_L'$, the motion in plane (x, y) is a sum of two rotations: one with frequency $\omega_0 + \omega_L'$ and another with frequency $\omega_0 - \omega_L'$. Magnetic field which, in the case under discussion, is directed along axis z does not affect the component of oscillations along this axis.

If we analyze the emission of this oscillator similarly to what we had done in § 25 (and here we can assume $\Gamma = 0$), we shall find that in addition to the spectral line with frequency ω_0 generated by oscillator vibrations along axis z, there appears radiation with frequencies $\omega_0 - \omega_L'$ and $\omega_0 + \omega_L'$. Hence, rotation of the oscillator in the applied magnetic field results in splitting of the emitted spectral line into three components. This phenomenon is called the *normal Zeeman effect*. This simple model is not valid for an emitting atom; consistent description of the emission of radiation in a magnetic field is given only in the framework of quantum theory.

26.4*. Let us turn now to the relativistic equation (8.3), that is to

$$m_0 \frac{du^k}{d\tau} = \frac{q}{c} uF_l^{\ lk}$$

If all components of field strength tensor F_{ik} are constant, we can draw some general conclusions concerning the character of the solution. Let us seek the solution in the form

$$x^l = x_0^l \exp\left(\frac{q}{m_0 c} \lambda\tau\right)$$

where x_0^l are real and λ is a complex constant. Substitution into (8.3) yields an algebraic system of equations

$$\lambda x_{0i} = x_{0l} F^l_{\cdot i}$$

This system has a solution if

$$\det(F^l_{\cdot i} - \lambda\delta_i^l) = 0 \tag{26.16}$$

that is if λ is a solution of the characteristic equation (the eigenvalue of matrix $\|F^l_{\cdot i}\|$). If λ is a root of equation (26.16), we can write

$$\det(F^l_{\cdot i} + \lambda\delta_i^l) = \det(F_i^{\cdot l} + \lambda\delta_i^l)$$
$$= \det(-F^l_{\cdot i} + \lambda\delta_i^l) = +\det(F^l_{\cdot i} - \lambda\delta_{li}) = 0$$

In the first of these equalities we have used the constancy of the determinant under permutation of rows and columns, and in the second, the antisymmetry of tensor $F^l_{\cdot i}$.[3] Hence, $-\lambda$ is also a root of (26.16). Consequently, equation (26.16) must have the form $\lambda^4 + \alpha\lambda^2 + \beta = 0$; as this equation is invariant with respect to the Lorentz transformations, constants α and β must be given in terms of quantities I_1 and I_2 found from equation (7.10). The easiest way to find α and β is to calculate the determinant directly.

By denoting $a^l_{\cdot i} \equiv F^l_{\cdot i} - \lambda\delta_i^l$ and using a unit pseudoscalar ε_{iklm} defined in Appendix A, we make use of formula

$$\det(a^l_{\cdot i}) = \varepsilon_{i_1 i_2 i_3 i_4} a^{i_1}_{\cdot 1} a^{i_2}_{\cdot 2} a^{i_3}_{\cdot 3} a^{i_4}_{\cdot 4} \tag{26.17}$$

in which the right-hand side is in fact the definition of the left-hand side and which is related directly to the transformation law (A.9) of pseudoscalars. As tensor $F^l_{\cdot i}$ is antisymmetric, diagonal elements $a^l_{\cdot l}$ are equal simply to $-\lambda$. The term in (26.17) with coefficient ε_{1234} is therefore equal to λ^4. Coefficient α with λ^2 will be obtained if we single out all the nonzero terms of the sum in which only two of the factors are equal to diagonal elements $a^l_{\cdot l}$. For instance, one

---

[3] The third equality holds for a determinant of an even order (in the case under discussion, of the fourth order).

such term will be obtained if $i_1 = 1$, $i_2 = 2$, $i_3 = 4$, $i_4 = 3$; as follows from (7.6), it is

$$\lambda^2 \varepsilon_{1243} F^{4}_{\cdot 3} F^{3}_{\cdot 4} = -\lambda^2 \varepsilon_{1234} (F_{34})^2 = -\lambda^2 E_3^2$$

To simplify calculations, we have replaced index 0 for the time being by 4. It can be shown therefore that $\alpha = \mathbf{B}^2 - \mathbf{E}^2$.

Let us calculate now coefficient β when none of the elements $a^{i}_{\cdot l}$ is diagonal. In this case

$$\beta = \det (F^{l}_{\cdot i}) = \varepsilon_{i_1 i_2 i_3 i_4} F^{i_1}_{\cdot 1} F^{i_2}_{\cdot 2} F^{i_3}_{\cdot 3} F^{i_4}_{\cdot 4}$$

The properties of antisymmetry of tensor $F^{i}_{\cdot k}$ and of coefficient ε_{iklm} lead to the following combinations of indices i_1, i_2, i_3, i_4 giving nonzero result: 2143, 2341, 2413, 3142, 3421, 4123, 4312, 4321. The corresponding terms of the sum have, for example, the form $\varepsilon_{2341} F^{2}_{\cdot 1} F^{3}_{\cdot 2} F^{4}_{\cdot 3} F^{1}_{\cdot 4} = -E_1 E_3 B_1 B_3$ or $\varepsilon_{3412} F^{3}_{\cdot 1} F^{4}_{\cdot 2} F^{1}_{\cdot 3} F^{2}_{\cdot 4} = -E_2^2 B_2^2$. Finally, we obtain

$$\det (F^{l}_{\cdot i}) = -(\mathbf{E} \cdot \mathbf{B})^2 \tag{26.18}$$

Earlier we have demonstrated on the basis of general arguments that coefficients of λ and λ^3 equal zero; of course, this result could be obtained in a straightforward manner similar to the above calculations.

Equation (26.16) can be reduced to the form

$$\lambda^4 + (\mathbf{B}^2 - \mathbf{E}^2) \lambda^2 - (\mathbf{E} \cdot \mathbf{B})^2 = 0 \tag{26.19}$$

whence

$$\lambda = \pm [(1/2) (E^2 - B^2) \pm \sqrt{(1/4) (E^2 - B^2)^2 + (\mathbf{E} \cdot \mathbf{B})^2}]^{1/2} \tag{26.20}$$

The case in which both invariants of electromagnetic field equal zero is a singular case and has to be analyzed separately. Therefore, with the exception of this case, the expression in parentheses under the radical sign is positive. In the general case, for $I_1 \neq 0$ and $I_2 \neq 0$,

$$(1/2) \, |B^2 - E^2| < \left| \sqrt{(1/4) (B^2 - E^2)^2 + (\mathbf{E} \cdot \mathbf{B})^2} \right|$$

Both with $B^2 > E^2$ and with $E^2 > B^2$ the choice of the plus sign in front of the radical makes the expression in the brackets in (26.20) positive, and the choice of the minus sign makes it negative. In the first case we obtain two real values with equal magnitudes and opposite signs: $\lambda_1 = -\lambda_2$. In the second case we obtain two purely imaginary conjugate values: $i\omega$, $-i\omega$. The real eigenvalues correspond to aperiodic motions depending on $\exp (\pm \lambda \tau)$, and imaginary eigenvalues correspond to periodic motions of the type $\exp (\pm i \omega \tau)$.

On the basis of this analysis we can write the possible solution in the real form:

$$\vec{x}(\tau)=a(\vec{\xi}\cos\omega\tau-\vec{\eta}\sin\omega\tau)+b(\vec{\alpha}\cosh\mu\tau-\vec{\beta}\sinh\mu\tau)+\vec{x}_0 \quad (26.21)$$

Here a, b are real amplitudes and $\vec{\alpha}$, $\vec{\beta}$, $\vec{\xi}$, $\vec{\eta}$, $\vec{x}_0$ are constant vectors in the Minkowski space with $\vec{x}_0$ determined by the initial data. The choice of sign in the above formula was dictated by convenience of calculations. The following equations for vectors are obtained by substituting (26.21) into (8.3) and equating the terms in the right- and left-hand sides, which depend identically on τ:

$$\omega\xi^n=\frac{q}{cm_0}\eta^l F_l^{\cdot n},\qquad \omega\eta^n=-\frac{q}{cm_0}\xi^l F_l^{\cdot n} \quad (26.22_1)$$

$$\mu\alpha^n=-\frac{q}{cm_0}\beta^l F_l^{\cdot n},\quad \mu\beta^n=-\frac{q}{cm_0}\alpha^l F_l^{\cdot n} \quad (26.22_2)$$

In (26.21) we use the same eigenvalues multiplied by q/m_0c. An analysis of equations (26.22) shows that they are compatible, and that vectors $\vec{\xi}$, $\vec{\eta}$, $\vec{\alpha}$, and $\vec{\beta}$ can be considered mutually orthogonal. Indeed, (26.22_1) yields

$$\omega\xi^i\eta_i=\frac{q}{cm_0}F_l^{\cdot i}\eta^l\eta_i=0$$

by virtue of antisymmetric nature of tensor $F_l^{\cdot i}$. By complete analogy (26.22_2) yields $\vec{\alpha}\vec{\beta}=0$. By using now the first equation of (26.22_1) and then the second equation of (26.22_2), we find

$$\omega\xi^i\alpha_i=\frac{q}{cm_0}F_l^{\cdot i}\eta^l\alpha_i=\mu\eta^l\beta_l$$

Similarly, from the first equation of (26.22_1) and the first equation of (26.22_2) we obtain

$$\omega\eta^i\beta_i=-\frac{q}{cm_0}\xi^l F_l^{\cdot i}\beta_i=-\mu\xi^l\alpha_l$$

Here we have used again the property of antisymmetry of $F_l^{\cdot i}$. The two preceding equations give

$$\omega\xi^l\alpha_l=-\frac{\mu^2}{\omega}\xi^l\alpha_l=\frac{\mu^3}{\omega^2}\eta^l\beta_l$$

This yields $(\omega^2+\mu^2)\,\xi^l\alpha_l=0$ and, since ω and μ are real, $\xi^l\alpha_l=0$ and therefore $\eta^l\beta_l=0$. Similarly, we can obtain $(\vec{\eta}\vec{\alpha})=(\vec{\xi}\vec{\beta})=0$. This proves that vectors $\vec{\xi}$, $\vec{\eta}$, $\vec{\alpha}$, $\vec{\beta}$ are mutually orthogonal. Equalities $\vec{\xi}^2=\vec{\eta}^2$ and $\vec{\alpha}^2=-\vec{\beta}^2$ are verified in an identical manner. The presence of coefficients a and b permits to consider vectors $\vec{\alpha}$, $\vec{\beta}$, $\vec{\xi}$, $\vec{\eta}$ as unit vectors. However, among the four mutually orthogonal vectors of the Minkowski space three must be spacelike and one timelike. The relations between lengths of these vectors show that only two possibilities are available: either vector $\vec{\alpha}$ or vector $\vec{\beta}$ must be timelike. But it follows from this that the component of motion in the timelike direction is, as could be expected, always aperiodic.[4] Four-dimensional velocity of motion must be timelike. Taking into account the indicated above properties of vector amplitudes, we obtain from (26.21) $(d\vec{x}/d\tau)^2 = -a^2\omega^2 \mp b^2\mu^2$, where the upper sign corresponds to timelike vector $\vec{\alpha}$ and the lower sign, to timelike vector $\vec{\beta}$. In the first case the condition for velocity cannot be satisfied. For this reason it is vector $\vec{\beta}$ that must be treated as timelike.

Further investigation with arbitrarily directed fields **B** and **E** cannot be carried out in the general case even if the fields are assumed, as before, to be static and uniform. However, we can carry out a complete analysis of several important particular cases. These cases are defined by constraints on the values of invariants of electromagnetic field. It will be useful to recall here the properties of field transformations presented in § 7.

26.5*. First of all, let us assume $\mathbf{E}\cdot\mathbf{B} = 0$, so that, according to (26.18), $\det(F_{li}) = 0$. We begin with the case $B^2 > E^2$. It was shown in § 7 that in these conditions there is a reference frame in which $\mathbf{E}' = 0$. Equation (26.20) shows that there are two nonzero eigenvalues: $i\omega = \pm i \frac{q}{m_0 c}(B^2 - E^2)^{1/2}$. Note that in this particular reference frame in which electromagnetic field is purely magnetic, $\mathbf{B}'$, the magnitude of these eigenvalues, coincides with the Larmor frequency (26.3). But one has to take into account additionally that, by virtue of the mentioned zero value of the determinant, there is a nonzero solution v^l of the system of algebraic equations

$$v^l F_l^{\cdot i} = 0 \tag{26.23}$$

and therefore equation (8.3) is satisfied not only by solutions of the type of (26.21) (where we must set $\mu = 0$) but also by a solution of type $x^l = v^l\tau + v_0^l$ in which the second term is constant. The general solution takes the form

$$\vec{x}(\tau) = a(\vec{\xi}\cos\omega\tau - \vec{\eta}\sin\omega\tau) + \vec{v}\tau \tag{26.24}$$

if the origin is chosen such that the constant term in (26.21) in the sum with $\vec{v}_0$ vanishes. Note that the argument in (26.24) is precisely the proper time τ and not time t in an arbitrary reference

[4] Obviously, a periodic motion along the time axis would mean violation of causality!

frame. In order to introduce t as an argument, one has to determine the functional dependence $\tau = \tau(t)$ and to resolve it with respect to t.

Denote by $\mathbf{v}$ the three-dimensional velocity corresponding to four-dimensional velocity $\vec{v}$. Then, in the three-dimensional notation, equation (26.23) can be rewritten in the form $\mathbf{v} \times \mathbf{B} = -c\mathbf{E}$ and $(\mathbf{v} \cdot \mathbf{E}) = 0$. We have found in § 7 that in this case $\mathbf{v}$ is the velocity of the reference frame in which $\mathbf{E}' = 0$. Forming a vector product of the above equality by $\mathbf{B}$ and using the formula for the triple vector product, we obtain that projection of velocity $\mathbf{v}$ onto a plane orthogonal to vector $\mathbf{B}$ is $\mathbf{v}_\perp = -\frac{c}{B^2}\mathbf{E} \times \mathbf{B}$. We have come again to the velocity of electric drift which is also found in the nonrelativistic equation (26.13). Velocity $\mathbf{v}$ is, therefore, the velocity of the leading center which moves uniformly along a straight line.

Let us change to a reference frame co-moving with the leading center, and denote the velocity of the particle in this reference frame by $\mathbf{v}$. As $\mathbf{E}' = 0$, the zero component of the equation of motion (8.3) of the particle takes the form $d(\gamma m_0 c)/d\tau = 0$, that is $\gamma = (1 - v^2/c^2)^{-1/2} = \text{const}$ and hence, $v = \text{const}$. The first two terms in (26.24) describe, therefore, a uniform Larmor rotation. And it follows from $d\tau/dt = \gamma^{-1}$ that $\tau = t\gamma^{-1}$ and the argument of trigonometric functions is written in the form $\omega\tau = \frac{qB'}{\gamma m_0 c} t$. Thus a relativistic particle in a reference frame co-moving with the leading center revolves just as in the nonrelativistic case (see Subsection 26.1). One only has to replace the rest mass m_0 in (26.3) by the mass of motion $m = m_0\gamma$.

The same result is obtained directly and in a still more simple manner from spatial components of equation (8.3) which in the co-moving reference frame of the leading center are quite analogous to equations (26.2), namely: $\frac{d\mathbf{v}_\perp}{dt} = \frac{q}{m_0 c\gamma}\mathbf{v}_\perp \times \mathbf{B}$; these components can be integrated as we have done already in Subsection 26.1. Consequently, the conclusions on the equivalent magnetic movement listed at the end of Subsection 26.1 remain valid when m_0 is replaced by m.

26.6. Still assuming $\mathbf{B} \cdot \mathbf{E} = 0$, we now specify $E^2 > B^2$. This case is analyzed best by making use of the results of § 7, and by performing calculations in the reference frame in which $\mathbf{B} = 0$. In this reference frame spatial components of equation (8.3) are written in the form $d\mathbf{p}/dt = q\mathbf{E}$, where $\mathbf{p} = m_0\gamma\mathbf{v}$ is the relativistic momentum of the particle. Let axis x be in the direction of vector $\mathbf{E}$. By projecting the equation onto axis x and onto a plane orthogonal to x we obtain $p_x = m_0\gamma v_x = qEt + p_{0x}$, $\mathbf{p}_\perp = m_0\gamma\mathbf{v}_\perp = \mathbf{p}_{0\perp} =$

= const. Here p_{0x} and $\mathbf{p}_{0\perp}$ are determined by the initial velocity. Now we can use the relation $m_0\gamma c = (\mathbf{p}^2 + m_0^2c^2)^{1/2} = (p_x^2 + w_0^2)^{1/2} = [(qEt + p_{0x})^2 + w_0^2]^{1/2}$ (cf. § 6), where $w_0^2 \equiv \mathbf{p}_\perp^2 + m_0^2c^2 = \text{const}$; this relation determines the time component of the momentum. In order to make the formulas clearer, in what follows we restrict the analysis to the case of $p_{0x} = 0$. Then

$$v_x = \frac{dx}{dt} = cqEt\,[w_0^2 + (qEt)^2]^{-1/2}$$

$$\mathbf{v}_\perp = \frac{d\mathbf{r}_\perp}{dt} = c\mathbf{p}_{0\perp}\,[w_0^2 + (qEt)^2]^{-1/2} \tag{26.25}$$

Obviously, one of the coordinate axes (let it be axis y) in the plane orthogonal to axis x can always be chosen in the direction of $\mathbf{p}_{0\perp}$. Integration of formulas (26.25) from the initial moment of time $t_0 = 0$ yields the following result:

$$x(t) = \frac{c}{qE}\,[w_0^2 + (qEt)^2]^{1/2}$$

$$y(t) = \frac{cp_{0y}}{qE}\,\text{Arcsinh}\,\frac{qEt}{w_0} \tag{26.26}$$

Here the origin in plane (x, y) is chosen for the constant terms in (26.26) to vanish. The path of the particle lies completely in this plane. We can eliminate time from (26.26) and obtain the equation of particle's trajectory in the form

$$x = \frac{cw_0}{qE}\cosh\left(\frac{qEy}{cp_{0y}}\right) \tag{26.27}$$

Assume in the nonrelativistic case $w_0 \simeq m_0c$ and $p_{0y} \simeq m_0v_{0y}$. Since in our case $p_{0x} = 0$, the initial kinetic energy of the particle is $(1/2)\,m_0v_{0y}^2 \equiv T_0$. The nonrelativistic approximation is valid only at sufficiently small distances from the point from which the particle starts its motion since in principle the acceleration imposed by the field can increase the velocity of the particle as near to the velocity of light c as necessary. By assuming the argument of cosh to be sufficiently small, we obtain

$$x \simeq \frac{1}{4}\,\frac{q}{m_0}\,\frac{E}{T_0}\,y^2 + \text{const} \tag{26.28}$$

This is the equation of a parabola in plane (x, y). An analysis of the trajectory in the general case is suggested for the reader as an exercise.

26.7. Consider the case of $\mathbf{E}\cdot\mathbf{B} = EB$, that is the case of parallel electric and magnetic fields. Axis z of spatial coordinates can be turned to lie in the common direction of these two vectors, so that $E_z = E$, $B_z = B$. The reader will easily find that this transforms equation (8.3) into the following set of equations:

$$\frac{d^2x}{d\tau^2}=\omega_L\frac{dy}{d\tau},\quad \frac{d^2y}{d\tau^2}=-\omega_L\frac{dx}{d\tau}$$

$$\frac{d^2z}{d\tau^2}=\frac{qE}{m_0}\frac{dt}{d\tau},\quad \frac{d^2t}{d\tau^2}=\frac{qE}{m_0c^2}\frac{dz}{d\tau}$$

Here ω_L is the nonrelativistic Larmor frequency (26.3). The first two equations describe periodic motion with frequency ω_L, and the second pair of equations describe aperiodic motion along axis z. This result can be compared with formula (26.21) but the motion itself will not be studied in more detail here.

26.8. To conclude this section, we consider the motion of a particle in the field of a plane wave. This means that both invariants of electromagnetic field equal zero. The analysis of motion in a field depending on time periodically (see Subsection 26.2) assumed that the field is uniform in space and, besides, is nonrelativistic. In this case formulas (18.7) are valid:

$$\mathbf{B}=\mathbf{n}\times\mathbf{E},\quad \mathbf{E}\cdot\mathbf{n}=0 \tag{26.29}$$

where $\mathbf{n}=\mathbf{k}/k$. The dependence on time has the form $\exp[i(\mathbf{k}\cdot\mathbf{r}-\omega t)]=\exp[-i\omega(t-\mathbf{n}\cdot\mathbf{r}/c)]$. The complex form of this function is unimportant here, and in fact we use the real part of the exponential function, assuming also that vectors **B** and **E** are real. With (26.29), spatial components of equation (8.3) take the form

$$\frac{d}{dt}(m\mathbf{v})=q\left[\mathbf{E}+\frac{\mathbf{v}}{c}\times(\mathbf{n}\times\mathbf{E})\right]=q\left[\left(1-\frac{\mathbf{v}\cdot\mathbf{n}}{c}\right)\mathbf{E}+\frac{1}{c}\mathbf{n}(\mathbf{E}\cdot\mathbf{v})\right] \tag{26.30}$$

As usual, the time component represents the energy conservation law, that is $\frac{d}{dt}(mc^2)=q\mathbf{E}\cdot\mathbf{v}$. By projecting (26.30) onto the constant vector **n** and taking into account (26.29), we obtain $\frac{d}{dt}(m\mathbf{v}\cdot\mathbf{n})=\frac{q}{c}\mathbf{E}\cdot\mathbf{v}=\frac{d}{dt}(mc)$. Consequently, $m(\mathbf{v}\cdot\mathbf{n}-c)=\text{const}$. We choose for the origin $t=0$ the moment of time in which $\mathbf{v}=0$ and, therefore, $m=m_0$. Then $m\mathbf{v}\cdot\mathbf{n}=(m-m_0)c$, that is $\mathbf{v}\cdot\mathbf{n}=\frac{\gamma-1}{\gamma}c$ or $\frac{d\tau}{dt}=\gamma^{-1}=1-\frac{\mathbf{v}\cdot\mathbf{n}}{c}=\frac{dt'}{dt}$, if we denote $t'\equiv t-\frac{\mathbf{n}\cdot\mathbf{r}}{c}$. Hence, $m\mathbf{v}=m_0\frac{dt}{dt'}\frac{d\mathbf{r}}{dt}=m_0\frac{d\mathbf{r}}{dt'}$. Multiply both sides of equation (26.30) by dt/dt'. Then we find from the above relations

$$\frac{d^2\mathbf{r}}{dt'^2}=\frac{q}{m_0}\mathbf{E}+\frac{q}{m_0c}\mathbf{n}\left(\mathbf{E}\cdot\frac{d\mathbf{r}}{dt'}\right) \tag{26.31}$$

It must be assumed that vector **E** can be given as a function of t'. Let axis x be directed along vector **n** of the plane wave propagation.

By projecting (26.31) on this vector and on the plane orthogonal to it, we find

$$\frac{d^2x}{dt'^2}=\frac{q}{m_0c}\left(\mathbf{E}\cdot\frac{d\mathbf{r}}{dt'}\right)=\frac{q}{m_0c}\left(\mathbf{E}\cdot\frac{d\mathbf{r}_\perp}{dt'}\right),\quad \frac{d^2\mathbf{r}_\perp}{dt'^2}=\frac{q}{m_0}\mathbf{E} \qquad (26.32)$$

From the second equation of (26.32) we have

$$\mathbf{r}_\perp(t')=\frac{q}{m_0}\int_0^{t'}dt''\int_0^{t''}\mathbf{E}(t''')\,dt'''$$

The first equation then yields

$$\frac{d^2x}{dt'^2}=\frac{q^2}{m_0^2c}\mathbf{E}(t')\int_0^{t'}\mathbf{E}(t'')\,dt''=\frac{q^2}{2m_0^2c}\frac{d}{dt'}\left[\int_0^{t'}\mathbf{E}(t'')\,dt''\right]^2$$

whence

$$x(t')=\frac{q^2}{2m_0^2c}\int_0^{t'}dt''\left[\int_0^{t''}\mathbf{E}(t''')\,dt'''\right]^2,\quad \mathbf{r}(t')=x(t')\,\mathbf{n}+\mathbf{r}_\perp(t') \qquad (26.33)$$

It is important to emphasize again that the above formulas give $\mathbf{r}$ as a function of auxiliary variable t' or, and this is the same in our case, a function of proper time τ. In order to find the motion of a particle as a function of time t, we should eliminate t' from equation $t = t' + \mathbf{r}(t')\cdot\mathbf{n}/c = t' + x(t')\,c$, in which $x(t')$ is found from (26.33). Integration in this formula is readily carried out since the real component of the field is written in the form $\mathbf{E}(t''') = \mathbf{E}_0\cos\omega t'''$; however, equation for t' is transcendental, and therefore can be solved only numerically.

The cases analyzed above do not exhaust all the possibilities of exact solution of the equations of motion. For example, we may discuss some static but nonuniform magnetic fields. The relevant results can be found in special literature[5].

§ 27. Theory of drift in nonuniform electromagnetic fields

27.1. Various approximate techniques are used to analyze the behaviour of a particle in nonuniform and time-dependent fields. Here we shall discuss one method of the perturbation theory widely used in such problems, and shall introduce the simplest assumptions. Roughly speaking, the physical picture on which the method is based is the motion in the form of sufficiently rapid Larmor oscillations superimposed on a comparatively slow displacement of the "leading center". This description of motion is exact in uniform

[5] See, for example: B. Lehnert, *Dynamics of Charged Particles*, North-Holland, Amsterdam, 1964.

static electric and magnetic fields (see Subsection 26.2). And under certain conditions it gives an adequate description of particle's behaviour in nonuniform fields as the first approximation.

All calculations will be carried out in the nonrelativistic limit. Assume first of all that magnetic field varies sufficiently slowly both in space and with time. This condition is expressed by the inequalities

$$r_{\mathrm{L}}\left|\frac{\partial B_j}{\partial x^k}\right| \ll |B_j|, \quad \frac{1}{\omega_{\mathrm{L}}}\left|\frac{\partial B_j}{\partial t}\right| \ll |B_j|$$

Here ω_{L} and r_{L} are the Larmor frequency and radius found from formulas of Subsection 26.1. Derivatives with respect to coordinates and time are found in a reference frame co-moving with the particle. Similar inequalities must hold for all the remaining external forces applied to the particle. Among such external forces we shall explicitly take into account hereafter only electric force $q\mathbf{E}$. If forces $\mathbf{f}$ of other origin are also applied, all the subsequent formulas will be valid if $q\mathbf{E}$ is replaced by $q\mathbf{E} + \mathbf{f} = \mathbf{F}$. We only have to assume that components $\mathbf{F}_{\|} = \mathbf{B}\,(\mathbf{F}\cdot\mathbf{B})/B^2$ parallel to magnetic field are small, and components $\mathbf{F}_{\perp}$ are nearly independent of the Larmor radius r_{L}.

Application of the perturbation theory requires that the equation of motion (26.1) be reduced to dimensionless form. For this we denote the characteristic size of the region in the nonuniform field in which we study the motion by L, the magnetic field strength at moment t_0 by $\mathbf{B}_0$, and the velocity of motion in the Larmor circle in this field by $\mathbf{v}_0$. Dimensionless variables can then be introduced as follows:

$$\boldsymbol{\rho} = 1/L\mathbf{r}, \quad \widetilde{T} = v_0 t/L, \quad \widetilde{\mathbf{B}} = \mathbf{B}\,(t)/B_0$$

If in addition we define a dimensionless quantity $\widetilde{\mathbf{E}} = \dfrac{c}{v_0 B_0}\,\mathbf{E}$, then after substituting all the functions into equation (26.1) we obtain a dimensionless expression

$$\frac{r_{\mathrm{L}}}{L}\,\frac{d^2\boldsymbol{\rho}}{d\widetilde{T}^2} = \widetilde{\mathbf{E}} + \frac{d\boldsymbol{\rho}}{d\widetilde{T}} \times \widetilde{\mathbf{B}} \tag{27.1}$$

Here, as in Subsection 26.1, $r_{\mathrm{L}} = m_0 c v_0/qB_0$. However, formally equation (27.1) is identical to (26.1) if r_{L}/L is replaced by $m_0 c/q$, and $\widetilde{\mathbf{E}}$ by $c\mathbf{E}$. There is, therefore, a one-to-one correspondence between the solutions of these two equations. The perturbation theory can thus be applied directly to equation (26.1). As a small parameter in (27.1) we choose r_{L}/L. The small parameter for an analysis of equation (26.1) is then $\varepsilon \equiv m_0 c/q = B/\omega_{\mathrm{L}}$:

$$\varepsilon\,\frac{d\mathbf{v}}{dt} = c\mathbf{E} + \mathbf{v} \times \mathbf{B} \tag{27.2}$$

We shall introduce now the basic assumption of the perturbation theory which will be considered here only to within the first-order terms in parameter ε. Namely, assume that radius vector of a particle can be written in the form

$$\mathbf{r}(t) = \mathbf{r}_c(t) + \mathbf{r}_L(t) = \mathbf{r}_c(t) + \varepsilon(\mathbf{e}_1 \cos \omega_L t + \mathbf{e}_2 \sin \omega_L t) \quad (27.3)$$

Here, in analogy to formula (26.6), vectors $\mathbf{e}_1$ and $\mathbf{e}_2$ are mutually orthogonal, and

$$\varepsilon \mathbf{e}_1 = \frac{\mathbf{B}_1 \times \mathbf{v}_0}{\omega_L}, \quad \varepsilon \mathbf{e}_2 = \frac{\mathbf{v}_0}{\omega_L} \quad (27.4)$$

Radius vector $\mathbf{r}_c(t)$ describes the motion of the "leading center" and $\mathbf{r}_L(t)$ describes the Larmor rotation mentioned at the beginning of this section. In what follows we shall also use the notation $\mathbf{v}_c = d\mathbf{r}_c/dt$ and $\mathbf{v}_L = d\mathbf{r}_L/dt$, so that $\mathbf{v} = \mathbf{v}_c + \mathbf{v}_L$.

By using the condition of slow variation of the field, we are justified in operating with expansions of the field in the neighbourhood of point $\mathbf{r}_c$ in the form

$$\begin{aligned} \mathbf{B}(\mathbf{r}) &= \mathbf{B}_c + \delta\mathbf{B} = \mathbf{B}_c + (\mathbf{r}\cdot \mathrm{grad})\,\mathbf{B}_c \\ \mathbf{E}(\mathbf{r}) &= \mathbf{E}_c + \delta\mathbf{E} = \mathbf{E}_c + (\mathbf{r}\cdot \mathrm{grad})\,\mathbf{E}_c \end{aligned} \quad (27.5)$$

Subscript "c" means calculation of each term at point $\mathbf{r} = \mathbf{r}_c$. Substitute (27.3) and (27.5) into (27.2), using instead of parameter ε its explicit expression. This gives

$$m_0 \frac{d\mathbf{v}_c}{dt} - q\left(\mathbf{E}_c + \frac{1}{c}\mathbf{v}_c \times \mathbf{B}_c\right) - \frac{q}{c}\mathbf{v}_L \times \delta\mathbf{B}$$
$$= q\delta\mathbf{E} + \frac{q}{c}\mathbf{v}_c \times \delta\mathbf{B} + \left(\frac{q}{c}\mathbf{v}_L \times \mathbf{B}_c - m_0 \frac{d\mathbf{v}_L}{dt}\right) \quad (27.6)$$

The last two terms in the right-hand side implicitly contain factor ε. The same factor is also contained in the last term in the left-hand side, but its description calls for very careful analysis and will be carried out somewhat later. We can assume, therefore, that difference $m_0 \frac{d\mathbf{v}_L}{dt} - \frac{q}{c}\mathbf{v}_L \times \mathbf{B}_c$ is "nearly zero". In other words, the particle participates predominantly in a rotational motion with the Larmor frequency. As a result of this (sufficiently fast) rotation, fields $\mathbf{B}$ and $\mathbf{E}$ vary periodically with time with frequency ω_L in the comoving reference frame.

Let us average both sides of equation (27.6) over the Larmor period of rotation, $T_L = 2\pi/\omega_L$, assuming that the averaging leaves $\frac{d\mathbf{v}_c}{dt}$ almost unaltered. This averaging is an extremely important feature of the discussed method of the perturbation theory; it enables us to separate relatively fast motions from slow motions. Clearly, average values of vectors $\mathbf{r}_L$ and $\mathbf{v}_L$ equal zero. Moreover, mean values of "fluctuations" $\delta\mathbf{E}$ and $\delta\mathbf{B}$ of electromagnetic field also equal zero. It would seem at the first glance that term $\mathbf{v}_L \times \delta\mathbf{B}$ will also give zero contribution; we shall presently see, however, that this is not the case.

First we shall write

$$\frac{1}{\omega_L}\mathbf{v}_L \times \delta\mathbf{B} = (\mathbf{r}_L \times \mathbf{B}_1) \times ((\mathbf{r}_L \cdot \text{grad})\,\mathbf{B})$$

$$= -\mathbf{r}_L(\mathbf{B}_1 \cdot (\mathbf{r}_L \cdot \text{grad})\,\mathbf{B}) + \mathbf{B}_1(\mathbf{r}_L \cdot (\mathbf{r}_L \cdot \text{grad})\,\mathbf{B})$$

Let the direction of axis z of a local coordinate system coincide with vector $\mathbf{B}_1$ in the origin of this system. Consequently, projection of $\mathbf{r}_L$ onto axis z is zero. Hence,

$$\frac{1}{\omega_L}\mathbf{v}_L \times \delta\mathbf{B}\,|_x = -r_{Lx}(\mathbf{r}_L \cdot \text{grad})\,B_z = -r_{Lx}^2\frac{\partial B_z}{\partial x} - r_{Lx}r_{Ly}\frac{\partial B_z}{\partial y}$$

$$\frac{1}{\omega_L}\mathbf{v}_L \times \delta\mathbf{B}\,|_y = -r_{Ly}(\mathbf{r}_L \cdot \text{grad})\,B_z = -r_{Ly}r_{Lx}\frac{\partial B_x}{\partial x} - r_{Ly}^2\frac{\partial B_z}{\partial y}$$

$$\frac{1}{\omega_L}\mathbf{v}_L \times \delta\mathbf{B}\,|_z = r_{Lx}(\mathbf{r}_L \cdot \text{grad})\,B_x + r_{Ly}(\mathbf{r}_L \cdot \text{grad})\,B_y$$

$$= r_{Lx}^2\frac{\partial B_x}{\partial x} + r_{Ly}^2\frac{\partial B_y}{\partial y} + r_{Lx}r_{Ly}\left(\frac{\partial B_x}{\partial y} + \frac{\partial B_y}{\partial x}\right)$$

Mean values of r_{Lx}^2 and r_{Ly}^2 over a period are equal to $r_L^2/2$, and the mean value of $r_{Lx}r_{Ly}$ equals zero. We shall also use equation $\text{div}\,\mathbf{B} = 0$, that is $\partial B_x/\partial x + \partial B_y/\partial y = -\partial B_z/\partial z$. Denoting the result of averaging by angle brackets $\langle\ \rangle$, we then have

$$\langle \mathbf{v}_L \times \delta\mathbf{B}\rangle = -(1/2)\,\omega_L r_L^2\,\text{grad}\,B_z \tag{27.7}$$

But if we use assumptions on smallness of field components B_x and B_y compared with components B_z and an assumption on slow variation of these components, we obtain

$$\text{grad}\,B_z = \frac{1}{2B_z}\,\text{grag}\,B_z^2$$

$$= \frac{1}{2B_z}(\text{grad}\,B^2 - 2B_x\,\text{grad}\,B_x - 2B_y\,\text{grad}\,B_y) \simeq \text{grad}\,B$$

Substituting (27.7) into (27.6) and taking into account all the mentioned approximations, we obtain the equation

$$m_0\frac{d\mathbf{v}_c}{dt} = q\mathbf{E} + \frac{q}{c}\mathbf{v}_c \times \mathbf{B} - M\,\text{grad}\,B \tag{27.8}$$

Here

$$M = \frac{q\omega_L r_L^2}{2c} = \frac{q}{2c}v_L r_L = \frac{m_0}{2}\frac{v_0^2}{B}$$

Quantity M is thus the magnitude of the magnetic moment of the current, corresponding to the motion of the particle in the Larmor orbit described by equation (26.7). One can say that in this approximation a real particle is substituted by an "equivalent" particle whose motion coincides with the trajectory of the leading center, and that this particle possesses, in addition to its charge, a magnetic moment **M**. The physical meaning of this picture is quite clear; however, we have to emphasize again the decisive role of averaging over periodic Larmor motion in the derivation of the picture.

The discussed above first approximation is naturally a very rough representation of the actual situation. To substantiate the approximation and determine the limits of its applicability, it should be necessary to analyze expansion of $\mathbf{r}(t)$ and of fields **B** and **E** into infinite series in powers of ε, as well as to perform averaging with sufficient accuracy. This problem has been solved by a number of authors but it is by far too complicated to be presented here[6].

27.2. Let us proceed with further analysis of the drift motion equation (27.8). We can drop the subscript "c". By projecting this equation onto vector $\mathbf{B}_1$, we obtain drift in the direction of the magnetic field:

$$m_0\left(\frac{d\mathbf{v}}{dt}\right)_{||}=q\mathbf{E}_{||}-M\mathbf{B}_1(\mathbf{B}_1\cdot\mathrm{grad}\,B)$$

The derivation of equation for the transverse drift in a convenient form is a more difficult task. Denote a unit vector of velocity along magnetic lines of force by $\mathbf{v}_{||1}$. Then $\mathbf{v}=v_{||}\mathbf{v}_{||1}+\mathbf{v}_{\perp}$ and $\frac{d\mathbf{v}}{dt}=\mathbf{v}_{||1}\frac{dv_{||}}{dt}+\frac{d\mathbf{v}_{\perp}}{dt}+v_{||}\frac{d\mathbf{v}_{||1}}{dt}$. If we form a vector product of equation (27.8) by $\mathbf{B}_1$ and take into account that $B^{-2}[\mathbf{B}\times(\mathbf{v}\times\mathbf{B})]=\mathbf{v}-\mathbf{B}_1(\mathbf{v}\cdot\mathbf{B}_1)\equiv\mathbf{v}_{\perp}$, and also that $\mathbf{v}_{||1}\times\mathbf{B}_1=0$, we obtain

$$\mathbf{v}_{\perp}=\frac{c}{qB^2}\left[q\mathbf{E}-M\,\mathrm{grad}\,B-m_0\left(\frac{d\mathbf{v}_{\perp}}{dt}+v_{||}\frac{d\mathbf{v}_{||1}}{dt}\right)\right]\times\mathbf{B} \qquad (27.9)$$

However, vector $\mathbf{v}_{||1}$ is by definition invariably directed along field **B**, that is $d\mathbf{v}_{||1}=d\mathbf{B}_1=(\mathbf{B}_1\cdot\mathrm{grad})\,\mathbf{B}_1\,dl$, where $dl=v_{||}dt$, and $v_{||}$ is the value of velocity at the beginning of the infinitesimal increment in question. Hence, $d\mathbf{v}_{||1}/dt=v_{||}(\mathbf{B}_1\cdot\mathrm{grad})\,\mathbf{B}_1$. As follows from (B.17), for $\mathbf{a}=\mathbf{B}_1$, that is for $\mathrm{grad}\,a^2=0$, we have $(\mathbf{B}_1\cdot\mathrm{grad})\,\mathbf{B}_1=-\mathbf{B}_1\times\mathrm{curl}\,\mathbf{B}_1$. Then, according to (B.14$_3$), $\mathrm{curl}\,\mathbf{B}_1=\mathrm{curl}\,\frac{\mathbf{B}}{B}=\frac{1}{B}\mathrm{curl}\,\mathbf{B}+\left[\mathrm{grad}\,\frac{1}{B}\times\mathbf{B}\right]$. Denoting $(\mathrm{grad}\,B)_{\perp}\equiv\mathrm{grad}\,B-\mathbf{B}_1(\mathbf{B}_1\cdot\mathrm{grad}\,B)$, we obtain

$$(\mathbf{B}_1\cdot\mathrm{grad})\,\mathbf{B}_1=\frac{1}{B}((\mathrm{grad}\,B)_{\perp}-\mathbf{B}_1\times\mathrm{curl}\,\mathbf{B}) \qquad (27.10)$$

[6] See, for example, Lehnert's monograph cited on p. 206.

Now substitute (27.10) into the formula for derivative $d\mathbf{v}_{\|1}/dt$, and the result—into equation (27.9). In calculating the vector product by $\mathbf{B}$ in this equation, we take into account that $(\text{grad } B)_\perp \times \mathbf{B} = (\text{grad } \mathbf{B}) \times \mathbf{B}$, and also use an obvious notation $(\text{curl } \mathbf{B})_\perp \equiv \mathbf{B}_1 \times [\text{curl } \mathbf{B} \times \mathbf{B}_1]$. Finally we obtain from the Maxwell equation $(\text{curl } \mathbf{B})_\perp = \frac{1}{c} \frac{\partial \mathbf{E}}{\partial t}$ that $\mathbf{v}_\perp = \mathbf{v}_E + \mathbf{v}_B + \mathbf{v}_I + v_P$, where

$$\mathbf{v}_E = \frac{c}{B^2} \mathbf{E} \times \mathbf{B}, \quad \mathbf{v}_I = \frac{cm_0}{qB^2} \mathbf{B} \times \frac{d\mathbf{v}_\perp}{dt}$$

$$\mathbf{v}_B = \frac{Mc}{qB^2} \left(1 + \frac{2v_\|^2}{v_\text{L}^2}\right) [\mathbf{B} \times \text{grad } B], \quad \mathbf{v}_P = \frac{m_0 v_\|^2}{qB^2 c} \frac{\partial \mathbf{E}_\perp}{\partial t}.$$

Term $\mathbf{v}_E$ describes the *electric drift* similar to that discussed in Subsection 26.2. Velocity $\mathbf{v}_I$ corresponds to the so-called *transverse inertial drift*. It is proportional to mass m_0 of the particle and results as if from the centrifugal force applied to this particle as it moves in a curved magnetic line of force. Term $\mathbf{v}_P$ is called the *polarization drift*. It appears because the electric drift of the particles changes with time as the particle moves in a variable electric field. In this process the external electric field does a certain work on the particle, leading to additional acceleration. A uniformly accelerated motion at velocity $\mathbf{v}_E$, which would be realized if the electric field were constant, is supplemented by "falling" of the particle in the field from one equipotential surface to another.

And finally, term $\mathbf{v}_B$ proportional to grad B and magnetic moment M is called the *gradient drift*. It is caused by the nonuniformity of magnetic field and is directed along surface $B = \text{const}$. More detailed analysis shows that the first term of the gradient drift describes the effect of the field nonuniformity on the Larmor rotation, and the second term is connected with the curvature of magnetic lines of force.

Note that all terms of the drift, with an exception of $\mathbf{v}_E$, change sign when the sign of the charge is reversed. Consequently, the phenomenon of drift is used for a practical separation of opposite charges, for example, of electrons and ions in plasma.

27.3. From the point of view of mechanics, the motion of a particle in the approximation under discussion is *nearly periodic*: the periodic motion, namely the rotation in the Larmor orbit, is superposed with the relatively slow, compared with the rotation, displacement of the leading center. If the motion which corresponds to the generalized coordinate q and generalized momentum p is strictly periodic, it can be proved in analytical mechanics that $J = \oint p\, dq$, in which integration is carried out over one period of oscillations, is an invariant of motion. In the case of nearly periodic motion, when parameters of the system vary adiabatically, that is, sufficiently slowly compared with the period and over characteristic time incommensurable with the period (this condition excludes the possibility of resonance) the indicated integrals remain constant and therefore are referred to as *adiabatic invariants* of the considered system.

It was shown in Subsection 26.1 for the case of strictly periodic motions in static magnetic fields that the equivalent magnetic moment M of the current generated by such motions is invariant. It can be proved that for the motion with drift described by equation (27.8) in the framework of the approximation used we have again $dM/dt = 0$, that is M is an adiabatic invariant.

A scalar product of equation (27.8) by vector $\mathbf{v}_c$ yields

$$\frac{m_0}{2}\frac{d\mathbf{v}_c^2}{dt} = q\mathbf{E}\cdot\mathbf{v}_c - M\mathbf{v}_c\cdot\operatorname{grad} B \tag{27.11}$$

which describes energy changes in the course of drift. As usual, the complete law of energy conservation takes the form

$$\frac{m_0}{2}\frac{d}{dt}(\mathbf{v}_c + \mathbf{v}_L)^2 = q\mathbf{E}\cdot(\mathbf{v}_c + \mathbf{v}_L) \tag{27.12}$$

Subtract (27.11) term by term from (27.12). This gives

$$\frac{m_0}{2}\frac{d}{dt}(\mathbf{v}_L^2 + 2\mathbf{v}_L\cdot\mathbf{v}_c) = q\mathbf{E}\cdot\mathbf{v}_L + M\mathbf{v}_c\cdot\operatorname{grad} B \tag{27.13}$$

The second term in the left-hand side can be neglected as in our approximation $v_c \ll v_L$. After substituting electric field $\mathbf{E}$ in (27.13) in terms of potential,

$$\mathbf{E} = -\operatorname{grad}\varphi - \frac{1}{c}\frac{\partial \mathbf{A}}{\partial t}$$

we shall integrate both sides of (27.13) over one period of the Larmor rotation, $T_L = 2\pi/|\omega_L|$. If $\mathbf{v}_L^2$, $\mathbf{B}$, $\mathbf{E}$, and $\mathbf{v}_c$ (but not $\mathbf{v}_L$) depend on time only slightly, the integration results in

$$\frac{m_0}{2}T_L\frac{dv_L^2}{dt}$$

$$\simeq T_L M(\mathbf{v}_c\cdot\operatorname{grad})B - q\int_0^{T_L}\operatorname{grad}\varphi\cdot\mathbf{v}_L\,dt - \frac{q}{c}\int_0^{T_L}\frac{\partial \mathbf{A}}{\partial t}\cdot\mathbf{v}_L\,dt \tag{27.14}$$

But we know that $\mathbf{v}_L\,dt = d\mathbf{r}_L$, and therefore

$$\int_0^{T_L}\operatorname{grad}\varphi\cdot\mathbf{v}_L\,dt = \oint\operatorname{grad}\varphi\cdot d\mathbf{r}_L \simeq 0$$

$$\frac{q}{c}\int_0^{T_L}\frac{\partial \mathbf{A}}{\partial t}\cdot\mathbf{v}_L\,dt = \frac{q}{c}\oint\frac{\partial \mathbf{A}}{\partial t}\cdot d\mathbf{r}_L = -\frac{|q|}{c}\int_\sigma\frac{\partial B_n}{\partial t}\,d\sigma$$

$$\simeq -\pi r_L^2\frac{|q|}{c}\frac{\partial B}{\partial t} \tag{27.15}$$

The first of these equations gives only an approximate equality to zero because the particle follows a nearly closed but nevertheless not a closed curve. In the second equation we have used Stokes' theorem applied to the Larmor circle and the area it enclosed. The following arguments are used to choose the sign: rotation in a given magnetic field in a Larmor circle proceeds in opposite directions for particles with opposite signs of charge (counterclockwise for negatively charged particles, and clockwise for positively charged ones if field **B** is in the positive direction). Besides, we have taken into account that magnetic field is almost orthogonal to the plane of rotation and that it changes only slightly over distances of the order of r_L.

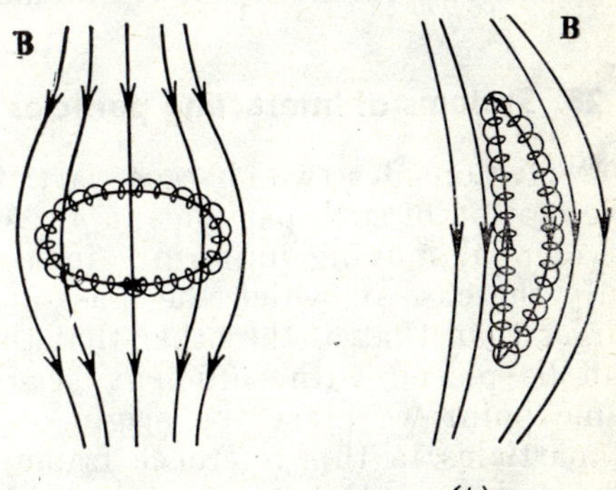

Fig. 29

Substitute (27.15) into (27.14) and take into account that according to (26.7), $M = \frac{|q|}{2c} v_L r_L$. Finally, we obtain

$$\frac{m_0}{2} \frac{d(v_L^2)}{dt} = M\left(\frac{\partial}{\partial t} + (\mathbf{v}_c \cdot \text{grad})\right) B$$

Here derivative with respect to time, d/dt, is "substantial", since it is calculated, as in the preceding formulas, in the co-moving reference frame. In the general case $d/dt = \partial/\partial t + (\mathbf{v} \cdot \text{grad})$, where $\mathbf{v}$ is the total velocity. In equation (27.8), however, which is averaged over periodic component $\mathbf{v}_L$, we must in fact assume that $d/dt = \partial/\partial t + (\mathbf{v}_c \cdot \text{grad})$. We can also write $M = m_0 v_L^2/2B$ (this has also been mentioned in (26.7)), so that in the approximation considered the preceding equation is transformed to

$$\frac{d}{dt}(MB) = M\frac{dB}{dt}, \text{ that is } \frac{dM}{dt} = 0$$

In certain configurations of a magnetic field the equivalent magnetic moment is not the only adiabatic invariant. Thus, the transverse adiabatic invariant of the type $\mathbf{p}_\perp^2/B$ may exist ($\mathbf{p}_\perp$ is the momentum of motion orthogonal to the field), as well as the longitudinal adiabatic invariant determined by the momentum of longitudinal drift. The first case corresponds to a nearly closed trajectory of the transverse drift motion (Fig. 29, *a*), and the second corresponds to a periodic drift between lines of force of the field and resembles consecutive reflections from the lines as from mirrors (Fig. 29, *b*). These modes will not be investigated here. As we have shown above, at each point of trajectories of this type a rapid Larmor rotation proceeds, and the equivalent magnetic moment is conserved.

§ 28. Systems of interacting particles

28.1. Consider two charged particles. It will be natural to analyze the motion of such particles in a reference frame fixed to their center of inertia, moving uniformly in a straight line. We shall consider only the case in which the mass of one of the particles is so much larger than that of the other that the center of inertia of the system can be placed with sufficient accuracy in the larger mass; at the same point we place the origin of spatial coordinates. Interaction of particles in this reference frame is described by Coulomb's law. The problem is then to determine the trajectory of the lighter particle in a given Coulomb field (Kepler's problem). We shall consider its relativistic solution.

The force of interaction is $\mathbf{f} = \frac{q_1 q_2}{4\pi r^3}\mathbf{r}$. The equations of motion are

$$\frac{d}{dt}(\gamma\mathbf{v}) = \frac{k}{r^3}\mathbf{r}, \quad \frac{d}{dt}(\gamma c^2) = \frac{k}{r^3}\mathbf{v}\cdot\mathbf{r} \tag{28.1}$$

where, as usual, $\gamma \equiv (1 - v^2/c^2)^{-1/2}$, and $k \equiv q_1 q_2/4\pi m_0$. In the case of attractive force $k < 0$, and in the case of repulsion $k > 0$. The right-hand side of the second of equations (28.1) equals $-\frac{d}{dt}\left(\frac{k}{r}\right)$, so that, as could be expected, this equation expresses energy conservation in the form

$$\gamma c^2 + k/r = W \tag{28.2}$$

where W is a constant. Form now the vector product of both sides of the first equation by $\mathbf{r}$. As

$$\mathbf{r}\times\frac{d}{dt}(\gamma\mathbf{v}) = \frac{d}{dt}[\gamma(\mathbf{r}\times\mathbf{v})]$$

we obtain

$$\gamma\mathbf{r}\times\mathbf{v} = \mathbf{A} \tag{28.3}$$

where $\mathbf{A}$ is a constant. This relation expresses conservation of angular momentum. It shows that motion is restricted to one plane. We introduce cylindrical coordinates with axis z along vector $\mathbf{r}\times\mathbf{v}$ normal to the plane of motion; r and φ denote polar coordinates in this plane. In these coordinates relation (28.3) takes the form

$$\gamma r^2 \frac{d\varphi}{dt} = A \tag{28.4}$$

Now, $v^2 = \dot{r}^2 + r^2\dot{\varphi}^2$ and, on the other hand, it is readily found that $\gamma^2 v^2 = c^2(\gamma^2 - 1)$, whence

$$\gamma^2\dot{r}^2 = c^2(\gamma^2-1) - \gamma^2 r^2\dot{\varphi}^2 = c^2(\gamma^2-1) - A^2/r^2$$

Here we have used formula (28.4). Quantity γ^2 which remains in the right-hand side will be expressed in terms of r by equations (28.2). This yields

$$\gamma\dot{r} = \left[\frac{1}{c^2}\left(W-\frac{k}{r}\right)^2 - \frac{A^2}{r^2} - c^2\right]^{1/2} \tag{28.5}$$

Equation (28.4) yields in its turn that $\gamma\dot{\varphi} = A/r^2$; since $d\varphi/dr = \dot{\varphi}/\dot{r}$, we obtain

$$\varphi = A\int\frac{r^{-2}\,dr}{\left[\frac{1}{c^2}\left(W-\frac{k}{r}\right)^2 - \frac{A^2}{r^2} - c^2\right]^{1/2}} + \varphi_0$$

$$= -A\int\frac{d\,(r^{-1})}{\left[\left(\frac{W}{c}-\frac{A_0}{r}\right)^2 - \frac{W_0^2}{c^2} - \frac{A^2}{r^2}\right]^{1/2}} + \varphi_0 \tag{28.6}$$

Here new notations are $W_0 \equiv c^2$, $A_0 \equiv k/c$. Energy W and angular momentum A introduced by (28.2) and (28.3) are referred to a unit mass of the moving particle. Let us take up again (28.5). Quantity $\gamma\dot{r}$, of course, assumes only real values. First we tend r in the right-hand side to infinity. When $|W| < c^2$, the radicand is negative, so that the particle cannot leave to infinity if $|W| < W_0$. Consider now the case $r \to 0$. In the radicand we retain only the terms proportional to r^{-2}. In this case the root becomes imaginary when $|A| > |k/c| = |A_0|$. Note that in the case of repulsion $A_0 > 0$, and in the case of attraction $A_0 < 0$. The above inequality must be interpreted as the condition which precludes the particle from falling onto the center.

Integration in formula (28.6) can be readily carried out. By introducing the notations

$$(A^2 - A_0^2)^{1/2} \equiv \alpha, \quad \frac{\alpha}{r} + \frac{WA_0}{c\alpha} \equiv \xi, \quad b \equiv \left[\left(\frac{WA}{c\alpha}\right)^2 - \frac{W_0^2}{c^2}\right]^{1/2} \tag{28.7}$$

where b is a real quantity if the motion is finite, and α is real if the falling onto the center is forbidden, (28.6) is transformed to

$$\varphi - \varphi_0 = -\frac{A}{\alpha}\int\frac{d\xi}{(b^2-\xi^2)^{1/2}} = \frac{A}{\alpha}\arccos\frac{\xi}{b} = \frac{A}{\alpha}\arccos\frac{1}{b}\left(\frac{\alpha}{r} + \frac{WA_0}{c\alpha}\right)$$

From this we derive

$$r = p\,[1 + \varepsilon\cos\zeta\,(\varphi - \varphi_0)]^{-1} \tag{28.8}$$

Here

$$p \equiv -\frac{c\alpha^2}{A_0 W} = \frac{A_0^2 - A^2}{A_0 W}\, c, \quad \varepsilon \equiv -\left[1 + \frac{(A^2 - A_0^2)}{A_0^2}\,\frac{(W^2 - W_0^2)}{W^2}\right]^{1/2}$$

$$\zeta \equiv \frac{\alpha}{A} = \left(1 - \frac{A_0^2}{A^2}\right)^{1/2} \tag{28.9}$$

In investigating equation of the trajectory, (28.8), it is necessary to take into account that, as follows from (28.2), attraction corresponds to $W < \gamma W_0$; a case of particular interest is that of $W < W_0$, that is the case of finite motions. Contrary to this, repulsive forces require that $W > W_0$. If we had $\zeta = 1$ (this would be possible only in the nonrelativistic case which is obtained formally for $c \to \infty$ and with $A_0 \to 0$), equation (28.8) would describe an ellipse (with one of the foci at the origin of coordinates) for $|\varepsilon| < 1$, a hyperbola for $|\varepsilon| > 1$, and a parabola for $|\varepsilon| = 1$.[7] Actually the motion is more complicated.

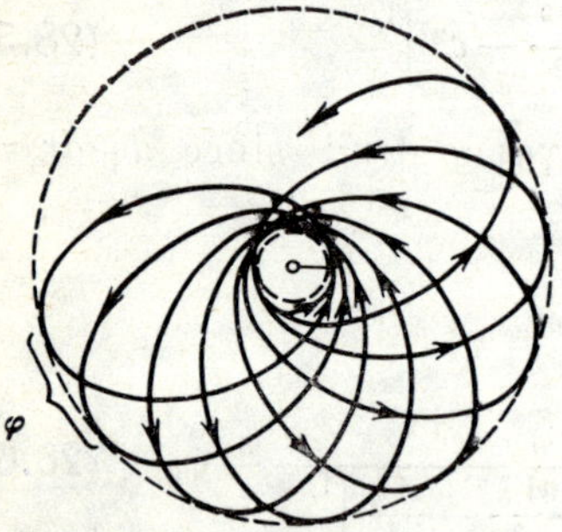

Fig. 30

Let us consider one important particular case. Let it be that of attraction, with $W < W_0$ and $A^2 > A_0^2$. Parameter ε in (28.9) is real and $|\varepsilon| < 1$ if inequality

$$A^2 (W_0^2 - W)^2 < A_0^2 W_0^2 = k^2 c^2$$

holds. Besides, condition $p > 0$ must hold; since $A_0 < 0$, this requires $W > 0$. It follows from (28.2) that in this case $\frac{|k|}{r} < \gamma c^2$. We then see from (28.8) that if all the above conditions are satisfied, the motion is finite, so that r returns to its initial value when angle changes not by $\Delta\varphi = 2\pi$ but by $\Delta\varphi = 2\pi/\zeta = 2\pi\,(1 - A_0^2/A^2)^{-1/2}$. This trajectory is an ellipse which precesses around the origin (resorting to astronomical terminology, one often refers to it as the shift of perihelion). This case is illustrated in Fig. 30. We suggest that the reader analyze the remaining cases which could be encountered in investigating the relativistic Kepler problem (and in particular, the case of repulsive forces).

28.2. If the masses of interacting particles have comparable values, the relativistic problem of determining their motion becomes tremendously complicated. In this situation one has to consider

---

[7] An expression for ε can also be written with the plus sign since the initial angle φ_0 is not fixed and can always be changed to $\varphi_0 + \pi$.

a system of equations of the type

$$m_{01}\frac{du_1^r}{d\tau}=\frac{q_1}{c}u_{1s}F^{sr}_{(2)},\quad m_{02}\frac{du_2^r}{d\tau}=\frac{q_2}{c}u_{2s}F^{sr}_{(1)}$$

In these equations $F^{sr}_{(2)}$ is determined by the field of the second particle acting on the first particle (retardation is taken into account), and $F^{sr}_{(1)}$ is determined by the field of the first particle acting on the second particle, again taking into account retardation. In mathematics these equations are classified as differential-difference equations. Not much progress has been achieved in their analysis in the case which interests us here.

Useful approximate results concerning the properties of the system consisting of N particles can be obtained under simplifying assumptions which are mostly reduced to considering motion of the particles slow compared to the velocity of light. In order to derive these results, it will be convenient to consider first retarding potentials produced by a continuous distribution of charges and currents given by formulas (13.11) (we shall assume $\varepsilon = 1$ and $\mu = 1$). We begin with the formula for scalar potential, namely

$$\varphi(\mathbf{r}, t)=\frac{1}{4\pi}\int\frac{\rho(\mathbf{r}', t-R/c)}{R}dV',\quad R\equiv|\mathbf{r}-\mathbf{r}'| \tag{28.10}$$

Assume that $\rho(\mathbf{r}', t')$ corresponds to a "nearly pointlike" particle; the velocity of this particle and higher derivatives of its coordinates with respect to time t' are assumed to be sufficiently slowly varying functions of time. Then function ρ can be expanded in powers of c^{-1}; retaining only several first terms, we obtain

$$\rho\left(\mathbf{r}', t-\frac{R}{c}\right)=\rho(\mathbf{r}', t)=\frac{R}{c}\frac{\partial}{\partial t}\rho(\mathbf{r}', t)+\frac{1}{2c^2}R^2\frac{\partial^2}{\partial t^2}\rho(\mathbf{r}', t)+\dots, \tag{28.11}$$

Substitution of (28.11) into (28.10) gives, if we take into account that $\int\rho(\mathbf{r}', t)\,dV' = q$ is independent of time (q is the total charge of the particle)

$$4\pi\varphi(\mathbf{r}, t)=\int\frac{\rho(\mathbf{r}', t)\,dV'}{R}+\frac{1}{2c^2}\frac{d^2}{dt^2}\int\rho(\mathbf{r}', t)\,R\,dV'+\dots \tag{28.12}$$

Taking also into account that $\rho(\mathbf{r}', t) = q_a\delta(\mathbf{r}'-\mathbf{r}_a(t))$, we rewrite (28.12) in the form

$$4\pi\varphi(\mathbf{r}, t)=\frac{q_a}{|\mathbf{r}-\mathbf{r}_a(t)|}+\frac{q_a}{2c^2}\frac{\partial^2}{\partial t^2}|\mathbf{r}-\mathbf{r}_a(t)|$$
$$=\frac{q_a}{|\mathbf{r}-\mathbf{r}_a(t)|}-\frac{q_a}{2c^2}\frac{\partial}{\partial t}\left(\frac{\mathbf{v}_a\cdot(\mathbf{r}-\mathbf{r}_a)}{|\mathbf{r}-\mathbf{r}_a|}\right) \tag{28.13}$$

where $\mathbf{v}_a\equiv d\mathbf{r}_a/dt$. Likewise, assuming $\mathbf{j} = \rho\mathbf{v}$, we can obtain the first term of the expansion of vector potential; this potential coincides with the derived earlier potential of static magnetic field

$$4\pi \mathbf{A}(\mathbf{r}, t) = \frac{q_a \mathbf{v}_a}{c\,|\mathbf{r} - \mathbf{r}_a|} \tag{28.14}$$

We neglect further terms in the expansion of vector potential since the terms of the equations of motion, containing vector potential, have an additional factor c^{-1}.

Now we shall use the gauge transformation (2.4) and (2.3) and set

$$4\pi\psi = -\frac{q_a}{2c}\frac{\mathbf{v}_a\cdot(\mathbf{r} - \mathbf{r}_a)}{|\mathbf{r} - \mathbf{r}_a|} \tag{28.15}$$

New scalar potential is reduced to the static Coulomb potential, and vector potential gains an additional term which is readily found from formula (B.18):

$$\operatorname{grad}\frac{\mathbf{v}\cdot\mathbf{r}}{r} = (\mathbf{v}\cdot\operatorname{grad})\frac{\mathbf{r}}{r}$$

Thus, if the new potentials are written with the same notations as the initial ones, we arrive at

$$4\pi\varphi = \frac{q_a}{|\mathbf{r} - \mathbf{r}_a|}, \quad 4\pi\mathbf{A} = \frac{q_a}{2c}\left[\frac{\mathbf{v}_a}{|\mathbf{r} - \mathbf{r}_a|} + \frac{(\mathbf{r} - \mathbf{r}_a)\,\mathbf{v}_a\cdot(\mathbf{r} - \mathbf{r}_a)}{|\mathbf{r} - \mathbf{r}_a|^3}\right] \tag{28.16}$$

A comparison with formulas (2.8) and (2.9) shows that in the course of calculations we changed to Coulomb gauge potentials. It is readily verified by means of (28.16) that equation $\operatorname{div}\mathbf{A} = 0$ characterizing this gauge is satisfied.

Consider the Lagrangian (8.19) describing the behaviour of a particle with charge q and radius vector $\mathbf{r}$ in a given field; we assume that this field is described by potentials (28.16) and is generated by another particle Obviously, terms $-q\varphi + \frac{q}{c}\mathbf{A}\cdot\mathbf{v}$ in the Lagrangian are symmetrical with respect to permutation of charges q and q_a and of velocities $\mathbf{v}$ and $\mathbf{v}_a$ of the particles. These terms can be interpreted as expressing the instantaneous long-range interaction between particles although, as we have seen above, their structure involves an approximate analysis of retardation. If we pass now to a consideration of a system consisting of N particles, the corresponding Lagrangian can be given by the formula

$$\mathcal{L} = -\sum_a m_a c^2\left(1 - \frac{v_a^2}{c^2}\right)^{1/2} - \frac{1}{8\pi}\sum_{\substack{a,\,b\\ a\neq b}}\frac{q_a q_b}{|\mathbf{r}_a - \mathbf{r}_b|}$$

$$+\frac{1}{16c^2\pi}\sum_{\substack{a,\,b\\ a\neq b}} q_a q_b\left[\frac{\mathbf{v}_a\cdot\mathbf{v}_b}{|\mathbf{r}_a - \mathbf{r}_b|} + \frac{\mathbf{v}_a\cdot(\mathbf{r}_a - \mathbf{r}_b)\,\mathbf{v}_b\cdot(\mathbf{r}_a - \mathbf{r}_b)}{|\mathbf{r}_a - \mathbf{r}_b|^3}\right] \tag{28.17}$$

In the approximation under discussion

$$-\sum_a m_a c^2 \left(1-\frac{v_a^2}{c^2}\right)^{1/2} \simeq -\sum_a m_a c^2 + \sum_a \frac{1}{2} m_a v_a^2 \qquad (28.18)$$

Formula (28.17) is known as the *Darwin Lagrangian*. The quantum-mechanical analogue of the discussed interaction is called the *Darwin-Breit interaction*.

Let us use equations (28.17) and (28.18) and find the expression for the Hamiltonian of a system of particles.[8] To contract the notation, we introduce the symbols: $\mathbf{r}_{ab} \equiv \mathbf{r}_a - \mathbf{r}_b$ and $\mathbf{n}_{ab} \equiv \mathbf{r}_{ab}/r_{ab}$. We can write

$$\mathbf{p}_a = \frac{\partial \mathscr{L}}{\partial \mathbf{v}_a} = m_a \mathbf{v}_a + \frac{1}{8\pi c^2} \sum_b \frac{q_a q_b}{r_{ab}} [\mathbf{v}_b + \mathbf{n}_{ab} (\mathbf{v}_b \cdot \mathbf{n}_{ab})]$$

Velocities can be written in terms of momenta by means of successive approximations. If the second term is ignored, $\mathbf{v}_a = \mathbf{p}_a/m_a$. Substitution of this approximation for velocity into the second term yields

$$m_a \mathbf{v}_a = \mathbf{p}_a - \frac{1}{8\pi c^2} \sum_b \frac{q_a q_b}{m_b r_{ab}} [\mathbf{p}_b + \mathbf{n}_{ab} (\mathbf{p}_b \cdot \mathbf{n}_{ab})]$$

A case of practical importance is that of all masses m_a equal to one and the same value m_0 (for instance, to the electron mass). By defining the Hamiltonian in the usual manner, it will be easy to obtain from the preceding formulas, to within the terms of the second order in (v/c):

$$\mathscr{H} = \sum_a (\mathbf{p}_a \cdot \mathbf{v}_a) - \mathscr{L}$$

$$\simeq \sum_a \left(m_0 c^2 + \frac{p_a^2}{2m_0}\right) + \frac{1}{8\pi} \sum_{a \neq b} \frac{q_a q_b}{r_{ab}} \left[1 - \frac{\mathbf{p}_a \mathbf{p}_b - (\mathbf{n} \cdot \mathbf{p}_a)(\mathbf{b} \cdot \mathbf{p}_b)}{2m_0^2 c^2}\right] \qquad (28.19)$$

where $\mathbf{n} \equiv \mathbf{n}_{ab}$. It is this formula that is of interest for the quantum theory of atom. If a system of interacting particles is placed in a given external field, φ in the initial expression for the Lagrangian must be replaced by $\varphi + \varphi'$, and $\mathbf{A}$ by $\mathbf{A} + \mathbf{A}'$ if φ' and $\mathbf{A}'$ are the potentials of this external field.

28.3. If there exists a reference frame in which all velocities of the charges are exactly zero and if we take the difference $U = \mathscr{H} - \sum m_0 c^2$ as a nonrelativistic definition of energy, we derive

[8] See also in: L. D. Landau and E. M. Lifshitz, *The Classical Theory of Fields* (Course of Theoretical Physics, vol. 2), Pergamon Press, Oxford, 1975.

from formula (28.19) the potential energy of this system of charges:

$$U = \frac{1}{8\pi} \sum_{\substack{a,\, b \\ a \neq b}} \frac{q_a q_b}{r_{ab}} \tag{28.20}$$

Obviously, this expression for energy can also be obtained directly from the equations of electrostatics discussed in § 11. We shall give this derivation because of the importance of formula (28.20). The general expression for the energy of electrostatic field in vacuo can be transformed as follows:

$$\int \mathbf{E}^2 \, dV = -\int (\mathbf{E} \cdot \operatorname{grad} \varphi) \, dV = -\int \operatorname{div} (\varphi \mathbf{E}) \, dV + \int \varphi \operatorname{div} \mathbf{E} \, dV$$

$$= -\oint_{\sigma} \varphi E_n \, d\sigma + \int \rho\varphi \, dV$$

Here we have used successively the relation $\mathbf{E} = -\operatorname{grad} \varphi$ valid in electrostatics, formula (B.14_2) of vector analysis, the Gauss theorem, and finally equation $\operatorname{div} \mathbf{E} = \rho$ (we still use the Gaussian system of units and therefore do not distinguish in vacuo between **D** and **E**). Assume now that all the charges are at finite distances from one another, namely within a closed surface σ chosen here. We can assume that when this surface recedes to infinity, the first of the obtained integrals tends to zero since the integrand is proportional to r^{-3}. If the charges are pointlike, their distribution is given by

$$\rho(\mathbf{r}) = \sum_a q_a \delta(\mathbf{r} - \mathbf{r}_a)$$

Therefore, the second integral takes the form $\sum\limits_a q_a \varphi(\mathbf{r}_a)$. Using the expression $\varphi(\mathbf{r}_a) = \frac{1}{4\pi} \sum\limits_{(a \neq b)} \frac{q_b}{r_{ab}}$ for the Coulomb potential produced at point $\mathbf{r}_a$ by all other charges, and taking account of factor 1/2 found in the formula for the energy of the field, we arrive at (28.20).[9]

Now let velocities and accelerations of charges be so small that despite their motion, electrostatic potential energy can be considered as a sufficiently exact expression of their instantaneous long-range interaction. The total energy of a system of charged particles is

[9] Actually $\varphi(\mathbf{r}_a)$ includes term $b = a$ which tends to infinity for pointlike charges, so that $\int E^2 \, dV$ comprises an infinite proper energy of the charges. Formula (28.20) gives finite energy only after "renormalization" in which the infinite proper energy is simply dropped and only the "mutual" energy of the charges is retained.

then $W = \sum_a \frac{m_a v_a^2}{2} + U$ and the motion of each particle is given by a classical mechanics equation

$$m_a \frac{d\mathbf{v}_a}{dt} = -\frac{\partial U}{\partial \mathbf{r}_a} \tag{28.21}$$

Multiplication of both sides of this equation scalarly by $\mathbf{r}_a$ enables us to rewrite it in the form

$$m_a \frac{d}{dt}(\mathbf{r}_a \cdot \mathbf{v}_a) - m_a \mathbf{v}_a^2 = -\mathbf{r}_a \frac{\partial U}{\partial \mathbf{r}_a}$$

Summing it over subscript a and substituting expression (28.20) for U into the right-hand side, we can write the result as a sum (over a and b for $a \neq b$) of the terms of the type

$$q_a q_b \left(\mathbf{r}_a \frac{\partial}{\partial \mathbf{r}_a} \frac{1}{\mathbf{r}_{ab}} + \mathbf{r}_b \frac{\partial}{\partial \mathbf{r}_b} \frac{1}{\mathbf{r}_{ab}} \right) = -q_a q_b \frac{1}{\mathbf{r}_{ab}^3} [\mathbf{r}_a \cdot (\mathbf{r}_a - \mathbf{r}_b) + \mathbf{r}_b \cdot (\mathbf{r}_b - \mathbf{r}_a)] = -\frac{q_a q_b}{\mathbf{r}_{ab}}$$

This gives $\sum_a \mathbf{r}_a \frac{\partial U}{\partial \mathbf{r}_a} = -U$. This result also follows from Euler's theorem on homogeneous functions. By denoting $Q \equiv \sum_a m_a (\mathbf{r}_a \cdot \mathbf{v}_a) = \frac{1}{2} \frac{d}{dt} \sum_a m_a \mathbf{r}_a^2$, we obtain

$$2K = -U + dQ/dt$$

where K is the total kinetic energy. If the particles of which the system is composed move almost periodically, then averaging (again denoted by angle brackets) over a sufficiently long time interval yields $\langle Q \rangle = 0$ and $\langle dQ/dt \rangle = 0$. The obtained result is known as the *virial theorem*:

$$2 \langle K \rangle = -\langle U \rangle \tag{28.22}$$

The derivation of this formula involved only the corollaries of classical mechanics. Formula (28.22) holds only when $\langle U \rangle < 0$, that is only if attractive forces dominate over repulsive forces. Further, we obtain for the total energy of a closed system

$$W = \langle K \rangle + \langle U \rangle = -\langle K \rangle$$

As the dimensions of the system in space diminish, its potential energy decreases but its kinetic energy grows (by an amount half that of the reduction in U).

28.4. *The Larmor theorem.* Let us assume, as in Subsection 28.3, that the effects of retardation and of magnetic fields produced by the system of charges are negligible, and formulate the problem as to the external electric field. We assume that it possesses cylindrical symmetry. Axis z of the coordinates will be directed along the symmetry axis. Charges making up the system are assumed equal to one another. With these assumptions the potential of external electric field applied to a charge a can be given in the form $\varphi(z_a, r_a)$, where $r_a \equiv \sqrt{x_a^2 + y_a^2}$. The equation of particle's motion is written in the form

$$m_a \ddot{\mathbf{r}}_a = -q\,\mathrm{grad}_a\,\Phi \tag{28.23}$$

where

$$\Phi \equiv \sum_b \varphi(z_b, r_b) + \frac{1}{q} U \tag{28.24}$$

and grad_a denotes differentiation with respect to coordinates $\mathbf{r}_a$.

The Larmor theorem which we are to derive indicates how the motion of such a system of charged particles is transformed by switching on an external uniform static magnetic field along axis z.

Let functions $x_a(t)$, $y_a(t)$, $z_a(t)$ describe the solution of the system of equations (28.23). Denote by $\mathbf{r}'_a(t)$ the coordinates describing the motion of the particle after magnetic field is switched on. Then electric interactions in the presence of a magnetic field will be determined by function Φ' which can be obtained from formula (28.24) by substituting $\mathbf{r}'_a$ for $\mathbf{r}_a$, that is $\Phi'(\mathbf{r}'_1, \ldots, \mathbf{r}'_N) = \Phi(\mathbf{r}'_1, \ldots, \mathbf{r}'_N)$ (here N denotes the number of particles). Equations of motion in the magnetic field take the form

$$m_a \ddot{\mathbf{r}}'_a = -q\,\mathrm{grad}'_a \Phi' + \frac{q}{c} \mathbf{v}'_a \times \mathbf{B} \tag{28.25}$$

Here grad'_a denotes the differentiation with respect to $\mathbf{r}'_a$, and $\mathbf{v}'_a \equiv d\mathbf{r}'_a/dt$. In our conditions the projection of (28.25) on axis z has the same form as the projection of equation (28.23) on the same axis. The point of main interest is therefore to compare the motion in plane (x, y) represented by equations (28.23) and (28.25). We assume the masses of all particles to be identical, that is $m_a = m_0$. By analogy with the procedure used in Subsection 26.1, we introduce complex variables: $\zeta_a(t) = x_a(t) + iy_a(t)$ and $\zeta'_a(t) = x'_a(t) + iy'_a(t)$. Then x- and y-components of equations (28.25) can be combined into a single equation

$$\ddot{\zeta}'_a = -\frac{q}{m_0}\left(\frac{\partial \Phi'}{\partial x'_a} + i\,\frac{\partial \Phi'}{\partial y'_a}\right) + 2i\omega'_{\mathrm{L}} \dot{\zeta}'_a \tag{28.26}$$

Here

$$\omega'_{\mathrm{L}} = -qB/(2m_0c) \tag{28.27}$$

This quantity is identical to the frequency introduced in Subsection 26.3 and denoted by the same symbol. It is important that this frequency is equal to one half of ω_{L} which corresponds to the motion of a charge interacting only with a static magnetic field.

We shall demonstrate that if the external magnetic field is sufficiently weak, the approximate solution of equations (28.26) takes the form

$$\zeta'_a = \zeta_a e^{i\omega'_{\mathrm{L}}t} \tag{28.28}$$

that is, in the real form,

$$x'_a = x_a \cos \omega'_{\mathrm{L}} t - y_a \sin \omega'_{\mathrm{L}} t, \quad y'_a = x_a \sin \omega'_{\mathrm{L}} t + y_a \cos \omega'_{\mathrm{L}} t$$

The analyzed motion is therefore a rotation in plane (x, y) with frequency ω'_{L}. This statement constitutes the Larmor theorem.

To prove the theorem, we make use of the fact that according to the assumption made, function Φ depends only on variables $\mathbf{r}_a$ and $(\mathbf{r}_a - \mathbf{r}_b)^2$. However, the indicated rotation does not change these quantities. Hence, if new radius vectors are found by (28.28), we find $\Phi'(\mathbf{r}'_1, \ldots, \mathbf{r}'_N) = \Phi(\mathbf{r}_1, \ldots, \mathbf{r}_N)$. On the other hand, transformation formulas for coordinates x_a and y_a yield

$$\left(\frac{\partial}{\partial x'} + i\frac{\partial}{\partial y'}\right)\Phi' = e^{i\omega'_{\mathrm{L}}t}\left(\frac{\partial}{\partial x} + i\frac{\partial}{\partial y}\right)\Phi$$

Substitute this formula and the expressions obtained from (28.28) by differentiating with respect to t into equations (28.26). It takes the form

$$\ddot{\zeta}_a = -\frac{q}{m_0}\left(\frac{\partial \Phi}{\partial x_a} + i\frac{\partial \Phi}{\partial y_a}\right) - \omega'^2_{\mathrm{L}}\zeta_a$$

Hence, if, as we assume, $\zeta_a(t)$ is a solution of (28.23), and the term proportional to ω'^2_{L} is negligibly small, then in the approximation under discussion functions (28.28) indeed represent a solution of equations (28.26).

In order to evaluate the chosen approximation we shall assume that the motion caused by electric forces only is described adequately as a nearly periodic process with a characteristic frequency ω_0. Then solution ζ_a in the absence of external magnetic field will be approximately proportional to $\exp(i\omega_0 t)$, and the assumption on the smallness of ω'_{L} can be written as $|\omega'_{\mathrm{L}}| = \left|\frac{qB}{2m_0 c}\right| \ll \omega_0$. If, for example, the characteristic frequency ω_0 is assumed to be of the order of optical vibrations frequency, and the value of ratio q/m_0 is chosen equal to that for the electron, we find that the Larmor theorem is applicable at very large values of field B (up to hundreds of millions of gausses). As for the process of switching-on the magnetic field if it was absent before, this aspect needs a special analysis since a variable magnetic field generates a vorticity electric field $\mathbf{E}$ for which curl $\mathbf{E} = -\frac{1}{c}\frac{\partial \mathbf{B}}{\partial t}$. Consequently, the Larmor theorem must naturally be regarded as a statement referring to a comparison of two different systems of charges one of which is permanently in a magnetic field and another is permanently free of it.

By using again the calculations similar to those carried out in Subsection 26.1, we easily find from (28.28) that the Larmor theorem can also be formulated as follows: if velocities of charged particles prior to switching-on of magnetic field were $\mathbf{v}_a^0$, then after switching the field on, the velocities become $\mathbf{v}_a = \mathbf{v}_a^0 + (\boldsymbol{\omega}'_{\mathrm{L}} \times \mathbf{r}_a)$, where $\boldsymbol{\omega}'_{\mathrm{L}} = \omega'_{\mathrm{L}}\, \mathbf{B}_1 = -\frac{q}{2m_0c}\mathbf{B}$. The arguments already used in Subsection 26.1 also show that the system of charges possesses an equivalent magnetic moment

$$\boldsymbol{\mu} = \frac{1}{2c}\sum_a q_a(\mathbf{r}_a \times \mathbf{v}_a) \tag{28.29}$$

This magnetic moment includes a contribution induced by external field; this contribution is given by the formula

$$\boldsymbol{\mu}(\mathbf{B}) = \frac{1}{2c}\sum_a q_a(\mathbf{r}_a \times (\boldsymbol{\omega}'_{\mathrm{L}} \times \mathbf{r}_a))$$

The observed value of this magnetic moment can be obtained by resorting to averaging over time. If we assume that all $q_a = q$, then the mean value of magnetic moment in the direction of magnetic field, that is along axis z, is

$$\langle \mu_z(\mathbf{B})\rangle = \frac{q\omega'_{\mathrm{L}}}{2c}\sum_a \langle r_a^2 - z_a^2\rangle = -\frac{q^2B}{4m_0c^2}\sum_a \langle x_a^2 + y_a^2\rangle$$

If the distribution of charges is spherically symmetric and the orbits of the charges can be assumed quasistationary, then $\langle x_a^2\rangle = \langle y_a^2\rangle = 1/3\ \langle r_a^2\rangle$, whence

$$\langle \mu_z(B)\rangle = -\frac{q^2B}{6m_0c}\sum_a \langle r_a^2\rangle \tag{28.30}$$

Coefficient of B is called the *diamagnetic susceptibility* of a system of charges. Under the same assumptions the mean values of the components of the induced magnetic moment, which are orthogonal to the field, vanish.

Another important conclusion obtained in Subsection 26.1 for a free charged particle in a magnetic field also has here its analogue. Namely, assuming again that the system consists of identical particles, we can write the angular momentum of this system in the form $\mathbf{M} = m_0 \sum_a \mathbf{r}_a \times \mathbf{v}_a$. A comparison with magnetic moment (28.29) yields

$$\mu/M = q/(2m_0c) \tag{28.31}$$

Thus, we again find that the ratio of magnetic moment of the system to its mechanical angular momentum is a universal constant.

CHAPTER 7

CONTINUOUS MEDIA IN ELECTRIC FIELD

§ 29. Introduction to electrodynamics of continuous media

29.1. We have mentioned at the beginning of the book (see § 1) that two approaches to the theory of electromagnetic phenomena must be distinguished, although these approaches are connected inseparably and cannot exist independently. One of them can be called the macroscopic electrodynamics. This approach was given its final form in the second half of the 19th century and is embodied in the Maxwell equations which describe electromagnetic properties of material media and treat electromagnetic field as a material medium of special kind. It can be said that Faraday's and Maxwell's studies resulted in a definition of what must be called electromagnetic phenomena from the macroscopic viewpoint. This definition agrees with all requirements of macroscopic physics, in particular with all experimental results, and is independent of the detailed theory of atomic properties of the matter. The other approach consists in explaining macroscopic properties of material media (and, in particular, of electromagnetic field itself) in terms of the concepts of their atomic structure and electromagnetic properties of molecules, atoms, electrons, and nuclei. This microscopic theory of electromagnetism was first developed by Lorentz at the turn of the 19th century and ever since is being developed and elaborated as our knowledge on the particles and their interactions is enriched. The modern microscopic theory must be based on quantum physics, and the explanation of macroscopic properties of materials must operate with quantum statistics.

Macroscopic quantities involved in the Maxwell equations are often obtained in the following manner. First one considers a classical system of charges and currents in vacuo and chooses it as a model of the distribution of microscopic field sources. Fields generated by these sources are also called microscopic. Then one assumes that measurements carried out by a macroscopic observer correspond to the mean values of these microscopic quantities. Averaging must be performed both over spatial coordinates and over time. The very idea of this averaging is absolutely reasonable; more than that, it is unavoidable in any attempt to substantiate the relationship between directly observed quantities and the theory of structure of the material. However, one should then keep to the modern theory and not to classical physics. In addition, the averaging operation must be based on a sufficiently consistent applications of the methods of the probability theory. The derivation of macroscopic Maxwell equations for material media, as given usually in textbooks on classical electrodynamics, does not satisfy these requirements. For this reason we restrict further presentation to the phenomenological aspect of electrodynamics of material media and do not attempt to substantiate it with the methods of classical physics. One must emphasize, to avoid misunderstandings, that it would be wrong to say that application of the abovementioned classical model in modern physics is always meaningless. However, the limits of its applicability must always be determined with respect to specific problems and not to the "electrodynamics in general". The reader will find a detailed presentation of the classical method of averaging in, for example, *Classical Electrodynamics* by J. D. Jackson, 2nd edition, Wiley, New York, 1975.

29.2. Let us itemize the fundamental relations which we have formulated in Chapter 1 and which must always be kept in mind by a reader beginning his study of nonrelativistic electrodynamics of continuous media. First of all, these are, of course, the Maxwell equations which take the form of (M.1)-(M.4) and permit the use of either the SI system or the Gaussian system of units. These equations must be supplemented by the definitions of electric polarization (1.11) and magnetization (1.17). As we have shown in § 1, with these definitions the system of the Maxwell equations can also be written in the form of (M.1′), (M.2), (M.3), and (M.4′). This form of equations will be used in this and subsequent chapters.

When interaction of charges and propagation of radiation in vacuo was analyzed in Chapters 4-6, it was most convenient to use the Gaussian system of units in which electric constant ε_0 and magnetic constant μ_0 are dimensionless and assumed equal to unity. On the opposite, properties of material media manifest themselves more clearly if the SI system is used. The difference in dimensions of field strengths and inductions in this system of units reflects a difference in physical meanings of these quantities; this aspect is not significant in the case of vacuum but remains not taken into account by the Gaussian system of units in the case of material media, that is precisely when it must constantly be kept in mind. We have shown in § 1 that in the case of the SI system, coefficient α in the Maxwell equations must be considered dimensionless and assumed equal to unity. Quantities ε_0 and μ_0 acquire dimensions, with their numerical values related via (1.25) in which c is the velocity of light in vacuo; for all practical purposes c can be set equal to 3×10^8 m/s. In SI the quantity of charge is measured in units independent of the fundamental units of mechanics. This unit of charge is called *coulomb*.

We shall not dwell on how to determine the units of basic electric and magnetic quantities and on the relations between these units in various systems. We shall only determine the numerical values of coefficients ε_0 and μ_0 in SI since these coefficients are encountered in many formulas of further sections. Clearly, if the value of one of them is known, the other is found via (1.25). We shall start with determining magnetic permeability μ_0. For this purpose we shall use equation (12.10) expressing the force of interaction between currents, and take into account that the fundamental unit in SI is a unit of current intensity, namely, *ampere* (cf. § 1). For our purposes we set in this equation $\alpha = 1$ and $\mu = \mu_0$. If mechanical quantities (force and distance) in (12.10) are measured in CGS units and we set $\mu_0 = 4\pi$ (still for $\alpha = 1$), then the current intensity unit can be expressed in terms of the units of mechanics. In this case its dimension is $M^{1/2}L^{1/2}T^{-1}$; it is known as the *electromagnetic unit of current* (CGSM unit). In SI 1 A is practically equal to 0.1 CGSM unit and, as we have mentioned above, is chosen as a fundamental unit of the system along with mechanical units. In the CGS system 1 dyne = = 1 (CGSM unit)^2. To pass from SI units to CGS units, one has to express a unit force in CGS, 1 dyne, via a unit force of SI, 1 newton, use the abovementioned relation between CGSM units and SI ampere, and take into account coefficient $\mu_0/4\pi$ in (12.10). By using also the definition $1\ A = 1\ C/s$, we obtain $10^{-5}\ N = 10^{-5}\ kg\cdot m/s^2 = \mu_0/4\pi \times 10^2\ C^2/s^2$, whence

$$\mu_0 = 4\pi\cdot 10^{-7}\ kg\cdot m/C^2 \simeq 12.57\times 10^{-7}\ kg\cdot m/C^2$$

Note that in the SI system a unit equal to 1 $kg\cdot m^2/C^2$ is called 1 henry (H). Thus, magnetic permeability is measured in H/m.

Relation (1.25) yields

$$\varepsilon_0 \simeq 8.854\times 10^{-12}\ C^2\cdot s^2/kg\cdot m^3$$

A unit 1 $C^2s^2/kg\cdot m^2$ in SI is called 1 farad (F), that is dielectric permittivity is measured in F/m.[1]

If the medium is linear, that is relations (1.27) are valid, it is often convenient to characterize the properties of the medium by dimensionless relative permittivities $\varepsilon' \equiv \varepsilon/\varepsilon_0$ and $\mu' \equiv \mu/\mu_0$. According to (1.28′) ε' and μ' are related, for constant ε and μ, by the following formula:

$$\sqrt{\varepsilon'\mu'} = c/v$$

Here v is the velocity of propagation of electromagnetic waves in the medium.

---

[1] Farad is the unit of capacitance (cf. § 30) and henry is the unit of inductance (§ 33).

§ 30. Ideal conductors in electrostatic field

30.1. We have shown in § 1 on the basis of the continuity equation for electric charge (1.31) that all material media can be classified into conductors and dielectrics. Although approximate, this classification is in many cases sufficiently well pronounced. Free transfer of electric charge within the volume of conductors can be regarded as their fundamental property. As a result (this has been mentioned in § 1), static electric field produces in a conductor a state with $\rho = 0$ and $\mathbf{E} = 0$ inside the material. If a conductor occupies a bounded volume and is surrounded with a dielectric medium, a distribution of charge with surface density λ can exist on its surface. In the present section we shall consider, as a rule, the properties of a system of N such bounded conductors. In addition, we assume that the dielectric into which these conductors are imbedded is unbounded (infinite), uniform, and isotropic, that is a relation of the type (1.27) holds, with $\varepsilon = \text{const}$. We also assume that there is no space charge in this dielectric, so that $\operatorname{div} \mathbf{D} = 0$ at each point within the dielectric. In practically important cases this condition is almost always satisfied, and so it somewhat simplifies the derivation of the results we are to obtain below.

Consider an electrostatic field in an immediate vicinity of surface σ of the conductor. We choose boundary conditions in the form of (4.11) and of the first of equations (4.14). Let medium I be a conductor, and medium II—a dielectric contiguous to its surface. It has been mentioned already that in this case $\mathbf{E}_1 = 0$ and $\mathbf{D}_1 = 0$. By making use of formula (11.2) which expresses electrostatic field $\mathbf{E}$ in terms of potential φ, we obtain from (4.11) (subscript II is dropped):

$$\left.\frac{\partial \varphi}{\partial s}\right|_{\sigma} = 0, \quad \text{i.e.} \quad \varphi(\mathbf{r})\,|_{\sigma} = \text{const} \tag{30.1}$$

In other words, the conductor surface is an equipotential surface of electrostatic field. The lines of force of this field are orthogonal to this surface at each point.

On the other hand, as follows from the boundary condition (4.14),

$$\left.\frac{\partial \varphi}{\partial n}\right|_{\sigma} = -\frac{\lambda}{\varepsilon}$$

whence

$$q = -\varepsilon \oint_{\sigma} \frac{\partial \varphi}{\partial n}\, d\sigma = \oint_{\sigma} D_n\, d\sigma \tag{30.2}$$

where q is the total charge on the surface of the conductor.[2]

---

[2] The direction of normal from conductor to dielectric is chosen as positive.

The equations derived above are fundamental for electrostatics of conductors and serve to obtain a number of necessary relations.

Denote by σ_i the surface of the ith conductor ($i = 1, 2, \ldots, N$) and by σ a closed surface drawn to enclose all the N charges (Fig. 31). Now choose two different distributions of surface charges on conductors σ_i such that in one case this distribution is represented by functions $\lambda_i(\mathbf{r})$, and in the other by $\lambda'_i(\mathbf{r})$. We assume that neither the shape nor the arrangement of the conductors change with time. In the first case the charges produce in the space around the conductors an electric field with potential $\varphi(\mathbf{r})$, and in the second—a field with a different potential $\varphi'(\mathbf{r})$. We shall make use of Green's formula (B.28) applying it to volume V external with respect to the conductors and enclosed within surface σ, and assuming $\psi = \varphi'$. Then $\Delta\varphi = \Delta\varphi' = 0$, so that the left-hand side of Green's formula vanishes, and the surface integral in the right-hand side is equal to the sum of integrals over all surfaces of the conductors and over surface σ. As for this last surface, the corresponding surface integral tends to zero as the surface recedes to infinity (the integrand is proportional to r^{-3}); after this limiting transition is performed, Green's formula yields the relation

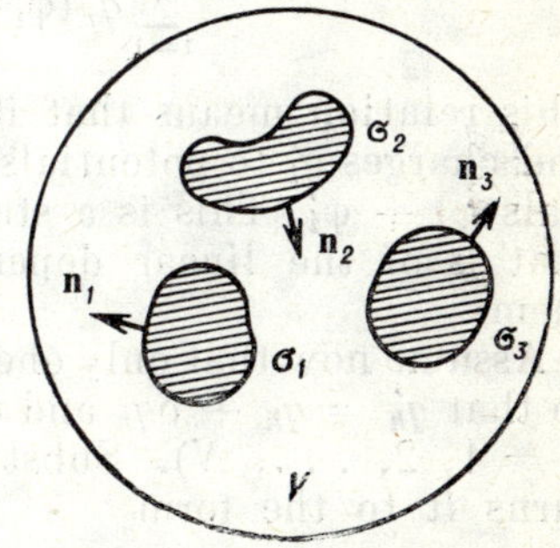

Fig. 31

$$\sum_{i=1}^{N} \oint_{\sigma_i} \varphi \frac{\partial \varphi'}{\partial n} d\sigma_i = \sum_{i=1}^{N} \oint_{\sigma_i} \varphi' \frac{\partial \varphi}{\partial n} d\sigma_i$$

But we have seen above that in any electrostatic field $\varphi|_{\sigma_i} = \varphi_i$, $\varphi'|_{ii} = \varphi'_i$, where φ'_i and φ_i are constant on the surface of the ith conductor. Factoring these constants out of the integral, we obtain in each term of the sum expressions of the type of (30.2) for the total charge q_i on the surface of the corresponding conductor. Thus,

$$\sum_{i=1}^{N} q_i \varphi'_i = \sum_{i=1}^{N} q'_i \varphi_i \tag{30.3}$$

This formula is called the *Green reciprocity relation*. It is a very important relation for resolving a number of practical problems of electrostatics; this will be demonstrated with several simplest examples.

Consider a particular case of Green's formula (30.3) in which $N = 2$, $q_1 = q'_2 = q$, and $q'_1 = q_2 = 0$. The formula yields $\varphi'_1 = \varphi_2$.

In other words, potential of noncharged conductor 1 produced by placing charge q on the surface of conductor 2 will be the same as that produced on noncharged conductor 2 by placing charge q on the surface of conductor 1.

Add a sum $\sum_{i=1}^{N} q_i\varphi_i$ to both parts of (30.3). This gives

$$\sum_{i=1}^{N} q_i(\varphi_i+\varphi_i')=\sum_{i=1}^{N}(q_i+q_i')\,\varphi_i$$

This relation means that if charges q_i correspond to potentials φ_i, and charges q_i' to potentials φ_i', then charges $q_i + q_i'$ produce potentials $\varphi_i + \varphi_i'$. This is a statement of the *principle of superposition*, that is of the linear dependence of potentials on charges producing them.

Assume now that only one of the charges is varied in formula (30.3), so that $q_k' = q_k + \delta q_k$ and $q_i' = q_i$ for $i \neq k$. Denote $\delta\varphi_i \equiv \varphi_i' - \varphi_i$ $(i = 1, 2, \ldots, N)$. Substitution of these notations into (30.3) turns it to the form

$$\varphi_k=\sum_{i=1}^{N} s_{ki}q_i \tag{30.4}$$

where we have defined $s_{ki} \equiv \delta\varphi_i/\delta q_k$. Quantities s_{ki} are called the *potential coefficients*. As follows from (30.4), the value of s_{ki} equals the potential which is produced on the kth conductor when a unit charge is placed on the ith conductor, with no charges on all other conductors. The conclusion drawn from the Green reciprocity relation for the above case of two conductors shows that $s_{ki} = s_{ik}$. It can be proved that this symmetry follows directly from (30.3) if this equation is rewritten in the form $\sum_i q_i\delta\varphi_i = \sum_i \delta q_i\varphi_i$. Coefficients s_{ik} are always positive, since positive charge placed on a conductor increases the potentials of the other conductors.

The system of equations (30.4) can be resolved with respect to charges q_k:

$$q_k=\sum_{i=1}^{N} c_{ki}\varphi_i \tag{30.5}$$

Here c_{ki} are elements of the matrix inverse with respect to the matrix composed of the potential coefficients. Coefficients c_{ki} are called the *capacitance coefficients* of the conductors involved. Note that $c_{ki} = c_{ik}$. Diagonal elements c_{kk} are called proper capacitances (or simply capacitances), and elements c_{ik} for $i \neq k$, mutual capacitances (or induction coefficients).

When two conductors are connected and thus have a common surface (by direct contact or through a conducting wire), a new value of potential, common for both conductors, sets up. A concept of considerable importance is that of grounding, that is connecting a conductor to the earth whose potential is assumed as the zero reference point. Owing to the tremendously large capacitance of the Earth, the potential of a grounded conductor is also practically zero. Using this concept, one can say that capacitance c_{kk} is equal to the ratio of the charge to the potential on the kth conductor when all other conductors are grounded.

30.2. We shall calculate now the *energy* of electrostatic field produced by surface charges on the conductors. We shall use formulas of § 3, and in particular, that for energy density (3.5). Relations ($B.14_2$) and (11.2) yield

$$2W = \int_V \mathbf{D}\cdot\mathbf{E}\,dV = -\int_V (\mathbf{D}\cdot\operatorname{grad}\varphi)\,dV$$

$$= -\int_V \operatorname{div}(\varphi\mathbf{D})\,dV + \int_V \varphi\operatorname{div}\mathbf{D}\,dV \qquad (30.6)$$

Here volume V is defined exactly as in the case of formula (30.3) (Fig. 31). We can assume that integration covers the whole region within surface σ since inside the conductors the integrand vanishes. Let us apply the Gauss theorem to the first term in (30.6):

$$\int_V \operatorname{div}(\varphi\mathbf{D})\,dV = \oint_\sigma \varphi D_n\,d\sigma - \sum_{i=1}^{N} \oint_{\sigma_i} \varphi\mathbf{D}_{n_i}\,d\sigma_i \qquad (30.7)$$

The integral over the outer surface σ vanishes as the surface recedes to infinity. In the remaining integrals $\varphi = \varphi_i = \text{const}$ on the surface of each of the conductors and, by virtue of the boundary condition, $D_{n_i} = \lambda_i$ (note that in accordance with the choice made in writing the Gauss theorem, the positive direction of the normal is the one going outward with respect to the range of integration, that is reversed compared with the choice for boundary conditions). Assuming (as an exception) that the dielectric contains space charge, so that $\operatorname{div}\mathbf{D} = \rho$, we obtain

$$W = \frac{1}{2}\sum_{i=1}^{N} q_i\varphi_i + \frac{1}{2}\int_V \rho\varphi\,dV \qquad (30.8)$$

If we again set $\rho = 0$, we find from (30.4) and (30.5) that the energy of a system of charged conductors can also be written as a bilinear form of charges and potentials:

$$W = \frac{1}{2}\sum_{i,\,k=1}^{N} s_{ik}q_iq_k = \frac{1}{2}\sum_{i,\,k=1}^{N} c_{ik}\varphi_i\varphi_k \qquad (30.9)$$

Energy in (30.8) takes the form typical for long-range interaction between charges located on conductors. In the case of a static field, that is with retardation being unimportant, the field itself is found to be eliminated from the expression describing interaction. Note also a formal analogy of the first term in (30.8) and relation (28.20) for the energy of interaction between pointlike charges. However, in this last case the energy of the charge acting on itself is infinite and has to be dropped, while in (30.8) the self-action, finite in this case, is also taken into account for each conductor.

Now let us prove the *Thompson theorem*: the surface distribution of charges on conductors is such that the energy of the field this distribution produces is minimal. This minimization holds with respect to such virtual variations of surface charge density which conserve total charges on the surface of each conductor.

Let $\mathbf{D}$ denote the induction field which is produced actually in the dielectric filling up the space around the conductors, and let $\mathbf{D}'$ denote the field resulting from variation of surface distributions. By using (30.2), we can write the conditions to which this variation is subjected in the form

$$\oint_{\sigma_i} (D'_{n_i} - D_{n_i})\, d\sigma_i = 0 \quad (i = 1, 2, \ldots, N) \tag{30.10}$$

As $\mathbf{D} = \varepsilon\mathbf{E}$ and $\mathbf{D}' = \varepsilon\mathbf{E}'$, we obtain the following identical transformation for the energy difference:

$$\begin{aligned} 2(W' - W) &= \int (\mathbf{E}'\cdot\mathbf{D}' - \mathbf{E}\cdot\mathbf{D})\, dV \\ &= \int (\mathbf{E}' - \mathbf{E})\cdot(\mathbf{D}' - \mathbf{D})\, dV - 2\int \mathbf{E}\cdot\mathbf{D}\, dV + \int \mathbf{E}\cdot\mathbf{D}'\, dV + \int \mathbf{E}'\cdot\mathbf{D}\, dV \\ &= \int (\mathbf{E}' - \mathbf{E})\cdot(\mathbf{D}' - \mathbf{D})\, dV + 2\int \mathbf{E}\cdot(\mathbf{D}' - \mathbf{D})\, dV \end{aligned} \tag{30.11}$$

The last term can be analyzed in the same way as in calculating energy. Namely, we substitute $\mathbf{E} = -\operatorname{grad}\varphi$, apply formula (B.$14_2$), and then the Gauss theorem. As before, the integral over the outer surface σ can be considered vanishing, so that

$$\begin{aligned} \int \mathbf{E}\cdot(\mathbf{D}' - \mathbf{D})\, dV &= \sum_{i=1}^{N} \oint_{\sigma_i} \varphi\, (D'_{n_i} - D_{n_i})\, d\sigma_i \\ &= \sum_{i=1}^{N} \varphi_i \oint (D'_{n_i} - D_{n_i})\, d\sigma_i = 0 \end{aligned}$$

Hence,

$$W' - W = \frac{\varepsilon}{2}\int (\mathbf{E}' - \mathbf{E})^2\, dV > 0 \tag{30.12}$$

for $\mathbf{E}' \neq \mathbf{E}$.

The solution of the electrostatics problem with fixed boundary conditions is unique (see §§ 4 and 11). The Thompson theorem must be interpreted, therefore, as expressing the extremal character of this solution with respect to the considered variations of sources.

It can readily be shown that both potential φ and its derivatives reach minimum or maximum only on the surface of the conductors. Indeed, let us assume the opposite, that is, let potential φ have a minimum at some point inside the dielectric surrounding the conductors (the arguments to follow are transformed trivially to the case of a maximum). This is realized if all three partial derivatives $\partial^2\varphi/\partial x^2$, $\partial^2\varphi/\partial y^2$, $\partial^2\varphi/\partial z^2$ are positive at this point. But this contradicts the assumption that in the dielectric $\Delta\varphi = 0$ (the Laplace equation). Hence, a pointlike charge cannot be in stable equilibrium at any point of the electrostatic field, unless it is subjected to forces of non-electric origin. This statement is a particular case of the *Earnshaw theorem*.

In analogy to the Thompson theorem, it can be proved that introduction of a noncharged conductor into the field of a given system of charged conductors diminishes the total energy of this field. Denote by σ_0 the surface of the noncharged conductor, and by V_0 the volume within this surface. As above, let V be the volume outer with respect to N charged conductors located within surface σ. If we introduce into the same surface an additional, noncharged, conductor, the volume outside them will be $V_1 = V - V_0$. The difference between the energy of the field before the introduction of the noncharged conductor, W, and energy after its introduction, W', is

$$2(W-W') = \int_V \mathbf{E}\cdot\mathbf{D}\,dV - \int_{V_1} \mathbf{E}'\cdot\mathbf{D}'\,dV$$

$$= \int_{V_0} \mathbf{E}\cdot\mathbf{D}\,dV + \int_{V_1} (\mathbf{E}\cdot\mathbf{D} - \mathbf{E}'\cdot\mathbf{D}')\,dV \quad (30.13)$$

The second term is transformed as in (30.11). It is equal again to (30.12) where integration must be carried out over volume V_1 since the total charge on each of the charged conductors remains unaltered. But the first term in (30.13) is also non-negative. Hence, $W > W'$, which was to be proved.

The energy of the field that finally sets up must, by virtue of the Thompson theorem, be minimal; hence, a noncharged conductor is as if pulled into the region occupied by charged conductors. The forces applied to this conductor appear because the field surrounding it causes its polarization: positive and negative charges are so redistributed on its surface that, remaining neutral on the whole, the conductor behaves as a system of dipoles. An excess of positive charge is accumulated at some points of its surface, and an excess of negative charge—at other points.

Assume now that a noncharged conductor is so far away from the charges that field **E** they produce in the neighbourhood of the conductor can be regarded as uniform. As a result of charge separation on the surface of this conductor, it can be replaced by an equivalent dipole with dipole momentum **p**. Potential energy of dipole, U, is given by formula (11.21). Let us assume that dipole moment **p** is a linear function of field strength **E**, namely $p_i = V\alpha_{ik}E_k$.[3] Here V is the volume of the conductor, and coefficients α_{ik} form the so-called polarizability tensor.[4] This gives

$$U = (-1/2)\, V\alpha_{ik}\, E^i E^k \tag{30.14}$$

In order to derive this formula, we have to calculate the energy expended to form the dipole moment when the conductor was moved into the field. Indeed, $dU = -\mathbf{p}\, d\mathbf{E}$ and $U = -V\alpha_{ik}\int_0^{\mathbf{E}} E^i\, dE^k$, so that by the symmetry of tensor α_{ik} we immediately obtain (30.14).

30.3. The general formula determining the forces acting on the surface of the conductor is (3.18), with the Maxwell stress tensor taken in the form of (3.9). We must take into account that field strength at the conductor surface has only a normal component. Hence, $\mathbf{E} = \pm E\mathbf{n}$, and the surface density of forces is

$$\varphi_i = T^{(e)}_{ik} n^k = \varepsilon E_i E_k n^k - \frac{\varepsilon E^2}{2}\, n^i = \frac{\varepsilon E^2}{2}\, n^i \tag{30.15}$$

Here the positive normal is directed from the conductor into the surrounding dielectric. We see that forces (30.15) are tensile, that is they tend to increase the volume of the conductor to whose surface they are applied. The total force is given by the integral $\int (1/2)\, \varepsilon E^2 \mathbf{n}\, d\sigma$. The effect of the change of a body volume (in the present case, of a conductor) in an electric field is known as *electrostriction*; we shall return to its analysis in the next section.

Calculation of forces acting on conductors is sometimes simplified by using the property of minimization of energy given by (30.9), and by equating the work done by the forces to the variation of energy under virtual displacements of the conductors; this approach is typical for mechanics. Assume that the arrangement of conductors can be determined by fixing a certain number of generalized coordinates ξ_α corresponding to the available mechanical degrees of free-

---

[3] Earlier we used Greek letters to indicate the components of 3-dimensional tensors. Here and in subsequent chapters we operate only with such tensors, and Latin letters will be more convenient.

[4] Polarizability tensor depends on the shape of the conductor. The tensor is symmetric; see, for example, books cited on p. 250.

dom. Two types of virtual displacements of the system, corresponding to the two forms of representing energy given in (30.9), are of interest: displacements with conserved charge on each conductor, and displacements with conserved potential of the conductor surface. Denote the generalized force corresponding to virtual variation of coordinate ξ_α by F_α. If the first of the indicated conditions holds, we obtain for the work done by the forces

$$\delta A = -(\delta W)_{\text{all } q_i=\text{const}} = -\sum_\alpha \left(\frac{\partial W}{\partial \xi_\alpha}\right)_{q_i} \delta\xi_\alpha$$

On the other hand, we have the usual relation $\delta A = \sum\limits_\alpha F_\alpha \delta\xi^\alpha$. Comparing these two expressions and using the first of equations (30.9), we obtain

$$F_\alpha = -\frac{1}{2}\sum_{i,\,k} \frac{\partial s_{ik}}{\partial \xi_\alpha} q_i q_k \tag{30.16}$$

If the conductors are now displaced so that their charges are unchanged, their potentials cannot be conserved. Consequently, constancy of potentials under a virtual displacement requires, in contrast to the assumptions used to obtain the preceding formula, that the charges on each of the conductors be changed. Imagine that in the process of displacement the conductors are connected to some charge "reservoir" which can supply needed charge or serve as a sink for surplus charge. Additional work $\delta A'$, which will have to be included into the net balance of energy, will be done to transfer these compensating charges between conductors, so that $\delta A = \delta A' - \delta W$. If we demand that potentials be constant, it follows from (30.5) that

$$dq_i = \sum_k \varphi_k \, dc_{ik} = \sum_k \varphi_k \sum_\alpha \frac{\partial c_{ik}}{\partial \xi_\alpha} d\xi_\alpha$$

Let the charge reservoir have potential $\varphi_0 = 0$. Then the work done to transfer charge dq_i from the reservoir to the ith conductor (in electrostatic field this work is independent of the path) is found from

$$A_i' = -(dq_i)\int \mathbf{E}\cdot d\mathbf{s} = (dq_i)\int_{\varphi_0}^{\varphi_i} d\varphi = \varphi_i\,(dq_i)$$

The minus sign is chosen because we calculate work against the forces of the field produced by the sources. Hence, the total work on charging the conductors in the course of virtual displacement is equal, under the condition of constancy of potentials, to

$$\delta A' = \sum_{i=1}^{N} \varphi_i \, dq_i = \sum_{i,\,k=1}^{N} \varphi_i \varphi_k \sum_{\alpha} \frac{\partial c_{ik}}{\partial \xi_\alpha} d\xi_\alpha$$

As we see from the second relation in (30.9),

$$\sum_{i,\,k=1}^{N} \varphi_i \varphi_k \frac{\partial c_{ik}}{\partial \xi_\alpha} = 2 \left(\frac{\partial W}{\partial \xi_\alpha} \right)_{\text{all } \varphi_i = \text{const}}$$

Therefore.

$$\delta A = \delta A' - \delta W = \sum_{\alpha} \left(\frac{\partial W}{\partial \xi_\alpha} \right)_{\varphi_i} d\xi_\alpha$$

that is

$$F_\alpha = \frac{1}{2} \sum_{i,\,k=1}^{N} \frac{\partial c_{ik}}{\partial \xi_\alpha} \varphi_i \varphi_k \tag{30.17}$$

Note that if work $\delta A'$ were ignored, the expression for F_α would have an opposite sign.

§ 31. Dielectrics in electrostatic field. Isotropic dielectrics

31.1. We have shown in the preceding section that the inner volume of ideal conductors placed in an electrostatic field contains neither charges nor fields. In contrast to this, electrostatic field pervades the whole volume of the dielectric and, therefore, can affect its structure. For example, even if material particles, of which a dielectric is composed, remain neutral, electric field may induce a dipole electric moment in each of them. This is manifested as polarization of the medium, and one has to distinguish between electric induction and electric field strength (this has already been pointed out in § 1).

A simple experiment (in fact, it was first conducted by Faraday) is very instructive and clarifies the meaning of the concepts of induction and polarization. Take two conductors with surface charges of opposite signs. Let us specify them to be metal plates parallel to each other (a plane capacitor). If the plates are in vacuo, the charges will produce in the capacitor a field $\mathbf{E}_0$ which, directly at the plate surface, is given the boundary condition (4.14): $E_{0n} = \lambda/\varepsilon_0$, in complete analogy to the general case analyzed at the beginning of § 30. If potential of one of the planes is φ_1^0, and that of the other is $\varphi_2^0 > \varphi_1^0$, then it is easy to find from $\mathbf{E}_0 = -\text{grad}\,\varphi$ that $E_0 = (\varphi_2^0 - \varphi_1^0)/d$, where d is the spacing between the plates.

Now we insert into the gap between the plates a layer of a dielectric. The experiment shows that the potentials of the plates will thereby be changed, as well as the electric field strength in the capacitor. This occurs because an additional internal electric field is produced in the dielectric. The new field strength is found from the condition $\mathbf{E} = -\text{grad}\ \varphi$, where φ is the new potential. In a particular case of a uniform and isotropic dielectric, $E = (\varphi_2 - \varphi_1)/d$. This is the force acting on a unit electric charge if this probe charge were inserted into the dielectric. Field strength $\mathbf{E}$ does not satisfy the condition $E_n = \lambda/\varepsilon_0$ but one can find vector $\mathbf{P}$ depending on the structure of the dielectric and such that $E_n = (\lambda - P_n)/\varepsilon_0$. Term P_n appears because surface "bound" charges, screening "true" charges on the capacitor plates, appear on the boundary surfaces of the dielectric layer.

Assume now that capacitor plates are connected to a source of electric charges (electric cell) which maintains on the plates the same potential difference both with the dielectric in the interplate gap and in the absence of such a layer. In this case the electric field strength inside the capacitor remains constant, $\mathbf{E} = \mathbf{E}_0$, but the quantity of charge on the capacitor plates is changed by the introduction of the dielectric; the charge density changes from λ to λ'. This change is governed by the following relation:

$$E_{0n} = \frac{1}{\varepsilon_0}(\lambda' - P_n') = \frac{\lambda}{\varepsilon_0}$$

We have taken into account all such effects when we formulated the Maxwell equations for the electric induction vector (see § 1).

Let us undertake a general investigation of equation (M.1). Substitution of $\mathbf{E} = -\text{grad}\ \varphi$ transforms (M.1) to

$$\Delta\varphi = -\frac{1}{\varepsilon_0}(\rho + \rho') \tag{31.1}$$

where $\rho' = -\text{div}\ \mathbf{P}$ is a quantity sometimes called the bulk density of bound charges. The term $(\mathbf{P}^{\text{I}} - \mathbf{P}^{\text{II}})\cdot\mathbf{n}$ in (4.13) can be called the surface density of bound charges at the interface between two dielectrics. The solution of the Poisson equation (31.1), taking into account a potential which can be produced by bound charges appearing at the interface between different media, can be written immediately by analogy to formulas (11.4) and (11.10):

$$\varphi(\mathbf{r}) = \frac{1}{4\pi\varepsilon_0}\int \frac{\rho - \text{div}'\ \mathbf{P}}{R}\,dV' + \frac{1}{4\pi\varepsilon_0}\int \frac{\lambda + (\mathbf{P}^{\text{I}} - \mathbf{P}^{\text{II}})\cdot\mathbf{n}}{R}\,d\sigma' \tag{31.2}$$

Here, as usual, $R \equiv |\mathbf{r} - \mathbf{r}'|$, and differentiation with respect to source coordinates is marked by primes. Apply now formula (B.14_2) to the first term:

$$\frac{1}{R}\,\text{div}'\ \mathbf{P} = \text{div}\left(\frac{\mathbf{P}}{R}\right) - \left(\text{grad}'\frac{1}{R}\cdot\mathbf{P}\right)$$

and then make use of the Gauss theorem. This yields

$$\int \operatorname{div}'\left(\frac{\mathbf{P}}{R}\right) dV' = \oint_{\sigma} \frac{P_n}{R}\, d\sigma' + \sum_{i=1}^{N} \oint_{\sigma_i} \frac{P_n}{R}\, d\sigma_i'$$

The first of the above integrals covers a sufficiently removed closed surface which encloses the volume of interest here. If the volume occupied by the dielectric is bounded, then this surface can always be drawn in such a manner that polarization be zero at each point of this surface. As for the second term, that is the sum over boundary surfaces making interfaces between dielectrics with different properties and over the outer boundary of the dielectric, we must take into account that each boundary enters the sum twice (Fig. 32), with oppositely directed normals (for example, in integrating "on the side of volume V_1" and "on the side of volume V_2"). Each time the outward normal is assumed to be positive. By substituting this result into

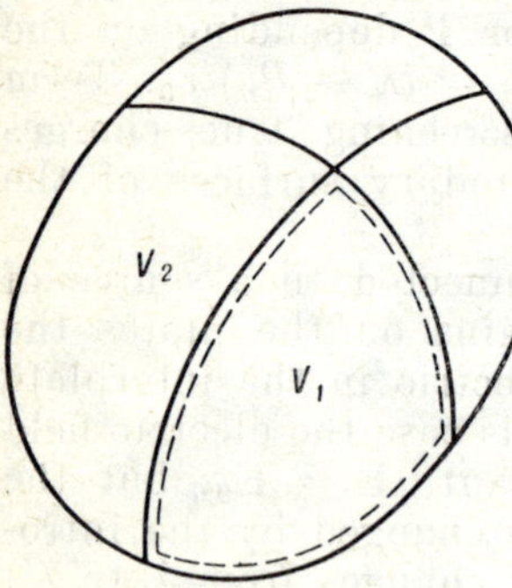

Fig. 32

(31.2) and taking into account that in (31.2) the positive direction of the normal follows from the boundary conditions (4.13), we find that surface integrals containing P_n cancel out. If we additionally assume $\lambda = 0$, the formula for $\varphi(\mathbf{r})$ takes the form

$$\varphi(\mathbf{r}) = \frac{1}{4\pi\varepsilon_0} \int \frac{\rho}{R}\, dV' + \frac{1}{4\pi\varepsilon_0} \int \left(\mathbf{P} \cdot \operatorname{grad}' \frac{1}{R}\right) dV' \qquad (31.3)$$

A comparison with formula (11.17) shows that the second term appears as a potential produced by a spatial distribution of dipoles with electric moment density $\mathbf{P}$ $\left(\text{recall that } \operatorname{grad}' \frac{1}{R} = -\operatorname{grad} \frac{1}{R}\right)$.

Creation of a spatial dipole moment in a dielectric requires an expenditure of energy. The corresponding calculation must take account of thermodynamical conditions in the dielectric medium. The energy necessary to produce a specific polarization depends on these thermodynamic conditions.

31.2. An analysis of energy conservation, conducted in § 3 directly on the basis of the Maxwell equations in their general form, has shown that regardless of the structure of the material (i.e. with no specific assumptions on the relation between **D** and **E**) the change in the field energy in the medium is given by $\mathbf{E} \cdot d\mathbf{D}$. Consequently, with the effect of field on the medium taken into account, we must formulate the first law of thermodynamics as

$$dU = đQ + \zeta\, d\varkappa + \mathbf{E}\cdot d\mathbf{D} \tag{31.4}$$

where all the quantities are referred to a unit of volume, U is the internal energy, $đQ$ is the quantity of heat, $\varkappa$ is the mass density, and ζ is the chemical potential. Symbol $đ$ is used to emphasize the basic fact of thermodynamics, namely, that the quantity of heat is not the total differential but depends on the sort of the process affecting the state of the medium. According to the second law of thermodynamics,

$$đQ = T\, dS \tag{31.5}$$

where T is temperature, and S is entropy which, along with internal energy U, is a function of state, that is a single-valued function of the parameters required to define the state. Here we choose for such parameters $\varkappa$ and $\mathbf{D}$, and consider reversible (quasistatic) processes.

Free energy, that is a function of state defined as

$$F = U - TS \tag{31.6}$$

will play an important role in further discussion. From (31.6) it follows that

$$dF = -S\, dT + \zeta\, d\varkappa + \mathbf{E}\cdot d\mathbf{D} \tag{31.7}$$

Hence, the electric field strength is

$$\mathbf{E} = \left(\frac{\partial U}{\partial \mathbf{D}}\right)_{S,\,\varkappa} = \left(\frac{\partial F}{\partial \mathbf{D}}\right)_{T,\,\varkappa} \tag{31.8}$$

We shall also use a thermodynamic function

$$\tilde{F} = F - \mathbf{E}\cdot\mathbf{D} \tag{31.9}$$

such that

$$d\tilde{F} = -S\, dT + \zeta\, d\varkappa - \mathbf{D}\cdot d\mathbf{E} \tag{31.10}$$

and therefore induction is written as

$$\mathbf{D} = -\left(\frac{\partial \tilde{F}}{\partial \mathbf{E}}\right)_{T,\,\varkappa} \tag{31.11}$$

Let us show that $\frac{1}{2}(\varepsilon E^2)$, that is relation (3.5) for the energy density in an isotropic dielectric, is directly connected to the thermodynamic free energy of the dielectric. We shall not need to assume that the dielectric is homogeneous, so that ε may be a function of coordinates. Experimental data show that dielectric permittivity may depend on the density and temperature: $\varepsilon = \varepsilon(\varkappa, T)$. Let us assume for the moment that density $\varkappa$ is constant. Then $d\mathbf{D} = \varepsilon\, d\mathbf{E} + \mathbf{E}\frac{\partial \varepsilon}{\partial T}\, dT$, so that with (31.5) taken into account, equation (31.4) is transformed to

$$dU = T\,dS + \frac{\varepsilon}{2}\,d(\mathbf{E}^2) + \mathbf{E}^2\frac{\partial\varepsilon}{\partial T}\,dT \tag{31.12}$$

The state of the medium is conveniently described by T and $\mathbf{E}^2$. Both U and S are functions of state, and we can write

$$dU = \frac{\partial U}{\partial T}\,dT + \frac{\partial U}{\partial(\mathbf{E}^2)}\,d(\mathbf{E}^2),\quad dS = \frac{\partial S}{\partial T}\,dT + \frac{\partial S}{\partial(\mathbf{E}^2)}\,d(\mathbf{E}^2) \tag{31.13}$$

By substituting these last formulas into (31.12) and comparing coefficients of dT and $d(\mathbf{E}^2)$, we obtain

$$\frac{\partial S}{\partial T} = \frac{1}{T}\left(\frac{\partial U}{\partial T} - \mathbf{E}^2\frac{\partial\varepsilon}{\partial T}\right),\quad \frac{\partial S}{\partial(E^2)} = \frac{1}{T}\left(\frac{\partial U}{\partial(E^2)} - \frac{\varepsilon}{2}\right) \tag{31.14}$$

But since dS is a total differential, the second derivatives of functions S are insensitive to the order in which the derivatives of S are taken. The same is true for U. Taking this into account and calculating

$$\frac{\partial^2 S}{\partial T\,\partial(E^2)} = \frac{\partial^2 S}{\partial(E^2)\,\partial T}$$

in a complete form by using (31.14), we find the equation

$$\frac{\partial U}{\partial(E^2)} = \frac{1}{2}\left(\varepsilon + T\frac{\partial\varepsilon}{\partial T}\right) \tag{31.15}$$

Substitution of (31.15) into the second of relations (31.14) yields

$$\frac{\partial S}{\partial(E^2)} = \frac{1}{2}\frac{\partial\varepsilon}{\partial T} \tag{31.16}$$

Now it follows from (31.15) and (31.16) that

$$U = U_0(T) + \frac{E^2}{2}\left(\varepsilon + T\frac{\partial\varepsilon}{\partial T}\right),\quad S = S_0(T) + \frac{E^2}{2}\frac{\partial\varepsilon}{\partial T} \tag{31.17}$$

Here U_0 and S_0 are independent of electric parameters ε and E^2. By using (31.17), we rewrite free energy (31.16) in the form

$$F = F_0(T) + (1/2)\,\varepsilon E^2 \tag{31.18}$$

Thus, we conclude that relation (3.5) gives one term in the expression for the free energy; as we know from thermodynamics, an increment of the free energy equals the maximum mechanical work which could be done on the system in an isothermal process. And the term $F_0 = U_0 - TS_0$ is equal to the free energy of the medium in the absence of electric field.

It must be remarked that in the general case two facts are responsible for polarization of isotropic dielectrics. In some dielectrics molecules possess dipole moments even when external field is absent.

The directions of these moments are spread randomly (dipole dielectrics). In a certain degree the external field orders molecular dipole moments thus producing a macroscopic polarization effect. In other media an external field polarizes the zero-dipole moment molecules and with them the dielectric as a whole. In many dipole dielectrics (gaseous and liquid) the following formula gives ε as a function of temperature:

$$\varepsilon = \varepsilon_0 + \frac{\text{const}}{T} \qquad (31.19)$$

The first formula in (31.17) shows that the field produces the change in energy equal to $(1/2)\,\varepsilon_0 E^2$, that is the temperature-dependent term in (31.19) makes no contribution to the energy.[5]

Formula (31.17) also shows that entropy increases with E if $\partial\varepsilon/\partial T > 0$ and diminishes if $\partial\varepsilon/\partial T < 0$. The latter case signifies that electric field increases ordering of the medium; this is the case, in particular, when (31.19) holds. The former case, realized in some solids, is characterized by certain ordering of molecular dipoles even when no external field is applied. An external field rotates these dipoles and may perturb the initial ordering.

As we found from (31.5) and (31.17), the heat absorbed in a unit volume when the field is switched on is given by

$$đQ = T\,dS = \mathbf{E}\cdot d\mathbf{D}\cdot\frac{T}{\varepsilon}\,\frac{d\varepsilon}{dT}$$

Thus, if formula (31.19) is valid, then $đQ < 0$ if $d\mathbf{D} > 0$, that is, heat is released when the field is applied and is absorbed when this field is removed.

31.3. Now we shall calculate the energy of a dielectric solid in an external field. Assume that field $\mathbf{E}_1$ is produced in a dielectric medium with permittivity ε_1, and we introduce into this medium a "foreign" dielectric with permittivity ε_2.[6] Sourcos of field $\mathbf{E}_1$ will be considered time-independent. The process of introduction of a foreign body will be assumed isothermal. The electrostatic energy before this process is initiated is

$$W_1 = \frac{1}{2}\int \mathbf{E}_1\cdot\mathbf{D}_1\,dV$$

where integration covers the entire space. After the body is inserted, new field $\mathbf{E}$ sets up. Consider a sufficiently large part of the space bounded by surface σ enclosing the introduced dielectric (Fig. 33). The volume of this dielectric will be denoted by V_2, and the rest

[5] Indeed, the components of ε depending on T cancel out when (31.19) is substituted into (31.17).

[6] To be precise, a certain volume of the dielectric with permittivity ε_1 is replaced with a dielectric with permittivity ε_2.

of the volume inside σ filled with the original dielectric, by V_1. What we want to know is the difference between field energies in the new and old states:

$$\Delta W = \frac{1}{2}\int\limits_{V_1+V_2} \mathbf{E}\cdot\mathbf{D}\,dV - \frac{1}{2}\int\limits_{V_1+V_2} \mathbf{E}_1\cdot\mathbf{D}_1\,dV \qquad (31.20)$$

which must be equal to the work done in the isothermal process of the introduction of the dielectric into the field; hence, it is equal to the potential energy of this body. Assume that neither space charges nor surface charges at the interfaces between two dielectrics are present. An identity transformation enables us to rewrite (31.20) in the form

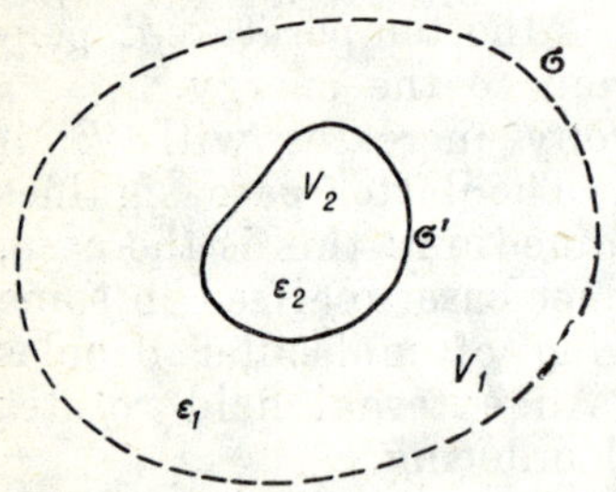

Fig. 33

$$2\Delta W = \int \mathbf{E}\cdot(\mathbf{D}-\mathbf{D}_1)\,dV + \\ + \int (\mathbf{E}-\mathbf{E}_1)\cdot\mathbf{D}_1\,dV \qquad (31.21)$$

As in a number of already familiar cases, we transform the first term by substituting $\mathbf{E} = -\text{grad}\,\varphi$ and using formula (B.14$_2$) and the Gauss theorem. Recalling that $\text{div}\,\mathbf{D} = \text{div}\,\mathbf{D}_1 = 0$, we obtain

$$\int\limits_{V_1+V_2} \mathbf{E}\cdot(\mathbf{D}-\mathbf{D}_1)\,dV = -\int\limits_{V_1+V_2} \text{div}\,[\varphi(\mathbf{D}-\mathbf{D}_1)]\,dV$$

If σ' denotes the interface between two dielectrics, we have

$$\int\limits_{V_1+V_2} \text{div}\,[\varphi(\mathbf{D}-\mathbf{D}_1)]\,dV = \int\limits_{V_1} \text{div}\,[\varphi(\mathbf{D}-\mathbf{D}_1)]\,dV \\ + \int\limits_{V_2} \text{div}\,[\varphi(\mathbf{D}-\mathbf{D}_1)]\,dV = \oint\limits_{\sigma} \varphi(D_n - D_{1n})\,d\sigma \\ + \oint\limits_{\sigma'} \varphi(D_n^{\mathrm{I}} - D_{1n}^{\mathrm{I}})\,d\sigma' - \oint\limits_{\sigma'} \varphi(D_n^{\mathrm{II}} - D_{1n}^{\mathrm{II}})\,d\sigma'$$

Assuming that the first of the above integrals on the right vanishes as surface σ recedes to infinity, and performing this limiting transition, we transform the right-hand side of the above formula to

$$\oint\limits_{\sigma'} \varphi(D_n^{\mathrm{I}} - D_n^{\mathrm{II}})\,d\sigma' - \oint\limits_{\sigma'} \varphi(D_{1n}^{\mathrm{I}} - D_{1n}^{\mathrm{II}})\,d\sigma' = 0$$

which is valid in the chosen case of no surface charges. Indeed, according to (4.9), here $D_n^{\mathrm{I}} = D_n^{\mathrm{II}}$ and $D_{1n}^{\mathrm{I}} = D_{1n}^{\mathrm{II}}$.

Now we take up the second term in (31.21). As $\mathbf{D} = \varepsilon_1 \mathbf{E}$ in volume V_1 and $\mathbf{D}_1 = \varepsilon_1 \mathbf{E}_1$, we can write

$$\int_{V_1} (\mathbf{E} - \mathbf{E}_1) \cdot \mathbf{D}_1 \, dV = \int_{V_1} (\mathbf{D} - \mathbf{D}_1) \cdot \mathbf{E}_1 \, dV$$

It can be shown, in complete analogy to the preceding paragraph, that

$$\int_{V_1+V_2} (\mathbf{D} - \mathbf{D}_1) \cdot \mathbf{E}_1 \, dV = \int_{V_1} \mathbf{E}_1 \cdot (\mathbf{D} - \mathbf{D}_1) \, dV + \int_{V_2} \mathbf{E}_1 \cdot (\mathbf{D} - \mathbf{D}_1) \, dV = 0$$

whence

$$2\Delta W = \int_{V_1+V_2} (\mathbf{E} - \mathbf{E}_1) \cdot \mathbf{D}_1 \, dV = \int_{V_2} (\mathbf{E} - \mathbf{E}_1) \cdot \mathbf{D}_1 \, dV - \int_{V_2} \mathbf{E}_1 \cdot (\mathbf{D} - \mathbf{D}_1) \, dV = \int_{V_2} (\mathbf{E} \cdot \mathbf{D}_1 - \mathbf{E}_1 \cdot \mathbf{D}) \, dV$$

Or, since $\mathbf{D} = \varepsilon_2 \mathbf{E}$ in volume V_2, and $\mathbf{D}_1 = \varepsilon_1 \mathbf{E}_1$

$$\Delta W = \frac{1}{2} \int_{V_2} (\varepsilon_1 - \varepsilon_2) \, \mathbf{E} \cdot \mathbf{E}_1 \, dV \qquad (31.22)$$

If $\varepsilon_1 = \varepsilon_0$, that is the dielectric is inserted into vacuum, and if we take into account that $\mathbf{P} = \mathbf{D} - \varepsilon_0 \mathbf{E} = (\varepsilon_2 - \varepsilon_0)\, \mathbf{E}$ formula (31.22) changes to

$$\Delta W = -\frac{1}{2} \int_{V_2} \mathbf{P} \cdot \mathbf{E}_1 \, dV \qquad (31.23)$$

As follows from (31.22), if dielectric permittivity in the dielectric-field volume undergoes an infinitesimal increment, so that $\varepsilon_2 = \varepsilon_1 + \delta\varepsilon$, where $\delta\varepsilon > 0$, and obviously $\mathbf{E} \simeq \mathbf{E}_1$ to within infinitesimal corrections, then

$$\Delta W = -\frac{1}{2} \int (\delta\varepsilon) \, \mathbf{E}^2 \, dV \qquad (31.24)$$

that is, the total energy of the field decreases. More intricate thermodynamic arguments which we omit here show that $\varepsilon > \varepsilon_0$ in any dielectric.

31.4. Now we are able to calculate *forces* occurring in a dielectric. Let us consider the forces which are applied to an infinitesimal plane within a dielectric medium. They are described by formula (3.18). Among all possible orientations of this plane we can single out those in which the vector of surface force coincides with the vector of the normal, that is such that $\boldsymbol{\varphi} = \lambda \mathbf{n}$ or $T_{ik} n^k = \lambda n_i$.

These are the so-called principal directions of the Maxwell stress tensor. They can be found from the condition of resolvability of the above equation, that is from equation $\det(T_{ik} - \lambda\delta_{ik}) = 0$. Note that symmetricity of the stress tensor ($T_{ik} = T_{ki}$) results in mutual orthogonality of the principal directions corresponding to different values of λ. Indeed, let $T_{ik}n_1^k = \lambda_1 n_{1i}$ and $T_{ik}n_2^k = \lambda_2 n_{2i}$. Multiply the first of these equations by n_{2i}, the second by n_{1i}, and sum them up over i. This gives

$$T_{ik}n_1^k n_2^i = \lambda_1(\mathbf{n}_1 \cdot \mathbf{n}_2), \quad T_{ik}n_2^k n_1^i = T_{ik}n_1^k n_2^i = \lambda_2 \mathbf{n}_1 \cdot \mathbf{n}_2$$

For $\lambda_1 \neq \lambda_2$ this gives $\mathbf{n}_1 \cdot \mathbf{n}_2 = 0$. If, however, two or all three of the possible values of λ coincide, then we can always orthogonalize the corresponding principal directions.

Thus, if the stress tensor has the form of (3.9), the calculation of the third-order determinant given above leads to the equation

$$8\lambda^3 + 4\lambda^2\varepsilon E^2 - 2\lambda\varepsilon^2 E^4 - \varepsilon^3 E^6 = 0$$

This equation has three roots:

$$\lambda_1 = \frac{\varepsilon}{2}E^2, \quad \lambda_2 = \lambda_3 = -\frac{\varepsilon}{2}E^2$$

Axis 1 is chosen along vector $\mathbf{E}$, and the other two axes lie in a plane orthogonal to this vector. The result is interpreted as a tensile force along the field and two contraction forces orthogonal to the direction of field $\mathbf{E}$.[7]

In calculating bulk forces in a dielectric we shall also take into account the dependence of ε on mass density $\varkappa$ and coordinates. According to the general principles of mechanics, the work $\delta\mathcal{A}$ of such forces $\mathbf{f}$ over an infinitesimal displacement $\delta\mathbf{r}$ of an element of volume dV with respect to the neighbouring volume elements is equal to $\delta\mathcal{A} = \mathbf{f} \cdot \delta\mathbf{r}$. The total work of forces in the chosen volume must be equal to the change in the free energy of this volume:

$$\Delta W = -\int \delta\mathcal{A}\, dV \tag{31.24'}$$

On the other hand, if we can find a tensor T_{ik} for which $f_i = \partial T_{ik}/\partial x_k$, it will be the stress tensor which in principle describes the effects of nonuniformity which were ignored in the derivation of the Maxwell stress tensor in § 3.

In the general case, an element of volume of the medium displaced in an electric field undergoes deformation. By denoting the initial volume of the element by dV_1 and its final volume containing the

[7] It must be emphasized that we mean tensor representation of the forces applied to bound charges by the electric field.

same mass of the matter by dV_2, we can write

$$dV_2 = \left[\det\left(\frac{\partial x^i_2}{\partial x^k_1}\right)\right] dV_1$$

where x^i_1 and x^i_2 are coordinates of the points within the volume in the initial and final positions. If $x^i_2 = x^i_1 + \delta x^i$, where the displacement is assumed infinitesimal, then in calculating the determinant to within infinitesimals of higher orders, we retain only its diagonal elements:

$$\det\left(\frac{\partial x^i_2}{\partial x^i_1}\right) \simeq 1 + \frac{\partial(\delta x^i)}{\partial x^i_1} = 1 + \operatorname{div} \mathbf{s}$$

where, to contract the formulas, we denote $\mathbf{s} \equiv \delta \mathbf{r}$.

The condition of mass conservation in the displaced element of volume shows that

$$\varkappa_1\, dV_1 = \varkappa_2\, dV_2 = \varkappa_2\, dV_1\,(1 + \operatorname{div} \mathbf{s})$$

that is $\varkappa_2 - \varkappa_1 \equiv \delta\varkappa \simeq -\varkappa \operatorname{div} \mathbf{s}$. If we go from point $\mathbf{r} - \mathbf{s}$ to point $\mathbf{r}$, the change in dielectric permittivity is, to the same accuracy,

$$\delta\varepsilon = -\mathbf{s}\cdot\operatorname{grad}\varepsilon + \frac{\partial\varepsilon}{\partial\varkappa}\,\delta\varkappa = -\mathbf{s}\cdot\operatorname{grad}\varepsilon - \varkappa\frac{\partial\varepsilon}{\partial\varkappa}\operatorname{div}\mathbf{s}$$

Now, in our approximation, formula (31.24) gives

$$\Delta W = \frac{1}{2}\int E^2\left(\mathbf{s}\cdot\operatorname{grad}\varepsilon + \varkappa\frac{\partial\varepsilon}{\partial\varkappa}\operatorname{div}\mathbf{s}\right) dV \tag{31.25}$$

By using formula (B.14_2), we transform the second term in the integrand to

$$E^2\varkappa\frac{\partial\varepsilon}{\partial\varkappa}\operatorname{div}\mathbf{s} = \operatorname{div}\left(E^2\varkappa\frac{\partial\varepsilon}{\partial\varkappa}\mathbf{s}\right) - \mathbf{s}\cdot\operatorname{grad}\left(E^2\varkappa\frac{\partial\varepsilon}{\partial\varkappa}\right)$$

From the Gauss theorem,

$$\int \operatorname{div}\left(E^2\varkappa\frac{\partial\varepsilon}{\partial\varkappa}\mathbf{s}\right) dV = \oint E^2\varkappa\frac{\partial\varepsilon}{\partial\varkappa} s_n\, d\sigma + \sum_{i=1}^{N}\oint E^2\varkappa\frac{\partial\varepsilon}{\partial\varkappa} s_n\, d\sigma$$

As before, we are to calculate the first integral in the right-hand side over a sufficiently remote surface and so we can assume it equal to zero. The second integral appears because N conductors may be imbedded in the dielectric, and this integral is to be taken over their surfaces. We assume these conductors rigid in the sense that displacement on their surfaces satisfies condition $s_n = 0$. Then

$$\Delta W = \frac{1}{2}\int\left[E^2\operatorname{grad}\varepsilon - \operatorname{grad}\left(E^2\varkappa\frac{\partial\varepsilon}{\partial\varkappa}\right)\right]\cdot\mathbf{s}\, dV$$

And finally we obtain from (31.24) and from the definition of work done by bulk forces

$$\mathbf{f} = -\frac{1}{2} E^2 \operatorname{grad} \varepsilon + \frac{1}{2} \operatorname{grad} \left(E^2 \varkappa \frac{\partial \varepsilon}{\partial \varkappa} \right) \tag{31.26}$$

The derivation of this formula for bulk forces shows that if a distribution of space charge ρ exists in the dielectric, one has also to take into account the work done by the forces applied to this charge (these forces have already been calculated in § 3 under the assumption of constant dielectric permittivity ε). These forces are equal to $\rho \mathbf{E}$ and must be added to the right-hand side of formula (31.26).

It is not difficult to verify by straightforward differentiation that, with term $\rho \mathbf{E}$ taken into account, we obtain $f_i = \partial T_{ik}/\partial x_k$, where

$$T_{ik} = E_i D_k - \frac{\delta_{ik}}{2} (1 - b) \mathbf{E} \cdot \mathbf{D} \tag{31.27}$$

and

$$b \equiv \frac{\varkappa}{\varepsilon} \frac{\partial \varepsilon}{\partial \varkappa} \tag{31.28}$$

Note that in calculating bulk forces we ignored the term $F_0 (T)$ independent of electric parameters and included into the complete expression (31.18) for free energy. However, this term also changes in a virtual displacement of an element of volume, and this change is responsible for mechanical pressure. We shall briefly outline the calculations necessary to find the total force in the dielectric. First of all,

$$\delta (F_0 \, dV) = F_0 \delta (dV) + \frac{\partial F_0}{\partial \varkappa} \delta \varkappa \, dV$$

But we have seen above that $\delta (dV) = \operatorname{div} \mathbf{s} \, dV$ and $\delta \varkappa = -\varkappa \operatorname{div} \mathbf{s}$. Hence

$$\delta \int F_0 \, dV = \int \left(F_0 - \varkappa \frac{\partial F_0}{\partial \varkappa} \right) \operatorname{div} \mathbf{s} \, dV \tag{31.29}$$

Recall that F_0 is referred to a unit volume. Free energy in a sufficiently small volume V enclosing mass M is $F_0 V = F_0 M/\varkappa$. Hence

$$\frac{\partial (F_0 V)}{\partial V} = \frac{\partial}{\partial (\varkappa^{-1})} (F_0 \varkappa^{-1})$$

This last expression coincides with the expression in the parentheses of (31.29). But if all thermodynamic parameters of state, with the exception of volume, are maintained constant, the left-hand side of the last expression equals $-p$. Calculations similar to those used to obtain (31.26) from (31.25) yield

$$\delta \int F_0 \, dV = \int \operatorname{grad} p \cdot \mathbf{s} \, dV$$

whence $\mathbf{f}_{\mathrm{pr}} = -\operatorname{grad} p$, where p is the pressure.

Collecting all the above arguments, we obtain the following general expression for bulk forces:

$$\mathbf{f} = -\operatorname{grad} p + \rho\mathbf{E} - \frac{E^2}{2}\operatorname{grad}\varepsilon + \frac{1}{2}\operatorname{grad}\left(E^2\varkappa\frac{\partial\varepsilon}{\partial\varkappa}\right) \quad (31.30)$$

Thermodynamic formulas (31.6) and (31.7) referred to quantities per unit volume. If one is interested in total values of the quantities in a known volume, this volume becomes one of the parameters describing the thermodynamic state. The first law of thermodynamics for these parameters (at this juncture we ignore the electric field energy) takes the form

$$d\mathcal{F}_0 = -\mathcal{S}\,dT - p\,dV \quad (31.31)$$

In fact, this relation has been used above in deriving the expression for hydrostatic pressure. Quantity ΔW given by formula (31.23) is the electric component of the free energy in the volume of the dielectric. Assume that electric field $\mathbf{E}$ in the considered volume is nearly uniform.[8] Then

$$\mathcal{F}_E \equiv \Delta W = (-1/2)\,\mathcal{P}\cdot\mathbf{E}$$

where $\mathcal{P}$ is the total polarization, that is the dipole moment of the volume as a whole. Further, let vector $\mathcal{P}$ be a linear function of $\mathbf{E}$: $\mathcal{P} = \varepsilon_0\chi_e\mathbf{E}V$, where χ_e is the scalar dielectric susceptibility. The variation of field $\mathbf{E}$ in constant volume V changes free energy $\mathcal{F}_E$ by $d\mathcal{F}_E = -\mathcal{P}\cdot d\mathbf{E}$. This term must be added to the former relation (31.31), and we denote $\mathcal{F} = \mathcal{F}_0 + \mathcal{F}_E$.

31.5. Let us analyze in more detail the thermodynamic meaning of $\mathcal{F}_E$. Recall first of all the first law of thermodynamics (31.4) for bulk densities, assuming for simplicity that $\varkappa$ is constant. By substituting (31.5) and making use of the relation $\mathbf{D} = \varepsilon_0\mathbf{E} + \mathbf{P}$, we obtain

$$d\left(U - \frac{\varepsilon_0 E^2}{2}\right) = T\,dS + \mathbf{E}\cdot d\mathbf{P} \quad (31.32)$$

The left-hand side expresses the change of internal energy per unit volume, minus energy of field $\mathbf{E}$ in vacuo. Hence

$$d\left(F - \frac{\varepsilon_0 E^2}{2}\right) = -S\,dT + \mathbf{E}\cdot d\mathbf{P}$$

We see that $F - (1/2)\,\varepsilon_0 E^2$ is a function of state if T and $\mathbf{P}$ are chosen as parameters. By analogy to a transformation from (31.6) to (31.9), we can change for parameters T and $\mathbf{E}$ if we introduce

$$F' = F - (1/2)\,\varepsilon_0 E^2 - \mathbf{P}\cdot\mathbf{E}$$

[8] In our analysis of the total free energy we assume that $\mathbf{E} \simeq \mathbf{E}_1$, that is, the introduction of the dielectric produces sufficiently small changes in the field.

Then

$$dF' = -S\,dT - \mathbf{P}\cdot d\mathbf{E} \tag{31.33}$$

If the process is isothermal ($dT = 0$) and if field $\mathbf{E}$, although varying, remains uniform within the volume under consideration, integration of the above equation over the volume gives us the value of $d\mathcal{F}_{\mathbf{E}}$. But the equation itself is much more general because no restrictions were imposed on the dependence of $\mathbf{P}$ on $\mathbf{E}$ in the course of the derivation.

It is often more convenient to choose pressure p as an independent parameter instead of V. For this purpose another function of state, namely thermodynamic potential $\Phi = \mathcal{F} + pV$, is introduced instead of free energy $\mathcal{F}$. We see from the above arguments that in isothermal processes $d\Phi = V\,dp - \mathcal{P}\cdot d\mathbf{E}$. And since $d\Phi$ is the total differential, we obtain

$$\left(\frac{\partial V}{\partial E_i}\right)_{T,\,p} = -\left(\frac{\partial \mathcal{P}_i}{\partial p}\right)_{T,\,\mathbf{E}} \tag{31.34}$$

This relation gives the volume of a dielectric as a function of the electric field applied (the effect of *electrostriction*). The dependence of polarization $\mathcal{P}$ on pressure p in the right-hand side is obviously determined by the structure of the specific dielectric. As a result, electric field may cause both expansion and contraction, depending on the dielectric. If equality (1.26) holds, relation (31.34) takes the form

$$V - V_0 = -\frac{\varepsilon_0 E^2}{2}\left(\frac{\partial \varkappa_e V}{\partial p}\right)_{T,\,\mathbf{E}} \tag{31.35}$$

Here V_0 is the dielectric's volume in the absence of external fields.

Let us return to formula (31.30) for bulk forces in isotropic dielectrics. Assuming $\rho = 0$ and $\operatorname{grad}\varepsilon = \frac{\partial \varepsilon}{\partial \varkappa}\operatorname{grad}\varkappa$, we can rewrite the condition of mechanical equilibrium of the dielectric, $\mathbf{f} = 0$, in the form

$$\operatorname{grad} p = \frac{\varkappa}{2}\operatorname{grad}\left(E^2 \frac{\partial \varepsilon}{\partial \varkappa}\right)$$

If the equation of state of a liquid, that is p as a function of $\varkappa$, is known we obtain

$$\int_{p_1}^{p_2} \frac{dp}{\varkappa} = \frac{E^2}{2}\,\frac{d\varepsilon}{d\varkappa}\Bigg|_{\mathbf{r}_1}^{\mathbf{r}_2} \tag{31.36}$$

where p_i is the pressure at point $\mathbf{r}_i$. Therefore, pressure depends only on electric field at a given point. It can often be assumed that incompressibility condition holds, that is, $\varkappa$ is independent of p. In a number of cases the dependence of ε on $\varkappa$ is approximated quite well by the following Clausius-Mosotti formula (also called the Lorentz-Lorenz formula):

$$\frac{\varepsilon'-1}{\varepsilon'+2}=C\varkappa$$

where $\varepsilon' \equiv \varepsilon/\varepsilon_0$ and C is a constant determined by the specific nature of the dielectric. We easily find then that

$$\varkappa\, d\varepsilon/d\varkappa = (1/3)\,\varepsilon_0\,(\varepsilon'-1)\,(\varepsilon'+2)$$

and (31.36) is rewritten in the form

$$p_2-p_1=\frac{\varepsilon_0 E^2}{2}\,\frac{(\varepsilon'-1)\,(\varepsilon'+2)}{3}\Bigg|_1^2$$

The following point should be noted in conclusion. If the stress tensor is known, we can derive a number of mechanical relations which should hold at the interface between two different media. Indeed, surface forces at this interface must be balanced due to the equality of action and reaction, namely $\varphi^{\mathrm{I}} = -\varphi^{\mathrm{II}}$, where φ^{I} is the surface force applied by the first medium to the second, and φ^{II} is the force applied to the first medium by the second. By using the stress tensor, we rewrite this equality in the form

$$T^{\mathrm{I}}_{ik}n^{\mathrm{I}k} = -T^{\mathrm{II}}_{ik}n^{\mathrm{II}k}$$

Here $\mathbf{n}^{\mathrm{I}}$ and $\mathbf{n}^{\mathrm{II}}$ point in opposite direction. Assuming $\mathbf{n} = \mathbf{n}^{\mathrm{I}}$, we obtain $(T^{\mathrm{I}}_{ik} - T^{\mathrm{II}}_{ik})\, n^k = 0$. If now T^{I}_{ik} is expressed in terms of quantities referring to the first medium, and T^{II}_{ik} in those referring to the second medium, then one can derive a relation for pressure difference at the interface from formula (31.27) (as we can see from the above arguments, we must add the term $-p\delta_{ik}$ to the right-hand side in order to take account of hydrostatic pressure). And as follows from boundary conditions, the equality of forces in tangential directions holds identically, while the relation for the normal direction is nontrivial.

§ 32. Anisotropic dielectrics

32.1. By definition dielectrics are called anisotropic if polarization $\mathbf{P}$ in the dielectric does not coincide in direction with electric field strength $\mathbf{E}$. Usually, anisotropy characterizes media whose particles form crystalline lattice. The lattice possesses a symmetry, that is regularity in the spatial arrangement of particles, and appears because interactions of particles in crystals are much stronger in some directions (these are specific features of chemical bonding between molecules). As a result, the directions of electric dipole moments induced in molecules by external fields are determined not only by these fields but also by the distribution of interactions in crystalline lattices. The simplest example is found in the so-called pyroelectric media in which polarization exists even in the absence of external field. For the dipole moments to give a nonzero net effect, certain regularity must exist in the arrangement of molecules. The total polarization vector $\mathbf{P}_0$ appearing in this case ("spontaneous" polarization) defines a chosen direction in the medium. As a result, the symmetry of a pyroelectric crystal must possess, among other symmetry elements, this chosen direction.

The theory of symmetry in crystals which treats the relationship between this symmetry and the physical properties of crystals, is too vast to be considered here[9]. However, some properties of anisotropic dielectrics can be discussed without resorting to specific results of this theory.

In what follows we assume the anisotropic medium to be linear. This means, as we have already mentioned in § 1, that polarization and electric field strength are related linearly:

$$P_i = \varepsilon_0 \chi_{ij} E^j + P_{0,\,i} \tag{32.1}$$

This formula takes into account the possibility of spontaneous polarization $\mathbf{P}_0$ mentioned above. Quantity χ_{ij} must be a tensor of rank 2 since $\mathbf{P}$ and $\mathbf{E}$ are vectors. It is called *dielectric susceptibility*. Field strength $\mathbf{E}$ and induction $\mathbf{D}$ are related by (1.29), where dielectric permittivity ε_{ij} is also a tensor of rank 2. Besides, we consider the medium as homogeneous. That is, consider components of tensors χ_{ij} and ε_{ij} as independent of coordinates in the medium. This means that physical properties of the medium measured along parallel directions at any two of its points are identical.

When thermodynamic properties of anisotropic dielectrics are investigated, special attention must be paid to the fact that the main thermodynamic relations discussed in the preceding section remain valid in the present case as well, since they were formulated without any reference to isotropy. This is true, in particular, for equations (31.4), (31.7), (31.10), as well as (31.33) and (31.32). The last of these equations is useful for proving the symmetricity of the dielectric susceptibility tensor χ_{ij}. In the adiabatic process, when $dS = 0$, this equation becomes $dW = \mathbf{E}\, d\mathbf{P}$, where W is a thermodynamic function of state interpreted as the internal energy of the medium. Let us substitute into it relation (32.1), assuming for the sake of simplicity that $\mathbf{P}_0 = 0$ and that χ_{ij} remains unaltered in

---

[9] See, for example, L. D. Landau and E. M. Lifshitz, *Electrodynamics of Continuous Media* (Course of Theoretical Physics, vol. 8), Pergamon Press, Oxford, 1960, or J. E. Nye, *Physical Properties of Crystals*, Clarendon Press, Oxford, 1957.

the thermodynamic process. Then

$$dW = \varepsilon_0 \chi_{ij} E^i \, dE^j$$

But since dW is a total differential, we must have

$$\frac{\partial^2 W}{\partial E_i \, \partial E_j} = \frac{\partial^2 W}{\partial E_j \, \partial E_i}$$

for any pair of indices i and j, whence $\chi_{ij} = \chi_{ji}$. This means at the same time that $\varepsilon_{ij} = \varepsilon_{ji}$. Consequently, a field in anisotropic medium possesses energy density w_e found from equation (3.5) in whose derivation symmetricity of tensor ε_{ij} was assumed without proof.

32.2. We shall consider now several problems concerning the relation between mechanical properties of solids and their electric properties. Assume that a surface force $\boldsymbol{\varphi}$ acts on the surface of a solid dielectric. In complete analogy to what we did in § 3 while considering the Maxwell stress tensor (where the isotropy condition was in fact used), we can define the stress tensor τ_{ij} by the relation

$$\varphi_i = \tau_{ij} n^j \tag{32.2}$$

where $\mathbf{n}$ is a unit normal to the surface. Assume that the stress tensor is symmetric: $\tau_{ij} = \tau_{ji}$. Here we are not interested in the relation between tensor components τ_{ij} and field strength and induction in anisotropic case.

In one type of solid crystals an application of forces to their surface produces electric polarization. This phenomenon is called the *direct piezoelectric effect*. The relation between polarization and mechanical stress is written in the form

$$P_i = d_{i,kl} \tau^{kl} \tag{32.3}$$

Piezoelectric coefficients $d_{i,kl}$ are components of a tensor of rank 3, and are symmetric in indices k, l. Later we shall also consider the *converse piezoelectric effect* consisting in the appearance of mechanical stress, and thus of a deformation of the crystal, due to the applied electric field. But first we shall refresh for the reader some results from the description of strains in the mechanics of continuous media.

Let the elements of the medium be displaced from their initial positions by the applied forces, so that the displacement of a point which initially had a radius vector $\mathbf{r}$ is $\mathbf{s}(\mathbf{r})$. Deformation, or strain, of the medium is characterized by the difference in displacements of the neighbouring points, that is, by vector $\delta\mathbf{s} = \mathbf{s}(\mathbf{r} + \delta\mathbf{r}) - \mathbf{s}(\mathbf{r}) \simeq (\delta\mathbf{r} \cdot \text{grad})\, \mathbf{s}$ (for infinitesimal $\delta\mathbf{r}$). This equation can be rewritten in the form $\delta s^i = a_{ik}\delta x^k + b_{ik}\delta x^k$, where

$$a_{ik} = \frac{1}{2}\left(\frac{\partial s_i}{\partial x^k} + \frac{\partial s_k}{\partial x^i}\right) \quad \text{and} \quad b_{ik} = \frac{1}{2}\left(\frac{\partial s_i}{\partial x^k} - \frac{\partial s_k}{\partial x^i}\right)$$

Tensor b_{ik} describes rotation of the considered infinitesimal volume as a whole. Indeed, formula (B.10) makes it possible to introduce $\overset{*}{\mathbf{b}} = \text{curl s}$, that is, $b_l^* = \varepsilon_{lki} \frac{\partial s_i}{\partial x^k}$, so that $b_{ik} = (1/2)\, \varepsilon_{kil} \overset{*}{b}_l$ and $b_{ik}\delta x^k = (1/2)\, \varepsilon_{ilk} \overset{*}{b}{}^l\, dx^k = (1/2)\, \text{curl s} \times \delta\mathbf{r}_i$. Therefore, this part of the displacement is of no significance. Tensor a_{ik} is symmetrical; it describes strain *per se*, namely, tensile and shear strain. Thus, pure tensile strain takes place if $a_{ik}\delta x^k = \lambda \delta x_i$, that is, it is realized when displacement goes along principal axes of tensor a_{ik}. An approximate calculation of the Jacobian easily verifies that the bulk expansion coefficient is $(\delta V' - \delta V)/\delta V = \text{div s}$ (compare with the calculation of forces in the isotropic dielectric in the preceding section).

In a number of cases the strain is described by the empirical *generalized Hook's law* which states that stress tensor τ_{ij} is linearly related to strain tensor a_{ij}:

$$\tau_{ij} = c_{ijkl} a^{kl}, \quad a_{ij} = s_{ijkl} \tau^{kl} \tag{32.4}$$

Coefficients c_{ijkl} are known as *elastic moduli*, and s_{ijkl} as *elasticity coefficients*. Matrices composed of c and s are mutually inverse.

Thermodynamic state of a solid can be specified by fixing temperature T, electric field $\mathbf{E}$, and, for example, components of stress tensor τ_{ij}. Then the following equations must hold:

$$da_{ij} = \left(\frac{\partial a_{ij}}{\partial \tau_{kl}}\right)_{\mathbf{E},\, T} d\tau_{kl} + \left(\frac{\partial a_{ij}}{\partial F_k}\right)_{\tau,\, T} dE_k + \left(\frac{\partial a_{ij}}{\partial T}\right)_{\mathbf{E},\, \tau} dT$$

$$dP_i = \left(\frac{\partial P_i}{\partial \tau_{kl}}\right)_{\mathbf{E},\, T} d\tau_{kl} + \left(\frac{\partial P_i}{\partial E_k}\right)_{\tau,\, T} dE_k + \left(\frac{\partial P_i}{\partial T}\right)_{\tau,\, \mathbf{E}} dT$$

$$dS = \left(\frac{\partial S}{\partial \tau_{kl}}\right)_{\mathbf{E},\, T} d\tau_{kl} + \left(\frac{\partial S}{\partial E_k}\right)_{\tau,\, T} dE_k + \left(\frac{\partial S}{\partial T}\right)_{\tau,\, \mathbf{E}} dT \tag{32.5}$$

Coefficients in the right-hand sides of these equations describe the following physical effects (below we enumerate these coefficients in the right-hand sides of each of equations (32.5) from left to right). In the first equation these are elasticity, converse piezoelectric effect, and thermal expansion; in the second equations these are the direct piezoelectric effect, electric polarization and pyroelectric effect; and finally, in the third equation these are the piezocaloric and electrocaloric effects and the heat capacitance divided by T. All these effects are connected through a number of relations which can be derived from the properties of the thermodynamic function of state.

The first law of thermodynamics can be written in the form

$$dW = \tau_{ij}\, da^{ij} + \mathbf{E} \cdot d\mathbf{P} + T\, dS$$

It can be shown in the theory of elasticity that the first term of the above equation describes the change in the elastic energy of the continuous medium. However, it is often more convenient to use another thermodynamic function, namely the Gibbs thermodynamic potential G defined as follows:

$$G = W - \tau_{ij} a^{ij} - \mathbf{E} \cdot \mathbf{P} - TS$$

so that

$$dG = -a_{ij}\, d\tau^{ij} - \mathbf{P} \cdot d\mathbf{E} - S\, dT$$

As dG is a total differential, we obtain for example, the equality

$$\frac{\partial^2 G}{\partial \tau_{ij}\, \partial E_k} = \frac{\partial^2 G}{\partial E_k\, \partial \tau_{ij}}, \quad \text{that is,} \quad \frac{\partial P_k}{\partial \tau_{ij}} = \frac{\partial a_{ij}}{\partial E_k}$$

The common value of this expression can also be denoted by $d_{k,ij}$, as shown by definition (32.3). But it then follows that the converse piezoelectric effect, that is the dependence of the stress tensor on the external field, is determined by the same piezoelectric coefficients as the direct effect.

A more detailed investigation makes it possible to calculate, for example, bulk forces produced in the anisotropic dielectric when it undergoes a strain in the applied electric field. This problem will not be discussed here[10], since it would call for a much more sophisticated analysis of the properties of stress and strain tensors.

---

[10] See, for example, the cited above monograph by Landau and Lifshitz, or J. A. Stratton, *Electromagnetic Theory*, McGraw-Hill, New York, 1941 (especially § 2.22).

CHAPTER 8

ELECTRIC CURRENT. MAGNETIC FIELD IN CONTINUOUS MEDIA

§ 33. Magnetic energy and forces in a system of direct-current loops. Quasistationary currents in linear circuits

33.1. We have already discussed in § 12 electric currents in isotropic homogeneous medium for which we have obtained relation (1.27) between magnetic induction and magnetic strength assuming magnetic permeability μ to be constant. Electric currents were studied as linear contours (loops) of direct current. The points of main interest were a magnetic field produced by currents and interaction between currents via this field. Thus we have derived formula (12.3) for vector potential, the Ampère law (12.7), and relation (12.10) for the total force between two linear contours. If the SI system of units is used, coefficient α in all these formulas must be set equal to unity. In the present section we shall add to the results of § 12, namely, we shall investigate energy properties of magnetic fields of dc currents. Later we shall discuss some properties of alternating currents (ac currents). In § 34 we shall resort to thermodynamic arguments and gain some knowledge of electric and magnetic effects produced by electric current and heat flux existing simultaneously in the medium.

We shall assume that current density satisfies the generalized Ohm law (1.30) which in isotropic media takes the form

$$\mathbf{j} = \sigma \mathbf{E} + \mathbf{j}_{\text{ext}} \tag{33.1}$$

Current density $\mathbf{j}_{\text{ext}}$ is generated by "external" forces making charges move through the medium. It has been mentioned in § 1 that such forces can be produced at the expense of chemical energy (electrochemical cells and batteries of cells), and by other means. In order to describe external forces, it is convenient to define vector $\mathbf{E}'$ as follows:

$$\mathbf{j}_{\text{ext}} = \sigma \mathbf{E}' \tag{33.2}$$

so that

$$\mathbf{j} = \sigma (\mathbf{E} + \mathbf{E}')$$

The total electromotive force U in a closed contour of sufficiently small cross section with electric current (we shall refer to such contours as quasilinear[1]) is found as an integral over this contour:

$$U = \oint (\mathbf{E} + \mathbf{E}')\, d\mathbf{s} \tag{33.3}$$

If magnetic field is independent of time, we can set $\mathbf{E} = -\mathrm{grad}\,\varphi$, so that the first term in (33.3), determined by the electrostatic field, vanishes. From (33.2) and (33.3) we obtain

$$U = \oint \mathbf{E}'\, d\mathbf{s} = \oint \frac{\mathbf{j}\, d\mathbf{s}}{\sigma} \tag{33.4}$$

Now recall relation (12.5) which can be regarded simply as a definition of current intensity I. The direction of current at each point of a quasilinear loop coincides with the direction of $d\mathbf{s}$ at this point. At the same time, an element of volume of the contour may be written in the form $dV = ds \cdot \Delta\Sigma$, where $\Delta\Sigma$ is the cross-sectional area of the contour. Using (12.5), we rewrite the integrand in (33.4) in the form

$$\mathbf{j} \cdot d\mathbf{s} = j\, ds = \frac{I\, ds}{\Delta\Sigma}$$

In the case of direct current, I is the same in all sections of the contour. Then substitution of the above relation into (33.4) yields

$$U = IR \tag{33.5}$$

where $R \equiv \oint \frac{ds}{\sigma\,\Delta\Sigma}$. Quantity R is called the total *resistance* of the contour; evidently, if electric conductivity σ and contour cross section area $\Delta\Sigma \equiv S$ are constant along the contour, then $R = l/\sigma S$, where l is the contour length. The dimension of σ is easily found if we know the dimensions of current density j and electric field strength E; this will give the dimension to resistance R and then, from formula (33.5), of electromotive force U. A unit of resistance in SI is 1 ohm, and the unit of electromotive force (also called the voltage across the contour) is 1 volt (V); electric current is then measured in amperes (cf. § 1 and § 29).

Assume now that the contour is open, so that no dc current is possible and $\mathbf{j} = 0$, but external electromotive forces in the contour exist. Then, integrating (33.3) along the contour between its ends 1 and 2, we obtain

$$-\int_1^2 \mathbf{E} \cdot d\mathbf{s} = \int_1^2 \mathbf{E}' \cdot d\mathbf{s} = \oint \mathbf{E}' \cdot d\mathbf{s} = U \tag{33.6}$$

In writing the integral over the closed loop, we take into account that $\mathbf{E}' = 0$ on the open segment of the circuit. The left-hand side

[1] To be precise, a quasilinear loop is defined as a loop (a conductor) in which electric conduction σ and cross section $\Delta\Sigma$ are functions of a single linear coordinate s along the contour.

of formula (33.6) characterizes the electrostatic voltage in the open circuit and is equal to potential difference $\varphi(1) - \varphi(2)$. The meaning of this formula is that the effect of external electromotive forces in an open circuit is balanced by the produced electrostatic field of the charges; in other words, e.m.f. is measured as potential difference across an open circuit.

33.2. Let a homogeneous isotropic medium with magnetic permeability μ contain N quasilinear contours s_α ($\alpha = 1, 2, \ldots, N$) with constant currents I_α. We should analyze an expression for energy W of magnetostatic field generated by these currents. By using the definition of magnetic energy density discussed at the beginning of § 3 and applying successively formula (2.1), relation (B.19), and the Maxwell equation (1.21) for $\partial\mathbf{D}/\partial t = 0$, we can write

$$2W = \int \mathbf{H}\cdot\mathbf{B}\,dV = \int \mathbf{H}\cdot\operatorname{curl}\mathbf{A}\,dV = \int \operatorname{div}(\mathbf{A}\times\mathbf{H})\,dV + \int \mathbf{j}\cdot\mathbf{A}\,dV \tag{33.7}$$

The first term in this formula can be transformed, by the Gauss theorem, to integral $\oint (\mathbf{A}\times\mathbf{H})\cdot\mathbf{n}\,d\sigma$ over a closed surface enclosing all current contours. As the surface is removed to infinity, this integral tends to zero since its integrand is proportional to r^{-3}. The second term can be transformed in two ways. First, we can immediately take account of the fact that integration should be carried out only in the region where $\mathbf{j} \neq 0$, that is only along current contours. Then (12.5) gives

$$\int \mathbf{j}\cdot\mathbf{A}\cdot dV = \sum_{\alpha=1}^{N} I_\alpha \oint_{s_\alpha} \mathbf{A}\cdot d\mathbf{s}_\alpha = \sum_{\alpha=1}^{N} I_\alpha \Phi_\alpha \tag{33.8}$$

and according to Stokes' theorem,

$$\oint_{s_\alpha} \mathbf{A}\cdot d\mathbf{s}_\alpha = \int \operatorname{curl}_n \mathbf{A}\,d\sigma_\alpha = \int B_n\,d\sigma_\alpha \equiv \Phi_\alpha \tag{33.9}$$

Here, as usual, σ_α is a two-dimensional surface bounded by contour s_α, and Φ_α is the magnetic flux across the area bounded by contour s_α.

On the other hand, let us use relation (12.3) for vector potential derived from the equations of magnetostatics:

$$\int \mathbf{j}\cdot\mathbf{A}\,dV = \frac{\mu}{4\pi}\int\int \frac{\mathbf{j}(\mathbf{r})\cdot\mathbf{j}(\mathbf{r}')}{|\mathbf{r}-\mathbf{r}'|}\,dV\,dV'$$

As before, integration covers only the quasilinear contours. One has to distinguish between two cases: volumes dV and dV' may belong to one contour or to different contours. Correspondingly, the double integral can be written in the following form:

$$\iint \frac{\mathbf{j}(\mathbf{r})\cdot\mathbf{j}(\mathbf{r}')}{|\mathbf{r}-\mathbf{r}'|}\,dV\,dV' = \sum_{\alpha=1}^{N} \iint \frac{\mathbf{j}(\mathbf{r}_\alpha)\,\mathbf{j}(\mathbf{r}'_\alpha)}{|\mathbf{r}_\alpha-\mathbf{r}'_\alpha|}\,dV_\alpha\,dV'_\alpha + \sum_{\substack{\alpha,\,\beta\\(\alpha\neq\beta)}} \iint \frac{\mathbf{j}(\mathbf{r}_\alpha)\cdot\mathbf{j}(\mathbf{r}_\beta)}{|\mathbf{r}_\alpha-\mathbf{r}_\beta|}\,dV_\alpha\,dV_\beta$$

Recalling that currents are constant (direct), we rewrite the right-hand side in the form

$$\sum_{\alpha=1}^{N} L_{\alpha\alpha}I_\alpha^2 + \sum_{\alpha\neq\beta}\sum L_{\alpha\beta}I_\alpha I_\beta = \sum_{\alpha,\,\beta} L_{\alpha\beta}I_\alpha I_\beta \tag{33.10}$$

where we have denoted

$$L_{\alpha\alpha} = \frac{\mu}{4\pi}\,\frac{1}{I_\alpha^2}\iint_{(V_\alpha)} \frac{\mathbf{j}(\mathbf{r}_\alpha)\cdot\mathbf{j}(\mathbf{r}'_\alpha)}{|\mathbf{r}_\alpha-\mathbf{r}'_\alpha|}\,dV_\alpha\,dV'_\alpha$$

$$L_{\alpha\beta} = \frac{\mu}{4\pi}\,\frac{1}{I_\alpha I_\beta}\int_{V_\alpha}\int_{V_\beta} \frac{\mathbf{j}(\mathbf{r}_\alpha)\cdot\mathbf{j}(\mathbf{r}_\beta)}{|\mathbf{r}_\alpha-\mathbf{r}_\beta|}\,dV_\alpha\,dV_\beta \tag{33.11}$$

These quantities are independent of the intensity of current. Applying relation (12.15) to the right-hand side of (33.11) for $L_{\alpha\beta}$, we obtain

$$L_{\alpha\beta} = \frac{\mu}{4\pi}\oint_{s_\alpha}\oint_{s_\beta} \frac{d\mathbf{s}_\alpha\,d\mathbf{s}_\beta}{|\mathbf{r}_\alpha-\mathbf{r}_\beta|}\quad(\alpha\neq\beta) \tag{33.11$'$}$$

A similar transition in $L_{\alpha\alpha}$ would be meaningless since the corresponding integral for linear conductors is divergent. Calculation of $L_{\alpha\alpha}$ must use formula (33.11).

Quantities $L_{\alpha\beta}$ are called the *inductances*; if $\alpha = \beta$, they are called the *self-inductances*, and if $\alpha \neq \beta$, the *mutual inductances* of the system of quasilinear contours. Radius vectors $\mathbf{r}_\alpha$, $\mathbf{r}'_\alpha$ scan the points of the αth contour.

Elements ds_α and ds_β in formula (33.11′) are taken on different contours. Analyzing two contours in § 12, we considered only this last case. The expressions (33.11) show that inductances are determined only by the geometric shape of each contour and the mutual arrangement of these contours in space. It is also obvious that $L_{\alpha\beta} = L_{\beta\alpha}$.

With (33.10), relation (33.9) for the magnetic flux is rewritten as follows:

$$\Phi_\alpha = \sum_{\beta=1}^{N} L_{\alpha\beta}I_\beta \tag{33.12}$$

As follows from (33.7), (33.8) and (33.12), the magnetic energy of a system of contours is

$$W=\frac{1}{2}\sum_{\alpha=1}^{N} I_\alpha\Phi_\alpha=\frac{1}{2}\sum_{\alpha,\,\beta=1}^{N} L_{\alpha\beta}I_\alpha I_\beta \tag{33.13}$$

This expression for energy corresponds to the notion of instantaneous long-range interaction between currents. Terms with $\alpha = \beta$ describe the effect of a current on itself. The terms with $\alpha \neq \beta$ must be interpreted as the result of interaction between different contours s_α and s_β. As $L_{\alpha\beta}$ is symmetric, the energy of this interaction is

$$W_{\alpha\beta} = L_{\alpha\beta}I_\alpha I_\beta \tag{33.14}$$

33.3. Let loop s_α be displaced virtually as a rigid body, so that radius vector $\mathbf{r}_\alpha$ at any point of the loop is given the same infinitesimal increment. By grad_α we denote the corresponding operation of differentiation. Assume that all other loops s_β for $\beta \neq \alpha$ are fixed. Then formula (12.10) for interaction force can be applied to interaction between loop s_α and any of the loops s_β. If subscript 1 in this formula is replaced by β, and subscript 2 by α, and we use the notations (33.11), then the force applied to loop α by loop β takes the form

$$\mathbf{F}_{\beta\alpha} = (\text{grad}_\alpha\, L_{\alpha\beta})\, I_\alpha I_\beta \tag{33.15}$$

If we assume that one of the loops s_β is displaced, while the former loop s_α remains fixed, then $\text{grad}_\alpha\, L_{\alpha\beta} = -\text{grad}_\beta\, L_{\alpha\beta}$ for $\alpha \neq \beta$, since the dependence on coordinates is determined by function $\frac{1}{|\mathbf{r}_\alpha-\mathbf{r}_\beta|}$ in formula (33.11′). This again gives the law of equal action and reaction, $\mathbf{F}_{\alpha\beta} = -\mathbf{F}_{\beta\alpha}$, already discussed in § 12.

However, it will be useful to try and substantiate the expression for force (33.15) directly by resorting to the field energy (33.13). Namely, we can make use again of a virtual displacement of each of the loops, and analyze the change in the energy this produces. For instance, if loop s_β is displaced as a solid at a velocity $\mathbf{u}_\beta$, then mechanical work of force $\mathbf{F}_\beta$ applied to this loop by the magnetic field is equal to $\mathbf{F}_\beta\cdot\mathbf{u}_\beta$ (per unit time). So it would seem that the energy conservation law can be written in the form $\mathbf{F}_\beta\cdot\mathbf{u}_\beta + dW/dt = 0$, and force $\mathbf{F}_\beta$ could be found by substituting (33.13). The result will be, however, incorrect. The thing is that magnetic fluxes crossing all loops would be changed by such displacement of the loop. This will be caused by electromagnetic induction which will generate additional electric field strength $\mathbf{E}$ in each loop, and this will lead to the generation of induction currents. Consequently, currents in each loop will not be constant during this displacement, and this will invalidate formula (33.13). However, we can assume that external electromotive forces acting in each loop and producing the current are so changed in the course of the virtual displacement that their combined action with the induction electromotive force maintains all currents at their initial value.

Let us write out the formulas representing this assumption. The energy conservation law must be written according to § 3:

$$\mathbf{F}_\beta\cdot\mathbf{u}_\beta+\frac{dW}{dt}+\int \mathbf{j}\cdot\mathbf{E}\,dV=0 \tag{33.16}$$

The last term here is the work of electric forces. With external electromotive forces taken into account, current is given by (33.2), and so $\mathbf{j}\cdot\mathbf{E}=\sigma^{-1}j^2-\mathbf{j}\cdot\mathbf{E}'$. By analogy to what we did above, we change from integration over volume to integration over quasilinear loops by (12.5), assuming the currents in each loop constant, and thus obtain

$$\int \mathbf{j}\cdot\mathbf{E}'\,dV=\sum_{\alpha=1}^{N} I_\alpha\oint_{s_\alpha}\mathbf{E}'\cdot d\mathbf{s}_\alpha=\sum_{\alpha=1}^{N} I_\alpha U_\alpha$$

$$\int \frac{j^2}{\sigma}\,dV=\sum_{\alpha=1}^{N} I_\alpha^2\oint_{s_\alpha}\frac{ds_\alpha}{\sigma\Delta\Sigma_\alpha}=\sum_{\alpha=1}^{N} I_\alpha^2 R_\alpha \tag{33.17}$$

Equation (33.16) then takes the form

$$\mathbf{F}_\beta\cdot\mathbf{u}_\beta+\frac{dW}{dt}+\sum_{\alpha=1}^{N} I_\alpha^2 R_\alpha-\sum_{\alpha=1}^{N} I_\alpha U_\alpha=0 \tag{33.18}$$

On the other hand, Faraday's law (M.3) for an arbitrary loop s_α is similarly written after substituting $\mathbf{E}=\sigma^{-1}\mathbf{j}-\mathbf{E}'$ in the form

$$I_\alpha R_\alpha-U_\alpha=-\frac{d\Phi_\alpha}{dt} \tag{33.19}$$

Assuming again all the currents to be constant, we can make use of formula (33.13), recalling that

$$\frac{dW}{dt}=\frac{1}{2}\sum_\alpha I_\alpha\frac{d\Phi_\alpha}{dt}$$

Substituting this relation together with (33.19) into (33.18), we obtain

$$\mathbf{F}_\beta\cdot\mathbf{u}_\beta=\frac{1}{2}\sum_\alpha I_\alpha\frac{d\Phi_\alpha}{dt}=\left(\frac{dW}{dt}\right)_{\text{all } I_\alpha=\text{const}} \tag{33.20}$$

If the above arguments were ignored, the right-hand side of this formula would have a reversed sign. Only loop s_β is displaced, and so the right-hand side can also be written in the form $dW/dt=$

$= \text{grad}_\beta W \cdot \mathbf{u}_\beta$ [2]. Of course, we in fact differentiate inductances (33.11) in the expression for energy W, and namely those which are functions of $\mathbf{r}_\beta$, that is coefficients $L_{\beta\alpha}$ for all values of α. Consequently, the force applied to the chosen loop s_β by all other loops (including itself) is equal, owing to the arbitrary value of displacement velocity $\mathbf{u}_\beta$, to

$$\mathbf{F}_\beta = \text{grad}_\beta W \tag{33.21}$$

Equation (33.15) for the force of interaction between two loops is a straightforward corollary of (33.21), here we need not carry out this derivation.

33.4. Let us consider now time-dependent currents in nonbranching linear loops. The currents will be considered quasistationary. This means that current $I(t)$ at any moment of time t is the same at all points of the loop but may change with time. In this case formulas (33.17) obtained above remain valid. Denote the self-inductance of the loop by L. According to (33.13), the magnetic energy associated with the electric current in the loop is equal to $W_{(m)} = (1/2)\,LI^2$. Assume that the circuit also contains a capacitor. When a current passes through the circuit, time-dependent electric charges appear on the capacitor plates. Introduction of a capacitor into a circuit consisting of conductors makes the circuit open. Nevertheless, alternating currents can pass through such circuit. Let us analyze this statement. We have seen already in § 1 that in the general case electric current is made up of two terms: conduction current (due to motion of free charges) and displacement current $\partial\mathbf{D}/\partial t$. The condition of quasistationary current in fact means that the same current is considered to set up simultaneously at all points of the circuit. In other words, the velocity of propagation of electromagnetic perturbations is assumed practically infinite. The derivation of the wave equation for an electromagnetic field shows that this assumption is justified only if term $\partial\mathbf{D}/\partial t$ is negligibly small. The situation is quite opposite in the case of a capacitor. Conduction current across a capacitor must be assumed zero, and the main role is played precisely by the displacement current. By virtue of the continuity equation, the displacement current across the capacitor must be equal to the conduction current in the conductors connected to the capacitor plates. Hence, $dq/dt = I$ and, additionally, if charge $q(t)$ is formed on one plate of the capacitor, the charge on the other plate must be $-q(t)$ (since the circuit as a whole is considered neutral).

According to (3.5), the change of the electric field energy in the volume of a capacitor is given by integral $\int \mathbf{E}\cdot\frac{\partial\mathbf{D}}{\partial t}\,dV$, which depends directly on the displacement current. If the medium between the

[2] The definition of operation grad_β see earlier, on p. 258.

capacitor plates is linear, then we have (as in § 3) $dW_{(e)}/dt$, where $W_{(e)} = (1/2)\int \mathbf{E}\cdot\mathbf{D}\,dV$. Assume that term $\partial\mathbf{A}/\partial t$ in the expression of field $\mathbf{E}$ in terms of potentials can be ignored (this is equivalent to neglecting the "vortex" electric field due to the Faraday induction in the volume of the capacitor), and that no space charge is present between the capacitor plates (only plate surfaces contain charges). Then calculations of § 30, which yielded formula (30.9), are valid for $W_{(e)}$ at any moment of time.

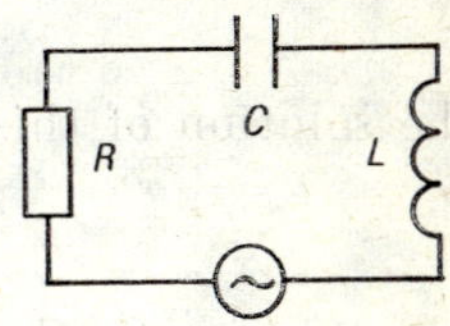

Fig. 34

In our case (a single capacitor) only one capacitance C and one potential coefficient are present, the two coefficients being mutually inverse. Therefore, electric energy can be written in the form $W_{(e)} = q^2/2C$. The law of conservation of total energy can be written in the form of (33.18), where we have to assume $\mathbf{u}_\beta = 0$, $W = W_{(m)} + W_{(e)}$ and consider a single loop:

$$\frac{dW}{dt} + I^2R = IU \tag{33.22}$$

In the general case coefficients L and C may be functions of the manner in which currents and charges vary with time. However, such dependence is often neglected in the theory of quasistationary currents. Consequently,

$$\frac{dW_{(m)}}{dt} = LI\frac{dI}{dt}, \quad \frac{dW_{(e)}}{dt} = \frac{1}{C}q\frac{dq}{dt} = \frac{1}{C}qI$$

The energy conservation law (33.22) thus takes the form

$$L\frac{dI}{dt} + \frac{1}{C}q + IR - U$$

In this form it is called the *second Kirchhoff law* for a nonbranching linear circuit. It can often be assumed for such linear contours (and in fact we did just that in our analysis) that thermal losses dominate on some segments of the loop (that is the "active" resistance R is predominant), while other segments are dominated by inductance (coils) or by capacitance (capacitors). The second Kirchhoff law can then be interpreted as the equality of the net voltage U produced in the circuit by external electromotive forces to the sum of voltages across the segments of ohmic resistance R and reactances L and C (Fig. 34). The first Kirchhoff law on branching of currents follows directly from the continuity equation. Together the first and second Kirchhoff laws make the foundation of the general theory of linear circuits in which it is assumed (as we did implicitly above) that inductance and capacitance of circuit elements are independent of current I.[3]

Since $I = \dot{q}$, $\dot{I} = \ddot{q}$, in our case the second Kirchhoff law takes the form

$$L\frac{d^2q}{dt^2}+R\frac{dq}{dt}+\frac{q}{C}=U \tag{33.23}$$

The solution of this equation for $U = 0$ can be written as follows:

$$q=C_1e^{k_1t}+C_2e^{k_2t} \tag{33.24}$$

Here constants C_1 and C_2 are determined by initial conditions (conditions of connecting the circuit to the power source). In this solution

$$k_{1,2}=-\frac{R}{2L}\pm\lambda,\quad \lambda=\left(\frac{R^2}{4L^2}-\frac{1}{LC}\right)^{1/2} \tag{33.25}$$

As follows from the last two formulas, for $\frac{R}{2L}>\frac{1}{\sqrt{LC}}$ the quantity λ is real and the process in the circuit is aperiodic. If, however, $\frac{R}{2L}<\frac{1}{\sqrt{LC}}$, then $\lambda = i\omega$, where $\omega\equiv\left(\frac{1}{LC}-\frac{R^2}{4L^2}\right)^{1/2}$ is a real quantity, and charge q on the capacitor (and consequently, current $I = dq/dt$ in the circuit) undergoes damped oscillations with frequency ω. In particular, for $R\to 0$ we arrive at the Thomson formula $\omega = (LC)^{-1/2}$, and the oscillations are undamped.

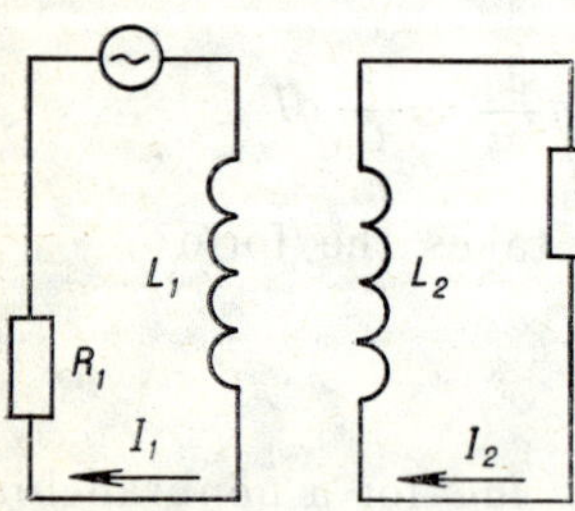

Fig. 35

Quite naturally, so far we were taking into account only self-inductance of the circuit. An inductive coupling between two alternating-current contours is an important case in which mutual inductances play a significant role. If the current in one of them is I_1, and in the other I_2, and capacitance effects can be ignored (Fig. 35), then the expression for magnetic energy $W_{(m)}$ which enters the energy conservation law for each contour must include the mutual inductance term $L_{12}I_1I_2$. The second Kirchhoff law for the set of such two contours is written as a system of two simultaneous differen-

[3] See textbooks on electronics, for example: A. P. Molchanov and P. N. Zanadvorov, *Electrical and Radio Engineering for Physicists*, Mir Publishers, Moscow, 1973.

tial equations

$$L_{11}\frac{dI_1}{dt}\pm L_{12}\frac{dI_2}{dt}+R_1I_1=U$$

$$L_{22}\frac{dI_2}{dt}\pm L_{12}\frac{dI_1}{dt}+R_2I_2=0$$

Currents I_1 and I_2 will be found by resolving this system. Note that in each of the contours the mutual induction generates an additional "external" electromotive force proportional to the time rate of the current variation in the other contour. Here we have assumed that the external e.m.f. U is included only into one of the contours.

The reader undoubtedly knows that oscillations corresponding, for example, to the Thomson formula may result in emission of electromagnetic waves to the surrounding space. Thus, a contour may be connected to an antenna which often can be described as an electric or magnetic dipole emitting waves. The process of the emission of radiation was already described in § 17. Energy dissipation via emission of radiation must be replenished by a power source included into the emitting contour.

§ 34. Eddy currents. Thermoelectric and thermomagnetic phenomena. Hall effect

34.1. So far we were studying only the properties of currents in linear circuits. Let us turn to more general problems originating with the electric current. First, we shall consider the problem of finding bulk distributions of currents in continuous media. Such currents are generated in conductors, for example, by alternating magnetic fields (eddy or the Foucault currents). The analysis requires that we turn to the general Maxwell equations. However, we shall assume that there is no displacement current, that is $\partial\mathbf{D}/\partial t = 0$. In fact this assumption is equivalent to neglecting the time lag effects (see § 33). Moreover, we assume, as before, that Ohm's law is valid, $\mathbf{j} = \sigma\mathbf{E}$, and that electric conductivity σ has the same value as in the stationary case. Microscopically this means that the frequency of the external field must be much lower than the inverse value of the mean free time of electrons in the conductor. Appropriate estimates lead to a conclusion that at the most the field may vary at infrared frequencies. Taking account of all these assumptions, we write the Maxwell equations in the form

$$\operatorname{curl}\mathbf{E}=-\frac{\partial\mathbf{B}}{\partial t},\ \operatorname{curl}\mathbf{B}=\mu\sigma\mathbf{E},\ \operatorname{div}\mathbf{B}=0 \tag{34.1}$$

The second equation yields div $\mathbf{E} = 0$ (the medium is assumed to be homogeneous). Besides, we assume that $\mathbf{B} = \mu\mathbf{H}$. Taking curl of the second equation in (34.1) and transforming curl curl in Cartesian coordinates via (B.21), we obtain

$$\Delta \mathbf{B} = \mu\sigma \frac{\partial \mathbf{B}}{\partial t} \tag{34.2}$$

This is the equation of the heat conduction (or diffusion) type. It must be solved under conditions (4.10) and (4.12) at the conductor's boundary, that is for $B_{n1} = B_{n2}$, $H_{t1} = H_{t2}$. Once the corresponding field distribution is found, the current distribution can be found from the formula $\mathbf{j} = \frac{1}{\mu}$ curl $\mathbf{B}$. However, we could derive the equation for current density directly. Indeed, taking curl of both sides of (34.2), we obtain

$$\Delta \mathbf{j} = \mu\sigma \frac{\partial \mathbf{j}}{\partial t}$$

Most often equation (34.2) is used to solve problems of the following types. Either one considers a case when the external field is switched off and the effect of interest is the damping out of the field in the conductor; or it is assumed that the external field is periodic with frequency ω; in this case one analyzes penetration into the conductor of magnetic field and of the electric field induced by magnetic field in this conductor, as well as the distribution of currents in the bulk of the conductor. Let us take the second problem. Let the conductor surface coincide with plane $z = 0$. Magnetic field is written in the form $\mathbf{B} = \mathbf{B}^{(0)} e^{-i\omega t}$. In addition, we assume that $\mathbf{B}^{(0)}$ depends only on z (and axis z points from the surface into the conductor). Equation (34.2) is then recast to

$$\frac{\partial^2 \mathbf{B}^{(0)}}{\partial z^2} + k^2 \mathbf{B}^{(0)} = 0$$

where

$$k \equiv (i\mu\sigma\omega)^{1/2} = (1+i)\sqrt{\mu\sigma\omega/2}$$

If we assume that $\mathbf{B} \to 0$ for $z \to \infty$, then $\mathbf{B}^{(0)} = \mathbf{b} e^{ikz}$, where $\mathbf{b}$ is a constant vector, that is $\mathbf{B}(0) = \mathbf{b} e^{-i\omega t}$. Denote $d \equiv \sqrt{2/\mu\sigma\omega}$. The corresponding solution is written in the form

$$\mathbf{B} = \mathbf{b} e^{-z/d} e^{i(z/d - \omega t)} \tag{34.3}$$

We thus find that field decreases exponentially into the conductor. Quantity d is known as the field penetration depth and determines the thickness of a surface layer in the conductor where magnetic field and currents must be taken into consideration. The existence of a magnetic field and currents in this surface layer is called the *skin effect*. Currents involved in the skin effect dissipate energy (the Joule heat is released). But here we are not going to calculate this dissipation. Later we shall analyze a problem of propagation of electromagnetic waves in a conducting medium with displacement currents $\partial\mathbf{D}/\partial t$ taken into account (see § 38).

34.2. Another generalization of the former approach to electric currents consists in taking account of thermodynamic arguments. Namely, we shall assume that temperature gradients, creating heat flux, exist in the continuous medium where conduction currents are generated. Electric and magnetic fields are also present. We shall begin with only electric field, and assume the medium to be isotropic. Now, two factors make a charged particle move in any direction: an electric field, and heat conduction. This brings an additional term proportional to grad T into the expression for the conduction current, and a term proportional to $\mathbf{E}$ into that of the heat flux. The two factors are as if superposed. The conduction current and heat flux densities, $\mathbf{j}$ and $\mathbf{q}$, will be written as follows:

$$\mathbf{j} = \sigma\mathbf{E} + \beta \operatorname{grad} T, \quad \mathbf{q} = \gamma\mathbf{E} + \zeta \operatorname{grad} T \qquad (34.4)$$

Transfer phenomena constitute one of the problems of the non-equilibrium (irreversible) thermodynamics. It can often be assumed that the Onsager reciprocity relation is valid, relating different transfer processes. Here we shall formulate this principle; its substantiation is given by the methods of statistical physics[4]. This principle establishes a relation between coefficients in formulas (34.4). Each current J_i in the medium ($\mathbf{j}$ and $\mathbf{q}$ in our case) is driven by a corresponding "generalized force" X_i. To find this generalized force we must take into account that transfer of the quantity of heat dQ to a volume of the medium changes its entropy S; namely, according to the second law of thermodynamics, $dS = dQ/T$. This gives the time rate of the variation of entropy $\dot{S} \equiv dS/dt$. Let us assume that $\dot{S}$ is a linear function of currents J_i: $\dot{S} = -\sum_i J_i X_i$. Coefficients of J_i represent the *generalized forces.* Assume now that currents too can be written as linear functions of generalized forces, namely

$$J_i = -\sum_i a_{ik} X_k \qquad (34.5)$$

The Onsager principle consists in stating that in the absence of magnetic field the matrix of coefficients a_{ik} is symmetrical[5]:

$$a_{ik} = a_{ki} \qquad (34.6)$$

---

[4] See, for example, L. D. Landau and E. M. Lifshitz, *Statistical Physics*, Course of Theoretical Physics, vol. 5 (3rd edition), Pergamon Press, Oxford, 1979.

[5] Linear relations of the type of (34.5) and (34.4) hold if the analysis is restricted to the effects which are small in a certain sense, that is, if currents are assumed to undergo sufficiently small variation when generalized forces are changed (see a discussion of this point in the monograph by Landau and Lifshitz, cited above).

Consider the change in entropy in the case we are discussing. The quantity of heat released per unit time in each element of volume is $-\text{div}\,\mathbf{q}$ (this is simply the definition of heat flux $\mathbf{q}$); in addition, the work done by electric field results in a dissipation of energy $\mathbf{j}\cdot\mathbf{E}$. Hence,

$$\dot{S} = -\int \frac{\text{div}\,\mathbf{q}}{T}\,dV + \int \frac{\mathbf{j}\cdot\mathbf{E}}{T}\,dV$$

The first integral can be transformed via formula (B.14$_2$) and the Gauss theorem. With obvious assumption, the above formula then takes the form

$$\dot{S} = \int \mathbf{q}\cdot\text{grad}\,\frac{1}{T}\,dV + \int \frac{\mathbf{j}\cdot\mathbf{E}}{T}\,dV$$

Evidently, the generalized forces here are $-\text{grad}\,(1/T)$ and $-(\mathbf{E}/T)$. Equations (34.4) are transformed to (34.5) as follows:

$$\mathbf{j} = -\sigma T\left(-\frac{\mathbf{E}}{T}\right) - \beta T^2\,\text{grad}\,\frac{1}{T}, \quad \mathbf{q} = -\gamma T\left(-\frac{\mathbf{E}}{T}\right) - \zeta T^2\,\text{grad}\,\frac{1}{T}$$

Relation (34.6) then is expressed by

$$\gamma = -\beta T, \quad \sigma = -\zeta T \tag{34.7}$$

It will be convenient to use equations (34.4) in the form resolved in $\mathbf{E}$, namely

$$\mathbf{E} = r\mathbf{j} + \eta\,\text{grad}\,T, \quad \mathbf{q} = \Pi\mathbf{j} - \varkappa\,\text{grad}\,T \tag{34.8}$$

Here

$$r \equiv \sigma^{-1}, \quad \eta = -\beta\sigma^{-1}, \quad \Pi = \gamma\sigma^{-1}, \quad \varkappa = \gamma\beta\sigma^{-1} - \zeta$$

From the first relation of (34.7) we obtain

$$\Pi = \eta T, \quad \varkappa = -\left(\frac{\beta^2 T}{\sigma} + \zeta\right) \tag{34.9}$$

In the case $\mathbf{j} = 0$ the first equation of (34.8) gives

$$\mathbf{E} = \eta\,\text{grad}\,T \tag{34.10}$$

that is, a distribution of temperatures in a conductor produces in it an electric field. By introducing a scalar potential, we can transform the above equation to $-d\varphi/dT = \eta$. Quantity η is called the *differential thermal electromotive force*; its value differs in different conductors. As a result, the thermal electromotive force can be employed as one of the external electromotive forces, that is, as a basis for a thermoelectric cell.

Consider a circuit composed of two different metals a and b soldered at points *1* and *2* (Fig. 36). Temperature at point *1* is denoted by T_1, and that at point *2*, by T_2. By analogy to the derivation of (33.6), determine the electromotive force in this circuit. We disconnect the circuit, for example, somewhere within the segment of metal a, and assume that the thus formed free ends have equal temperatures (see Fig. 36). Equations (33.6) and (34.10) yield

$$U=\oint \eta \frac{dT}{ds}\, ds=\int_{T_0}^{T_1} \eta_a\, dT+\int_{T_1}^{T_2} \eta_b\, dT+\int_{T_2}^{T_0} \eta_a\, dT=\int_{T_1}^{T_2} (\eta_b-\eta_a)\, dT$$

All other coefficients in (34.8) are, like η, determined by specific properties of the conductor. This produces a number of additional effects. Let current $\mathbf{j}$ pass through a junction of two different metals a and b. Temperature is assumed equal at all points of the circuit.

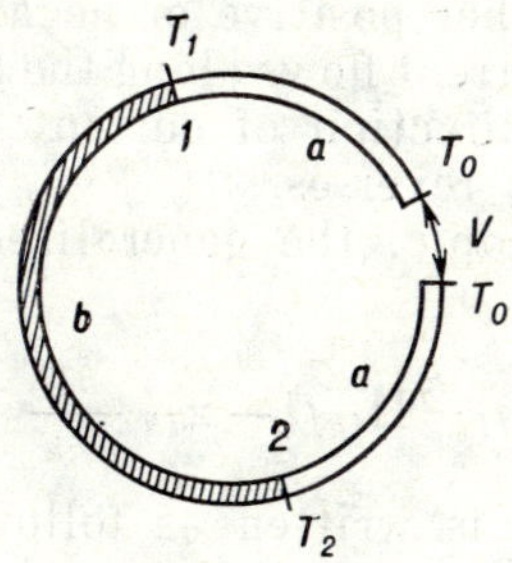

Fig. 36

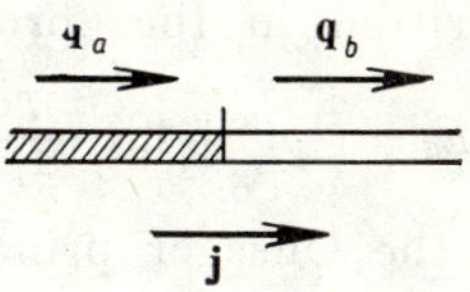

Fig. 37

According to the second equation of (34.8), current-induced heat fluxes $\mathbf{q}_a$ and $\mathbf{q}_b$ (Fig. 37) will appear on the two sides of the junction, with, in the general case, $q_a \neq q_b$. Consequently, there will be an excess heat W, released or absorbed in the junction, namely,

$$W = q_a - q_b = (\Pi_a - \Pi_b)\, j$$

The effect is called the *Peltier effect*, and coefficient Π is the Peltier coefficient. The direction of current through the junction determines whether the heat will be released or absorbed (i.e. it determines the sign of W).

Let us introduce a temperature gradient. We shall calculate in more detail the quantity of heat Q released in the conductor. We have obtained above, in entropy calculations, that

$$Q = \mathbf{j}\cdot\mathbf{E} - \operatorname{div} \mathbf{q}$$

per unit volume and unit time. Substitution of (34.8) yields

$$Q=rj^2+\eta\,(\mathbf{j}\cdot\operatorname{grad} T)-\frac{d\Pi}{dT}\,(\mathbf{j}\cdot\operatorname{grad} T)=rj^2-\tau\,(\mathbf{j}\cdot\operatorname{grad} T)$$

We also assume that $\operatorname{grad} \Pi = \frac{d\Pi}{dT} \operatorname{grad} T$, and that the term including the thermal conductivity in (34.8) makes a comparatively small contribution. The first term represents the Joule heat, and the second corresponds to the so-called *Thomson effect* (an additional quantity of heat released as a result of a nonuniform heating of the conductor). The Thomson coefficient τ is related to the Peltier coefficient by the following relation:

$$\tau = -\eta + \frac{d\Pi}{dT} = T\frac{d\eta}{dT}$$

Here we have used formula (34.9). The Thomson effect is easily distinguished from the Joule heat because the latter is proportional to j^2 and, in contrast to the former, is independent of the direction of current. Coefficient τ may be either positive or negative. In the first case heat is released when the current flows along the temperature gradient, and absorbed when the direction of current is reversed. In the second case the relationship reverses.

34.3. If the conductor is anisotropic, the generalized equations (34.8) are written in the form

$$E_i = r_{ik}j^k + \eta_{ik}\frac{\partial T}{\partial x_k}, \quad q_i = \Pi_{ik}j^k - \varkappa_{ik}\frac{\partial T}{\partial x_k} \tag{34.11}$$

In this case the Onsager principle is written as follows:

$$r_{ik} = r_{ki}, \quad \varkappa_{ik} = \varkappa_{ki}, \quad \Pi_{ik} = T\eta_{ki} \tag{34.12}$$

Assume now that the conducting medium is in an external magnetic field **H**. Both the experiments and the microscopic theory show that in this case coefficients in equations (34.11) are functions of this magnetic field. Then the Onsager principle must be written in the generalized form:

$$r_{ik}(\mathbf{H}) = r_{ki}(-\mathbf{H}), \quad \varkappa_{ik}(\mathbf{H}) = \varkappa_{ki}(-\mathbf{H})$$
$$\Pi_{ik}(\mathbf{H}) = T\eta_{ki}(-\mathbf{H}) \tag{34.13}$$

Even if the medium is isotropic in the absence of magnetic field, the field usually induces anisotropy. For example, if magnetic field is sufficiently small, we can expand function $r(\mathbf{H})$ in powers of its components, that is

$$r(\mathbf{H}) = r(0) + r_i^{(1)}H^i + r_{ik}^{(2)}H^iH^k + \ldots$$

Terms with coefficients $r_i^{(1)}$, $r_{ik}^{(2)}$, ... describe the field-induced dependence of properties of the medium on the field direction. Let us assume that the field is so weak that nonlinear effects in the expansion can be dropped. In this case we can, for example, find the additional terms in equations (34.8) in a straightforward manner, using the property of field transformation under rotations and reflections in space. It has been shown at the beginning of § 2 that field **E** is a polar vector; the same is true for **j**, **q** and grad T, while **H** is an axial vector. Thus, for example a correction to vector **E**, characterized by the required transformation properties of a polar vector and linear with respect to both $\mathbf{j}$ and $\mathbf{H}$, must be proportional to vector product $\mathbf{H} \times \mathbf{j}$. Thus, it can be expected that equations (34.8) must be replaced with the following:

$$\mathbf{E} = r\mathbf{j} + R\mathbf{H} \times \mathbf{j} + \eta \operatorname{grad} T + N\mathbf{H} \times \operatorname{grad} T \qquad (34.14_1)$$

$$\mathbf{q} = \Pi\mathbf{j} + \mathrm{B}\mathbf{H} \times \mathbf{j} - \varkappa \operatorname{grad} T + L\mathbf{H} \times \operatorname{grad} T \qquad (34.14_2)$$

Here R, N, B, L are constant coefficients. The principle of symmetry of kinetic coefficients yields $B = NT$.

The first of the above equations shows that with grad $T = 0$ the current flowing in the conductor in a magnetic field produces electric field $R\mathbf{H} \times \mathbf{j}$. Expansion of coefficient r in powers of field $\mathbf{H}$ must correspond to a similar expansion of electric conductivity σ. If we use for the currents an expression of the type of (34.4), this will produce a correction term in the expression for $\mathbf{j}$; it is clear from similar arguments concerning the transformation relation that this correction must be proportional to vector product $\mathbf{E} \times \mathbf{H}$. Thus, magnetic field generates an additional current in the direction normal to the electric field, and the magnitude of this current is proportional to the magnetic field. The phenomena outlined above constitute different aspects of the *Hall effect*.

34.4. Several other observable effects can be analyzed using equations (34.14). Let us choose axis z of Cartesian coordinates in the direction of field $\mathbf{H}$, and consider a particular case of current flowing along axis x: $j = j_x$.

First we assume that temperature gradient along axis x is zero, $\partial T/\partial x = 0$, and heat flux is zero along axis y, $q_y = 0$. By projecting equation (34.14_2) onto axis y we obtain $\partial T/\partial y = BHj/\varkappa$. This means that in a magnetic field applied along axis z the current flowing along x produces a temperature gradient along axis y. This is the so-called *Ettingshausen effect*.

Assume now that, as before, $q_y = 0$ but, in contrast to the preceding case, $j = 0$ and $\partial T/\partial x \neq 0$. Projecting again equation (34.14_2) onto axis y, we obtain a temperature gradient due to the magnetic field, directed along axis y: $\frac{\partial T}{\partial y} = \frac{L}{\varkappa} H \frac{\partial T}{\partial x}$. This formula gives the *Leduc-Righi effect*: magnetic field affects heat conduction.

Equation (34.14_1) also shows that the temperature gradient along axis x may produce in a magnetic field a thermo-e.m.f. along axis y. Indeed, if $j = 0$ and $\partial T/\partial y = 0$, then $E_y = NH\, \partial T/\partial x$. This effect of magnetic field on thermo-e.m.f. is called the *Nernst effect*.

The elementary information on thermoelectric and thermomagnetic phenomena given above makes up a certain, though incomplete, picture of the causes producing these phenomena. An investigation of transport processes in conductors placed in an electromagnetic field constitutes a considerable and very important part of the modern irreversible statistic thermodynamics. Note also that we have omitted from our analyses of factors generating electric current the transformation of chemical to electric energy, fundamental for functioning of electrolytic cells.

§ 35. Elements of magnetohydrodynamics

35.1. Magnetohydrodynamics studies the properties of conducting liquids or gases in electromagnetic fields. As these media are more or less easily deformable, the interaction of the electromagnetic field with the currents in these media may substantially affect their hydrodynamic properties. Here, as in all other sections of the book, we restrict the exposition to the phenomenological approach and do not consider problems concerning the microscopic structure of the medium. However, the liquid considered in magnetohydrodynamics is assumed consisting of negative and positive charged particles (for example, electrons and ions) freely moving with respect to one another. In most cases the medium can be thought of as neutral on the whole, that is the positive and negative charges in any microscopically small volume balance each other. This property is typical for molten metals and, which is especially important, for completely ionized gases (this state of the matter is called *plasma*).

At present the study of phenomenological properties of plasmalike media by the methods of magnetohydrodynamics, and the study of their microscopic properties by the methods of statistical physics, are rapidly progressing and play an extremely important part in modern astrophysics and in the theory of controlled thermonuclear reactions[6]. In both cases only completely ionized media are considered. Here we shall give only the most elementary exposition of magnetohydrodynamics meant to clarify the specific features of the subject.

In analyzing the moving media, it is necessary to distinguish between the laboratory and the co-moving (Lagrange) reference frame. Denote electromagnetic field in the laboratory reference frame by $\mathbf{E}$ and $\mathbf{B}$, and that in the Lagrange frame by $\mathbf{E}'$ and $\mathbf{B}'$. Let the medium move in the laboratory reference frame (at least, for a sufficiently small time interval) at a constant velocity $\mathbf{v}$. Then the co-moving reference frame is inertial, and fields $\mathbf{E}'$, $\mathbf{B}'$ and $\mathbf{E}$, $\mathbf{B}$ are related by the Lorentz transformation discussed in § 7. If $v \ll c$, we are to use formulas (7.13′). In the SI system of units one has to take into account the remark made immediately after

[6] A consistent study of magnetohydrodynamics was pioneered by H. Alfvén in 1942.

equation (7.21). Equations (7.13′) are then written in the form

$$\mathbf{E}' \simeq \mathbf{E} + \mathbf{v} \times \mathbf{B}, \quad \mathbf{B}' \simeq \mathbf{B} - \frac{1}{c^2} \mathbf{v} \times \mathbf{E} \simeq \mathbf{B} \tag{35.1}$$

It will be instructive to derive the first of the nonrelativistic relations (35.1) by resorting to the Faraday law of induction for a moving contour. If the contour were at rest in the laboratory reference frame, the law of induction should have been written in the form

$$\oint \mathbf{E} \cdot d\mathbf{s} = -\int \frac{\partial \mathbf{B}}{\partial t} \cdot \mathbf{n}\, d\sigma \tag{35.2}$$

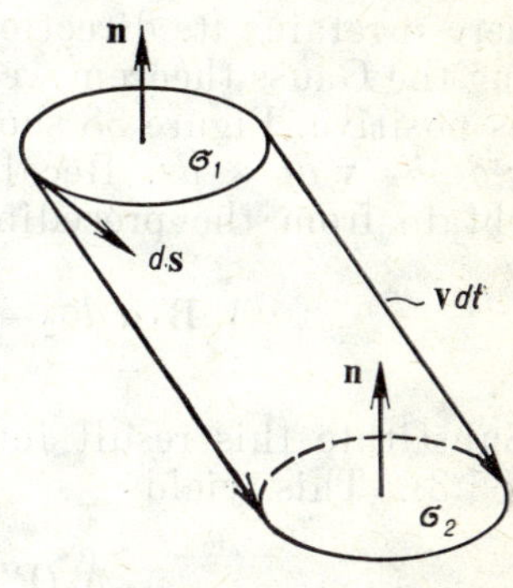

Fig. 38

In our approximation $\mathbf{B}' \simeq \mathbf{B}$, so that the motion of the medium can be treated as its translation with respect to the magnetic field. Consequently, the difference between $\mathbf{E}'$ (field strength determining e.m.f. in the moving contour) and $\mathbf{E}$ is stipulated, from the standpoint of the laboratory reference frame, by the need to apply the Faraday law to the contour moving together with the medium, namely

$$\oint \mathbf{E}' \cdot d\mathbf{s} = -\frac{d}{dt} \int \mathbf{B} \cdot \mathbf{n}\, d\sigma \tag{35.3}$$

because the right-hand side depends on time both in the integrand and owing to the motion of the contour, so that integration limits are variable (Fig. 38). The variation of the integral is

$$d \int \mathbf{B} \cdot \mathbf{n}\, d\sigma = \int\limits_{t+dt} \mathbf{B} \cdot \mathbf{n}\, d\sigma_2 - \int\limits_{t} \mathbf{B} \cdot \mathbf{n}\, d\sigma_1$$

$$\simeq \int\limits_{t} \mathbf{B} \cdot \mathbf{n}\, d\sigma_2 - \int\limits_{t} \mathbf{B} \cdot \mathbf{n}\, d\sigma_1 + \int \frac{\partial \mathbf{B}}{\partial t} \cdot \mathbf{n}\, d\sigma_1\, dt \tag{35.4}$$

As can be seen from Fig. 38, we have introduced surface σ_1 covering the contour at time t, surface σ_2 at time $t + dt$, and a unit vector of normal $\mathbf{n}$. In the second part of (35.4) we have used the expansion into Taylor series $\mathbf{B}(t + dt) \simeq \mathbf{B}(t) + \left.\frac{\partial \mathbf{B}}{\partial T}\right|_t dt$. Consequently, the integrand in the first term of the right-hand side is determined by the values taken by field $\mathbf{B}$ at time t at those points of space which will be occupied by surface σ_2 at time $t + dt$. Therefore we shall consider a cylindrical surface shown in Fig. 38 and take at all its points the values of vector $\mathbf{B}$ which it assumes at time t. Then, according to the Maxwell equation (M′.2), $\oint \mathbf{B} \cdot \mathbf{n}\, d\sigma = 0$. On the other hand,

$$\oint \mathbf{B}\cdot\mathbf{n}\, d\sigma = -\int \mathbf{B}\cdot\mathbf{n}\, d\sigma_2 + \int \mathbf{B}\cdot\mathbf{n}\, d\sigma_1 + \int \mathbf{B}\cdot\boldsymbol{\nu}\, d\sigma$$

where the last integral is taken over the lateral surface of the cylinder, and $\boldsymbol{\nu}$ is the outward normal to this surface. Of course, the signs of the first two terms are opposite because in the arrangement chosen here $\mathbf{n}$ retains its direction when the contour moves, while in applying the Gauss theorem we choose the direction of the outward normal as positive. Figure 38 shows that we can write for the lateral surface $\boldsymbol{\nu}\, d\sigma = \mathbf{v}\, dt \times d\mathbf{s}$. Recalling that $\mathbf{B}\cdot(\mathbf{v} \times d\mathbf{s}) = -(\mathbf{v} \times \mathbf{B})\cdot d\mathbf{s}$, we obtain from the preceding equation that

$$\int_t \mathbf{B}\cdot\mathbf{n}\, d\sigma_2 - \int_t \mathbf{B}\cdot\mathbf{n}\, d\sigma_1 = -\oint (\mathbf{v}\times\mathbf{B})\cdot d\mathbf{s}\, dt$$

Substitute this result into (35.4) and take into account (35.2) and (35.3). This yields

$$\oint (\mathbf{E}' - \mathbf{E})\cdot d\mathbf{s} = \oint (\mathbf{v}\times\mathbf{B})\cdot d\mathbf{s} \qquad (35.3')$$

This equality is satisfied for (35.1). Term $\mathbf{v} \times \mathbf{B}$ in the expression for $\mathbf{E}'$ is, in the laboratory reference frame, precisely the Lorentz force applied to charges moving in a magnetic field. And the observer co-moving with the medium records vector $\mathbf{E}'$ as an "aggregate", as the e.m.f. in the contour.

35.2. Consider now the relation between conduction currents in the medium recorded in the two mentioned reference frames. It is given by the first of formulas (7.18). In our nonrelativistic limit, when $\gamma \approx 1$, we find $\mathbf{j} = \mathbf{j}' + \rho'\mathbf{v}$ (the second term is called the *convection current*). If the medium is assumed neutral, that is $\rho' = 0$, then $\mathbf{j} = \mathbf{j}'$. Assume now that Ohm's law, $\mathbf{j}' = \sigma\mathbf{E}'$, holds in a fixed medium. It then follows from the above arguments and from formula (35.1) that in the laboratory reference frame (i.e. for a moving medium) Ohm's law takes the form

$$\mathbf{j} = \sigma\,(\mathbf{E} + \mathbf{v} \times \mathbf{B}) \qquad (35.5)$$

Here we shall neglect such phenomena as the Hall effect.

Using (35.5), we can formulate a condition under which the liquid is an ideal conductor. In fact, we have to demand that $\sigma \to \infty$ for a finite current density $\mathbf{j}$. This requirement is met if

$$\mathbf{E} + \mathbf{v} \times \mathbf{B} = 0 \qquad (35.6)$$

For $\mathbf{v} = 0$ this condition reduces to the absence of electric field inside the ideal conductor. If $\mathbf{v} \neq 0$, condition (35.6) becomes nontrivial. From the point of view of transformation (35.1) this means $\mathbf{E}' = 0$, that is, there is no electromotive force in the moving contour. In other words, if Φ is the magnetic flux through the moving contour, then equation (35.3) yields $d\Phi/dt = 0$ if this contour moves together with the medium having infinite conductivity. The magnetic lines of force are thus said to be *frozen* into the medium. As follows from the given above derivation of the law of induction, the condition of freezing reflects the fact that matter in its motion carries with itself the magnetic lines of force. In other words, if a magnetic line of force passes at a given moment of time through a chosen element of matter, and this element is translated, then this line will pass through the same element after the translation. Clearly, the direction of translation velocity of this element need not coincide, in the general case, with the direction of magnetic lines of force. It is important that the condition of freezing was obtained for an absolutely arbitrary contour passing through the particles of medium and moving together with them. This point is important because for any fixed configuration of the magnetic field it could be possible to find such specially selected moving contours that the magnetic flux crossing them would remain constant. However, if the condition of freezing is violated, there should necessarily be such contours in which the corresponding magnetic flux varies.

The condition of freezing can also be obtained in the differential form. By using Stokes' theorem and equation curl $\mathbf{E} = -\partial\mathbf{B}/\partial t$ corresponding to formula (35.2), we obtain from (35.3′) that the law of induction for a moving contour can be rewritten in the form

$$\operatorname{curl} \mathbf{E}' = -\frac{\partial \mathbf{B}}{\partial t} + \operatorname{curl}(\mathbf{v} \times \mathbf{B}) \qquad (35.3'')$$

Hence, condition (35.6) is equivalent to

$$\frac{\partial \mathbf{B}}{\partial t} = \operatorname{curl}(\mathbf{v} \times \mathbf{B}) \qquad (35.6')$$

Either this equation or the original relation (35.6) can be used, the choice being dictated by convenience.

Denote by $\mathbf{v}_\infty$ the component of velocity $\mathbf{v}$ orthogonal to field $\mathbf{B}$ in the case $\sigma \to \infty$. As follows from (35.6),

$$\mathbf{v}_\infty = \frac{\mathbf{E} \times \mathbf{B}}{B^2} \qquad (35.7)$$

This motion of liquid at a velocity $\mathbf{v}_\infty$ is called the *electric drift* of the liquid. Formula (35.7) coincides with the expression for electric drift $\mathbf{v}_E$ (see p. 211) obtained in the analysis of motion of charged particles in an electromagnetic field.

35.3. It is clear that in the general case of an arbitrary electric conductivity the motion of a conducting liquid must be found simultaneously with the field acting in it. Indeed, the basic equations of

18–01150

hydrodynamics take in this case the following form:

$$\frac{\partial \varkappa}{\partial t} + \operatorname{div}(\varkappa \mathbf{v}) = 0 \tag{35.8}$$

$$\varkappa \frac{d\mathbf{v}}{dt} = \mathbf{j} \times \mathbf{B} + \mathbf{F} \tag{35.9}$$

Here $\varkappa$ is the mass density of the liquid. The first of the relations is the continuity equation. The second of them is the equation of motion, and its main feature is the presence of magnetic force $\mathbf{j} \times \mathbf{B}$ applied to electric currents in the liquid. Term $\mathbf{F}$ describes all other external forces affecting the motion, for example,

$$\mathbf{F} = -\operatorname{grad} p + \varkappa \mathbf{g} \tag{35.10}$$

where p is the pressure and $\mathbf{g}$ is the acceleration of free fall. Forces due to viscosity may, if necessary, be taken into account in the right-hand side of (35.10). The time derivative in (35.9) is substantial, that is

$$\frac{d}{dt} = \frac{\partial}{\partial t} + \mathbf{v} \cdot \operatorname{grad} \tag{35.11}$$

The presence of the magnetic field in the right-hand side of (35.9) constitutes the link between this equation and the Maxwell equations. In a good conductor the displacement current $\partial \mathbf{D}/\partial t$ can be considered negligibly small, so that the Maxwell equations can be used in the form

$$\operatorname{curl} \mathbf{E} + \frac{\partial \mathbf{B}}{\partial t} = 0, \quad \operatorname{curl} \mathbf{B} = \mu \mathbf{j} \tag{35.12}$$

We also assume that in the medium of interest $\mathbf{B} = \mu \mathbf{H}$ and μ is constant (in many cases $\mu \simeq \mu_0$). Hence, equations (35.8), (35.9), and (35.12) must be chosen as the basic equations of magnetohydrodynamics. In the general case they must be solved simultaneously for specific boundary and initial conditions. Besides, Ohm's law in its form (35.5) must be taken into account.

Specific effects which immediately follow from the system of equations of magnetohydrodynamics in a liquid conductor are the magnetic diffusion, magnetic viscosity, and magnetic pressure.

The *magnetic diffusion* characterizes the behavior of a magnetic field in a moving conductor. For its analysis we should use equations (35.12) and (35.5). They yield

$$\frac{\partial \mathbf{B}}{\partial t} = -\operatorname{curl}(\sigma^{-1}\mathbf{j} - \mathbf{v} \times \mathbf{B}) = -\frac{1}{\mu\sigma} \operatorname{curl} \operatorname{curl} \mathbf{B} + \operatorname{curl}(\mathbf{v} \times \mathbf{B})$$

In the Cartesian system of coordinates, $\operatorname{curl} \operatorname{curl} \mathbf{B} = -\Delta \mathbf{B}$ since $\operatorname{div} \mathbf{B} = 0$. As a result, the preceding equation is rewritten in the form

$$\frac{\partial \mathbf{B}}{\partial t} = \frac{1}{\mu\sigma} \Delta \mathbf{B} + \operatorname{curl}(\mathbf{v} \times \mathbf{B}) \tag{35.13}$$

Thus, equation (35.13) becomes an ordinary diffusion equation (for each component of vector $\mathbf{B}$) for an observer moving together with the liquid, that is for $\mathbf{v} = 0$. Therefore, if electric conductivity σ is finite, the value of magnetic field at each point of the conductor decreases with time. A clear illustration of this effect is a pattern of magnetic lines of force whose density decreases with time. The specific character of this diffusion can be found by integrating the above equation if the shape of conductor and the boundary and initial conditions are known. Coefficient $(\mu\sigma)^{-1}$ plays a role similar to that of the diffusion coefficient. If characteristic dimensions of the conductor specimen are denoted by l, the quantity $\tau \sim \mu\sigma l^2$ has the dimension of time and characterizes the relaxation time of the field in the given conductor. If $\sigma \to \infty$ but $\Delta\mathbf{B}$ remains finite, there is practically no diffusion; this corresponds to the discussed above condition of freezing.

Let us turn now to the equation of motion (35.9) of the liquid. If the velocity of motion is represented by the sum of components $\mathbf{v} = \mathbf{v}_\perp + \mathbf{v}_\parallel$, where $\mathbf{v}_\parallel$ is parallel to field $\mathbf{B}$ and $\mathbf{v}_\perp$ is orthogonal to it, the magnetic force takes the form

$$\mathbf{j} \times \mathbf{B} = \sigma\mathbf{E} \times \mathbf{B} - \sigma\mathbf{B} \times (\mathbf{v} \times \mathbf{B}) = \sigma B^2(\mathbf{v}_\infty - \mathbf{v}_\perp)$$

where $\mathbf{v}_\infty$ is the velocity of electric drift in the case of infinite electric conductivity, given by formula (35.7). Equation (35.9) can therefore be rewritten in the form of a system

$$\varkappa \frac{d\mathbf{v}_\perp}{dt} = \mathbf{F}_\perp + \sigma B^2 (\mathbf{v}_\infty - \mathbf{v}_\perp), \qquad \varkappa \frac{d\mathbf{v}_\parallel}{dt} = \mathbf{F}_\parallel$$

We see that a friction force proportional to the velocity of motion appears in the direction orthogonal to the field. This force is of purely electromagnetic origin. This effect which decelerates the flow is called the *magnetic viscosity* of the liquid conductor.

Magnetic forces can also be written in a somewhat different form if we use relations (3.11) and (3.12) between the forces and the magnetic component of the Maxwell stress tensor, namely,

$$\frac{\partial T^{ab}}{\partial x^b} = \mathbf{j} \times \mathbf{B}|_a, \quad T^{ab} = \frac{1}{\mu}\left(B^a B^b - \frac{\delta^{ab}}{2} B^2\right) \tag{35.14}$$

From this we obtain

$$\mu\mathbf{j} \times \mathbf{B}|_a = B^a \frac{\partial B^b}{\partial x^b} - \frac{\delta^{ab}}{2} \frac{\partial (B^2)}{\partial x^b} + \frac{\partial B^a}{\partial x^b} B^b$$

That is, since $\operatorname{div} \mathbf{B} = 0$,

$$\mu\mathbf{j} \times \mathbf{B} = (\mathbf{B} \cdot \operatorname{grad})\,\mathbf{B} - \frac{1}{2} \operatorname{grad}(B^2) \tag{35.15}$$

If external forces of nonelectromagnetic origin have the form of (35.10), where we can substitute $\mathbf{g} = -\operatorname{grad} \chi$ (χ is the gravitational potential), then we can derive the equation of motion from (35.9) and (35.15) for $\varkappa = \text{const}$:

$$\varkappa \frac{d\mathbf{v}}{dt} = -\operatorname{grad}(p + p_{\mathrm{m}} + \varkappa\chi) + \frac{1}{\mu}(\mathbf{B}\cdot\operatorname{grad})\,\mathbf{B} \tag{35.16}$$

The quantity

$$p_{\mathrm{m}} \equiv \frac{1}{2\mu}\mathbf{B}^2 = \frac{\mathbf{B}\cdot\mathbf{H}}{2} \tag{35.17}$$

is called the *magnetic pressure*. The physical meaning of this term is clear from equation (35.16) in which p_{m} is added to hydrostatic pressure p.

The physical meaning of the two terms in (35.16) for the bulk magnetic force becomes clear if we notice that surface forces applied to the boundary of the volume are given, according to § 3, by the stress tensor (35.14):

$$\varphi^i = T^{ij}n_j = \frac{1}{\mu}B^i\mathbf{B}\cdot\mathbf{n} - \frac{1}{2\mu}B^2n^i \tag{35.18}$$

This force $\boldsymbol{\varphi}$ applied to a unit surface area is composed of two components: one along field $\mathbf{B}$, and another along normal $\mathbf{n}$. These are the components corresponding to the addends in the bulk force. The first of them thus describes the tension along the lines of force, and the second describes the force compressing the volume normally to its surface (in the preceding formula $\mathbf{n}$ is the outward normal, and the corresponding term has the minus sign), that is, the force of pressure.

In many cases tension forces in (35.16) are negligibly small compared with pressure and can be ignored. This means that the condition of static equilibrium in the liquid takes the form

$$p + p_{\mathrm{m}} + \varkappa\chi = \text{const}$$

Clearly, hydrostatic pressure p can be balanced off by the appropriate choice of magnetic pressure p_{m}. In other words, a magnetic field of the appropriate configuration can be used to confine the liquid in a fixed volume ("magnetic mirror"). This phenomenon is called the *pinch effect*. However, the equilibrium thus achieved is unstable with respect to random deformations of the liquid and is easily destabilized by them.[7]

---

[7] A detailed exposition of magnetohydrodynamics can be found, for example, in: J. A. Shercliff, *A Textbook of Magnetohydrodynamics*, Pergamon Press, Oxford, 1965; the applications to astrophysics see in: S. B. Pikel'ner, *Foundations of Cosmical Electrodynamics*, NASA, 1964.

§ 36. Elementary properties of ferromagnetics

36.1. Macroscopic description of magnetic properties of materials is based on relation (1.17) which links three characteristics of the field and the medium: magnetic induction **B**, magnetic field strength **H**, and magnetization **M**. We have mentioned already in § 1 that magnetization is a function of field, **M** (**H**) (instead of **H**, field **B** can be chosen as the argument of this function, if this proves more convenient). In linear media relation (1.26) holds; magnetization can be maintained in the media only by an external field **H** and vanishes together with **H**. Magnetization of such media is directed opposite to the external field if $\chi_m < 0$ (diamagnetism), and along the external field if $\chi_m > 0$ (paramagnetism). In principle, the diamagnetic effect is generated in any substance; it appears because of the molecular currents induced by the external field; by the Lenz law, these induced magnetic moments tend to compensate this field. Paramagnetism takes place when particles of the matter possess their own magnetic moments which are oriented along the magnetic field. Evidently, this effect is observable only if it is sufficiently strong and "overcomes" the effect of diamagnetism. Quantum mechanics plays an especially important role in the theoretical interpretation of magnetic properties. Thus, for example, the so-called Van-Leeuwen-Terletsky theorem states that in the thermodynamically equilibrium state the magnetic moment of any classical system of moving charges in a constant external field is zero. A nonzero result (for instance, for magnetization of dia- or paramagnetics) is therefore obtained in the classical theory of magnetism only by implicitly taking into account the quantum-mechanical considerations.[8]

As for the phenomenological relations valid in linear magnetic media, they can be derived in a complete analogy to the theory of dielectrics discussed in §§ 31 and 32. One only has to substitute in the relevant sections **H** for **E**, **B** for **D**, and μ for ε. It is very rare in practice that nonuniformity or anisotropy of linear magnetic media is taken into account.

The nature of nonlinear magnetics is also explained on the basis of quantum mechanics. In the final analysis, one comes to intrinsic (spin) moments of elementary particles (electrons and nuclei) which are ordered owing to specific quantum interactions (exchange forces). However, it is also necessary to investigate the relationships which exist on a macroscopic level in a phenomenological description of matter.

The simplest of such relationships will make the subject of the present section. Among several types of nonlinear magnetics (ferro-

[8] This aspect is discussed in: S. V. Vonsovski, *Magnetism*, Wiley, New York, 1974.

magnetics, antiferromagnetics, ferrites), we shall specifically discuss only *ferromagnetics* (i.e. "permanent magnets" in which spin moments of electrons in ordered state are parallel) and shall analyze elementary properties of magnetic field they generate.

Let a ferromagnetic material be placed in a magnetic field produced, for example, by electric currents. If magnetization of the ferromagnetic is initially zero, and the field is slowly increased (by increasing the electric current), then curve **B** (**H**) at any point of the ferromagnetic varies according to the dashed curve in Fig. 39. Magnetization of the medium consumes work which is equal numerically to the area bounded by the curve within the rectangle in Fig. 39. Let us now decrease the external field. Function **B** (**H**) is not only nonlinear but also not single-valued, namely, when field diminishes to zero, **B** traces not the initial dashed curve but the solid curve going somewhat higher. As a result, vector **B** is nonzero at zero external field **H**, that is the material retains residual magnetization **M** (0). When the field is varied further, we can obtain the remaining part of the curve shown by dots in Fig. 39. The variation of **B** lags behind that of **H**; this effect is called *hysteresis*.

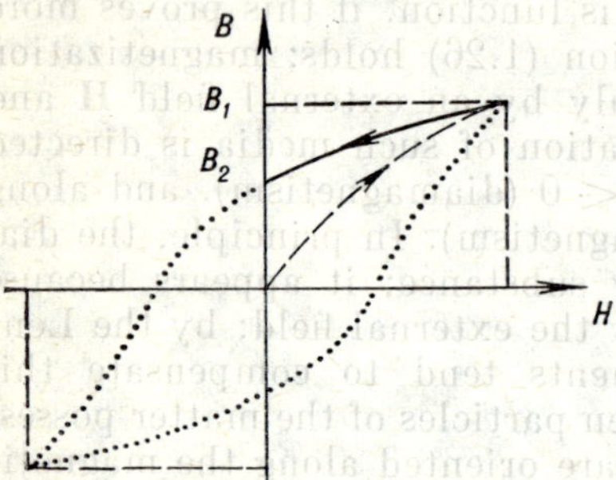

Fig. 39

Hysteresis is accompanied by dissipation of energy in the form of heat, which can be calculated as follows. When the field changes from $\mathbf{B}_1$ to **B**, the work done per unit volume of the ferromagnetic is

$$W=\int_{\mathbf{B}_1}^{\mathbf{B}} \mathbf{H}\cdot d\mathbf{B}=\mathbf{H}\cdot\mathbf{B}\,\Big|_{\mathbf{B}_1}^{\mathbf{B}}-\int_{\mathbf{H}_1}^{\mathbf{H}} \mathbf{B}\cdot d\mathbf{H}$$

Calculation for the hysteresis loop returning to point $\mathbf{B}_1$ yields

$$W=-\oint \mathbf{B}\cdot d\mathbf{H}$$

This is precisely the energy dissipated for hysteresis per unit volume.[9]

[9] The "limiting" closed hysteresis loop, similar to that shown in the figure, is obtained if each time H is increased, the ferromagnetic is magnetized to saturation. But if magnetization does not reach saturation, then the curve traced by B when H decreases is below the limiting curve. As a result, the curve may pass through any point (H, B) within the hysteresis loop by a proper choice of the magnetization path. This means that function B (H) is infinitely multivalued, and the actual values of B and M are determined by the "prehistory" of the specimen.

When the external field is removed, the "residual magnetization" $\mathbf{M}_0 \equiv \mathbf{M}(0)$ sets up in ferromagnetics. This residual magnetization is a source of magnetic field in the space surrounding the ferromagnetic. Indeed, let us turn to the Maxwell equation (M.4′). For $\mathbf{j} = 0$, $\mathbf{E} = 0$, $\mathbf{P} = 0$ this equation takes the form $\text{curl}\,\mathbf{B} = \mu_0\,\text{curl}\,\mathbf{M}_0$ if $\mathbf{M}_0$ is the only magnetization in the specimen. The right-hand side of this equation plays the role of a source with respect to field $\mathbf{B}$. This shows that we can set $\mathbf{B} - \mu_0\mathbf{M}_0 = -\mu_0\,\text{grad}\,\psi$, where ψ is a scalar function of space coordinates (both the sign and the factor are chosen for reasons of convenience), and a comparison with (1.17) shows that $\mathbf{H} = -\text{grad}\,\psi$. The equation for function ψ can be obtained from (M.2) which, after the substitution of (1.17), takes the form

$$\text{div}\,\mathbf{H} = -\text{div}\,\mathbf{M}_0 \tag{36.1}$$

We thus arrive at the Poisson equation for function ψ:

$$\Delta\psi = \text{div}\,\mathbf{M}_0 \tag{36.2}$$

Obviously, the same result follows from the Maxwell equation (M.4) for $\mathbf{j} = 0$. Moreover, we have already discussed it in § 12 but we came to a conclusion that the introduction of scalar magnetic potential ψ for the field generated by currents is not correct. However, the objectives given in § 12 are not valid in the present case if the magnetic field appears only owing to the residual magnetization $\mathbf{M}_0$.

Formally equation (36.2) coincides with the fundamental equation (11.1) for electrostatic potential if we introduce the "magnetic charge density" ρ_m defined by

$$\rho_m = -\text{div}\,\mathbf{M}_0 \tag{36.3}$$

Nevertheless, we shall presently demonstrate that the concept of magnetic charge is devoid of physical meaning, in complete accordance with the interpretation of the Maxwell equations given in § 1. Still, sometimes it is useful as a formal technique in the case of ferromagnetics.

Consider a solution of equation (36.2), paying special attention to the possible presence of discontinuity surfaces of the magnetization field. For example, magnetization is nonzero inside a permanent magnet placed in a nonmagnetic medium, and equals zero outside of the surface of the magnet. Let σ' be a closed surface bounding volume V_1. Introduce an auxiliary surface Σ enclosing surface σ' and denote the volume enclosed between these two surfaces by V_2 (Fig. 40). Assume that magnetization changes jumpwise across the surface σ'. In what follows we drop subscript 0, and denote the magnetization of the material within volume V_1 by $\mathbf{M}_1$ and that in volume V_2 by $\mathbf{M}_2$. We want to find the magnetic potential at point P (this point may be chosen both in volume V_1 and in volume V_2). In order to solve the Poisson equation taking into account the surface of discontinuity σ', we can apply Green's formula (B.28).

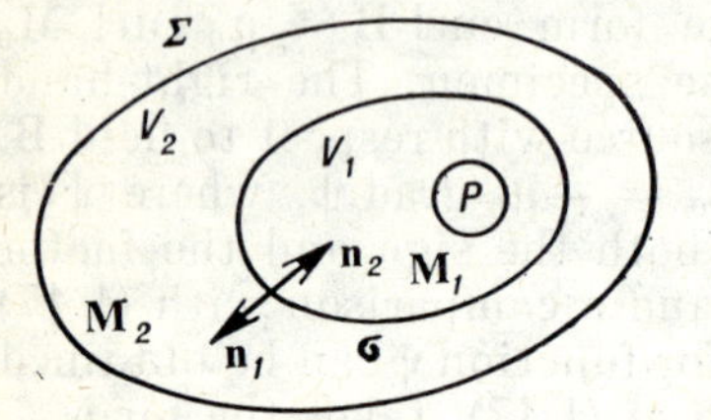

Fig. 40

The point of observation P must first be surrounded by a small sphere whose radius tends to zero. The surface integral in Green's formula (in which function ψ and its gradient must be assumed continuous) will be divided into four parts, namely, an integral over the mentioned above small sphere, an integral over the outer surface Σ, and integrals over the outer and inner (with respect to volume V_1) sides of surface σ'. Assuming, as usual, the integral over Σ vanishing, and evaluating the integral over the small sphere by analogy to the procedure used in § 13, we obtain

$$4\pi\psi(P)=\int\frac{\rho_m}{R}\,dV+\oint\limits_{\sigma^+}\left[\frac{1}{R}\frac{\partial\psi^+}{\partial n_2}-\psi^+\frac{\partial}{\partial n_2}\frac{1}{R}\right]d\sigma^+$$
$$+\oint\limits_{\sigma^-}\left[\frac{1}{R}\frac{\partial\psi^-}{\partial n_1}-\psi^-\frac{\partial}{\partial n_1}\frac{1}{R}\right]d\sigma^- \qquad (36.4)$$

Here σ^- is the inner side of surface σ', σ^+ is its outer side, and ψ^- and ψ^+ are the values of magnetic potential on these two sides, respectively. Let us take into account now that

$$\frac{\partial}{\partial n_1}=-\frac{\partial}{\partial n_2}$$

and denote

$$\frac{\partial}{\partial n_1}\equiv\frac{\partial}{\partial n}$$

Assume, in addition, that the magnetic potential *per se* has no discontinuity across surface σ'.[10] The preceding equation then takes the form

$$4\pi\psi(P)=\int\frac{\rho_m}{R}\,dV+\oint\left[\left(\frac{\partial\psi}{\partial n}\right)_1-\left(\frac{\partial\psi}{\partial n}\right)_2\right]\frac{d\sigma'}{R} \qquad (36.5)$$

It is consistent to refer to the quantity in the brackets as the *magnetic surface charge*; we denote it by λ_m. The meaning of this quantity will be elucidated if we recall the boundary condition for the magnetic field, that is

$$(\mathbf{H}_2-\mathbf{H}_1)\cdot\mathbf{n}=(\mathbf{M}_1-\mathbf{M}_2)\cdot\mathbf{n} \qquad (36.6)$$

[10] Possible discontinuities in potential across surface σ' will be taken into account at the end of this subsection.

Hence, λ_m is equal to $M_{1n} - M_{2n}$. As for the volume integral, we invoke definition (36.3) and transform the integrand to

$$-\frac{\operatorname{div}' \mathbf{M}}{R} = -\operatorname{div}'\left(\frac{\mathbf{M}}{R}\right) + \mathbf{M} \cdot \operatorname{grad}' \frac{1}{R}$$

The first term in the right-hand side enables us to apply the Gauss theorem. It is transformed exactly as the similar term in Subsection 31.1, and the resulting surface integrals cancel out when substituted into (36.5).

If, as is normally the case, magnetization on Σ is for some reason equal to zero, formula (36.5) takes the final form

$$4\pi\psi(P) = \int \mathbf{M} \cdot \operatorname{grad}' \frac{1}{R}\, dV \qquad (36.7)$$

Hence, the concept of a magnetic charge is definitely unacceptable. Formula (36.7) can be interpreted as a field generated by a system of dipoles distributed in space with bulk density $\mathbf{M}$ of the dipole moment. It is this magnetic moment that we have to regard as the primary concept in the theory of permanent magnets. If scalar potential ψ itself has a discontinuity on surface σ', then we see from (36.4) that an additional term $\oint (\Delta\psi) \frac{\partial}{\partial n} \frac{1}{R}\, d\sigma'$, where $\Delta\psi \equiv \psi^+ - \psi^-$, appears in the expression. This term describes the potential of a layer of magnetic dipoles distributed over surface σ' (double layer).

36.2. The case of a cylindrical permanent magnet (with constant magnetization inside it) placed in a medium with zero magnetization (for example, in vacuo) is simple but quite instructive. Here it will be convenient to calculate the field produced by the magnet via formula (36.5) in which the volume integral vanishes (because $\operatorname{div}' \mathbf{M} = 0$). Assume also that voctor $\mathbf{M}$ inside the magnet is parallel to its axis, so that $M_n = 0$ on its lateral surface; this leavos only the integrals over the cylinder bases σ_1 and σ_2:

$$4\pi\psi = \int \frac{M_n}{R}\, d\sigma_1 - \int \frac{M_n}{R}\, d\sigma_2$$

In this case quantities $\pm M_n$ play a role similar to that of surface charges of opposite sign deposited on the cylinder bases. In other words, this is the situation of "magnetic poles". Phenomenologically we can say that the charges of these poles generate inside the magnet a field $\mathbf{H}_\circ$ directed opposite to the external field $\mathbf{H}_{\text{ext}}$ and to magnetization $\mathbf{M}$. Consequently, field $\mathbf{H}_\circ$ is called the *demagnetizing field*. The net field inside the magnet is $\mathbf{H} = \mathbf{H}_{\text{ext}} + \mathbf{H}_\circ$. In frequent cases of proportionality $\mathbf{H}_\circ = -\nu\mathbf{M}$ (ν is known as the *demagnetization factor*) so that $\mathbf{H} = \mathbf{H}_{\text{ext}} - \nu\mathbf{M}$. If, in addition, $\mathbf{M} = \chi\mathbf{H}$, where $\chi = \chi(\mathbf{H})$, then $\mathbf{H} = (1 + \chi\nu)^{-1}\mathbf{H}_{\text{ext}}$ and $\mathbf{M} = \chi(1 + \chi\nu)^{-1}\mathbf{H}_{\text{ext}} = \chi_\circ\mathbf{H}_{\text{ext}}$. Coefficient χ is determined only by the structure of the magnetic material and is therefore called the *magnetic susceptibility of the matter*, and χ_∞ (magnetic susceptibility of the *specimen*) also depends on factor ν which in its turn is related to the geometric shape of the magnet. This is the customary terminology of the theory of magnetism. We shall return to the cylindrical magnet when considering Ampère's theory (see p. 285).

Equation (36.7) shows that magnetic field of a pointlike magnetic dipole is determined by the scalar potential:

$$\psi = \frac{1}{4\pi} \mathbf{m} \cdot \mathrm{grad}' \frac{1}{R} = -\frac{1}{4\pi} \mathbf{m} \cdot \mathrm{grad} \frac{1}{R} \tag{36.8}$$

where $\mathbf{m}$ is the magnetic moment of this dipole. Namely,

$$\mathbf{H} = \frac{1}{4\pi} \mathrm{grad} \left(\mathbf{m} \cdot \mathrm{grad} \frac{1}{R} \right) \tag{36.9}$$

With these equalities we can compare permanent magnets and electric currents as sources of magnetic field. The comparison will be based on *Ampère's theorem* which states that the magnetic field of a circular electric current at large distances from the current loop coincides with the magnetic field of a dipole if the magnetic moment of this dipole is defined by formula (12.13) (in the system of units used here $\alpha = 1$).

The assumption that the magnetic field is measured far from the current loop may be substituted by calculating the limit of infinitesimal contour, when the area enclosed by the contour tends to zero and $I \to \infty$, so that vector $\mathbf{m}$ given by formula (12.13) remains finite. In the first nonvanishing approximation the vector potential of the magnetic field of the current is given by (12.16). This means that

$$\mathbf{H} = \frac{1}{\mu} \mathbf{B} = -\frac{1}{4\pi} \mathrm{curl} \left(\mathbf{m} \times \mathrm{grad} \frac{1}{R} \right) \tag{36.10}$$

Ampère's theorem will be proved if the right-hand sides of the two preceding formulas are shown to be equal, that is if

$$-\mathrm{curl} \left(\mathbf{m} \times \mathrm{grad} \frac{1}{R} \right) = \mathrm{grad} \left(\mathbf{m} \cdot \mathrm{grad} \frac{1}{R} \right) \tag{36.11}$$

Let us transform the left-hand side of the equality to be proved by applying rule (B.20), and the right-hand side, by applying rule (B.18), assuming in these rules $\mathbf{a} = \mathbf{m}$ and $\mathbf{b} = \mathrm{grad} \frac{1}{R}$. Assuming that differential operations applied to constant vector $\mathbf{m}$ yield zero and, in addition, $\mathrm{curl}\, \mathbf{b} = \mathrm{curl\, grad} \frac{1}{R} \equiv 0$, we find that in our case relation (B.18) transforms to $\mathrm{grad}\, (\mathbf{a} \cdot \mathbf{b}) = (\mathbf{a} \cdot \mathrm{grad})\, \mathbf{b}$, and relation (B.20) to $\mathrm{curl}\, (\mathbf{a} \times \mathbf{b}) = -(\mathbf{a} \cdot \mathrm{grad})\, \mathbf{b} + \mathbf{a}\, \mathrm{div}\, \mathbf{b}$. However, $\mathrm{div}\, \mathbf{b} = \Delta \frac{1}{R} = 0$ because the field is analyzed under the condition

$R \neq 0$ (we have assumed above that the distance from the current loop is large). Therefore, relation (36.11) indeed holds, and Ampère's theorem is proved.

Ampère's theorem was a logical foundation for Ampère's hypothesis on the nature of the field produced by permanent magnets. This hypothesis assumes that the field is the result of "molecular currents" in the microscopic structure of the media. This hypothesis played an extremely important role in the development of the physics of ferromagnetism.

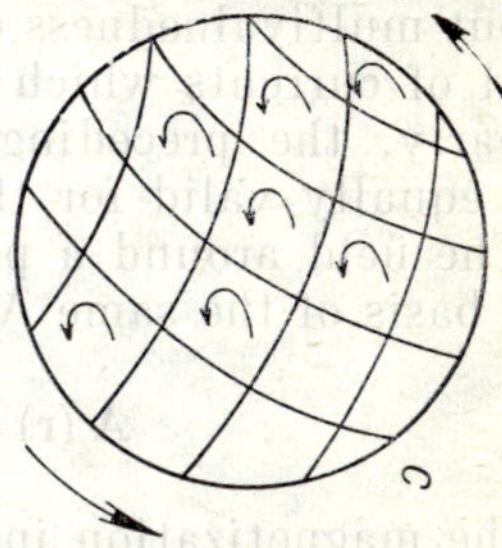

Fig. 41

By virtue of Ampère's theorem one can, if necessary, replace electric currents generating magnetostatic field by an equivalent distribution of magnetic moment, and vice versa. For example, if the field source is a linear current I, we can consider an arbitrary surface bounded by the contour C of this current and divide it into sufficiently small elementary cells (Fig. 41). Assume now that current I circulates along the contour of each of these cells in the same direction as the current flowing in the main contour. All the currents in contour segments inside the main current contour cancel out, so that only the original current remains. At the same time, when elementary cells tend to zero area, the current along each infinitesimal contour produces the magnetic field coinciding with that of a magnetic dipole with moment $d\mathbf{m} = I\mathbf{n}\, d\sigma$. Ultimately the field of a current flowing along contour C will coincide with the field produced by a distribution of elementary magnetic dipoles on the surface, with the distribution independent of the shape of this surface. The density of a double layer consisting of magnetic dipoles must have constant magnitude equal to $d\mathbf{m}/d\sigma \equiv \boldsymbol{\tau}$. The potential produced by such a double layer is

$$\psi(\mathbf{r}) = \frac{1}{4\pi} \int \boldsymbol{\tau}(\mathbf{r}') \cdot \operatorname{grad}' \frac{1}{R}\, d\sigma = -\frac{1}{4\pi} \int\limits_{\Omega} \tau\, d\Omega = -I\Omega \cdot \frac{1}{4\pi} \quad (36.12)$$

Here Ω is the solid angle at which the surface is seen from the observation point having radius vector $\mathbf{r}$. The second integral is negative because $d\Omega$ is positive if the layer is observed on the side of "negative magnetic charges" where the potential is negative.

Let us supplement surface σ in order to close it, assuming that the density of magnetic dipoles on the whole closed surface is, as before, equal to τ. Then the total solid angle Ω at which this surface is seen from the observation point equals 4π if this point is inside the surface, and equals zero if the point is outside the surface. Hence, there is a jump of potential at the boundary separating the outer and inner areas, equal to $\psi_+ - \psi_- = I$. However, this jump of potential is determined only by those surface dipoles which are located at the point on the surface for which the jump is calculated, so that the scalar magnetic potential ψ has a jump equal to

$$\psi_+ - \psi_- = \tau = I \tag{36.13}$$

across any double magnetic layer σ. This returns us to the conclusion about multivaluedness of scalar magnetic potential for the magnetic field of currents which was obtained in a different manner in § 12 (clearly, the preceding arguments concerning the potential jump are equally valid for the double layer of electrostatics).

The field around a permanent magnet can also be calculated, on the basis of the same Ampère theorem, by using vector potential:

$$\mathbf{A}(\mathbf{r}) = \frac{\mu}{4\pi} \int \mathbf{M} \times \operatorname{grad}' \frac{1}{R}\, dV' \tag{36.14}$$

if the magnetization inside the magnet is given by function $\mathbf{M}(\mathbf{r}')$. Formula (B.14$_3$) enables us to rewrite the integrand in the form

$$\mathbf{M} \times \operatorname{grad}' \frac{1}{R} = \frac{\operatorname{curl}' \mathbf{M}}{R} - \operatorname{curl}' \frac{\mathbf{M}}{R}$$

Now let us resort to a formula[11]

$$\int_V \operatorname{curl} \mathbf{a}\, dV = \oint_\sigma \mathbf{n} \times \mathbf{a}\, d\sigma \tag{36.15}$$

Then

$$\mathbf{A}(\mathbf{r}) = \frac{\mu}{4\pi} \int_V \frac{\operatorname{curl}' \mathbf{M}}{R}\, dV' + \frac{\mu}{4\pi} \oint_\sigma \frac{\mathbf{M} \times \mathbf{n}}{R}\, d\sigma' \tag{36.16}$$

The second of these integrals is calculated over the surface of the magnetized specimen. The volume integral can be interpreted by analogy to the expression for the vector potential of current discussed in § 12. Namely, it is logical to call curl′ **M** the *density of magnetization current* in volume V. Consequently, expression $\mathbf{M} \times \mathbf{n}$ should be interpreted as the density of the surface magnetization currents. Therefore, the field around a permanent magnet is modelled by the field of a distribution of currents. It should be emphasized that this modelling has no direct relation to the microscopic theory of ferromagnetism. However, the Maxwell equation (M.4′) written in the form $\operatorname{curl} \mathbf{B} = \mu_0 (\operatorname{curl} \mathbf{M} + \mathbf{j})$ can be interpreted as reflecting the fact that induction **B** of the magnetic field is produced only by currents, namely, by the conduction current **j** and magnetization current $\mathbf{j}_m = \operatorname{curl} \mathbf{M}$, and the latter of them is related to the microscopic structure of the ferromagnetic (in terms of the abovementioned

[11] Derivation of this relation can be found in any course on vector calculus.

Ampère hypothesis, to molecular currents). For this reason $\mathbf{j}_m = \operatorname{curl} \mathbf{M}$ is called the density of current of "bound" charges incorporated into the microscopic structure of the matter.

If magnetization of a ferromagnetic is uniform, all internal currents cancel out and the magnetic field outside such a ferromagnetic can be considered generated by surface currents. One example of this is the abovediscussed cylinder magnetized along its axis. The field of such a magnet coincides with the field of a solenoid with coils arranged on the surface of the cylinder in the planes perpendicular to the axis. This shows that the field must be determined, as has been shown above, by the ends of a cylindrical magnet. The field around this magnet can also be described as generated by "magnetic charges" (cf. an analysis above) and in terms of the vector potential, namely, via the second term in equation (36.16). Magnetization $\mathbf{M}$ differs from zero only inside the magnet and on its surface. Therefore vectors $(1/\mu)\,\mathbf{B}$ and $\mathbf{H}$ coincide outside the magnet. Inside the magnet vectors $\mathbf{B}$ and $\mathbf{H}$ are opposite to each other, because the former has a component normal to the surface, which is continuous across the interface of the ferromagnetic while the latter is described by the lines of force directed from positive "magnetic charges" to negative ones both outside and inside the magnet.

Note that in linear isotropic media the following system of equations holds for magnetic induction:

$$\oint \mathbf{H}\cdot d\mathbf{s} = I, \quad \operatorname{div} \mathbf{B} = 0, \quad \mathbf{B} = \mu \mathbf{H}$$

Formally this system is analogous to the equations for direct current in the presence of an external electromotive force U, analyzed in § 34:

$$\oint \mathbf{E}\cdot d\mathbf{s} = U, \quad \operatorname{div} \mathbf{j} = 0, \quad \mathbf{j} = \sigma \mathbf{E}$$

In this last case the conductor is characterized by resistance $R = \int \frac{ds}{\sigma\cdot\Delta\Sigma}$, where the integral is taken along the contour of a given conductor. This analogy makes it possible to introduce a concept of a "magnetic circuit", with total electric current $I = \int j_n \, d\Sigma$ being analogous to magnetic flux $\Phi = \int B_n \, d\sigma$. Similarly to the theory of electric currents, where we had $I = \mathscr{E}/R$, that is, Ohm's law for the magnetic flux, we obtain $\Phi = I/R_m$ with the "magnetic resistance" of a medium equal to

$$R_m = \int \frac{ds}{\mu\cdot\Delta\Sigma}$$

where integral is calculated in a specimen with cross-section $\Delta\Sigma$.

This equation of the magnetic flux makes possible the calculation of the magnetic field distribution in a medium. The solution thus obtained can, however, be only approximate because usually boundary conditions of a problem of current distribution differ from those in a problem of magnetic field. Nevertheless, this method can often be used to solve practical problems.

36.3. We shall consider some energy properties of media placed in a magnetic field. Assume that the sources of the magnetic field are electric currents. We choose a volume V filled with a matter possessing linear magnetic properties so that induction and strength of magnetic field are related by the equation $\mathbf{B}_1 = \mu_1 \mathbf{H}_1$. Magnetic permeability may be a function of coordinates; in other words, the material is not assumed to be homogeneous. Energy of the magnetic field in volume V is

$$W_1 = \frac{1}{2} \int \mathbf{H}_1 \cdot \mathbf{B}_1 \, dV$$

Now we switch off the field sources (this can be considered a "gedanken" experiment which makes possible the calculation of a physical quantity of interest in this section, that is, the energy of a magnetic in a given field) and insert into volume V a body with volume $V_1 < V$, so that $V = V_1 + V_2$. No assumption is made about the linearity of the substance of which this body is composed, but we can assume that the function $\mathbf{H}$ ($\mathbf{B}$) in volume V_1 is known, that is, we know the law governing the process of magnetization. Let us specify that initially magnetization in volume V_1 is zero. Now let the original intensity of magnetic field sources (i.e. current) be restored. This produces magnetization of the inserted body. As a result, magnetic induction in volume V will differ from the original value. Denote it by $\mathbf{B}$. Note that the portion V_2 of volume V remains occupied with the original matter.

Restoration of former sources requires that work W_2 be done:

$$W_2 = \int_{V_1+V_2} dV \int_0^{\mathbf{B}} \mathbf{H} \cdot d\mathbf{B} = \frac{1}{2} \int_{V_2} \mathbf{H} \cdot \mathbf{B} \, dV + \int_{V_1} dV \int_0^{\mathbf{B}} \mathbf{H} \cdot d\mathbf{B} \quad (36.17)$$

In the first term (integral over volume V_2) we use linearity of the medium occupying this volume. The difference in the calculated energies is

$$W = W_2 - W_1 = \frac{1}{2} \int_{V_2} (\mathbf{H} \cdot \mathbf{B} - \mathbf{H}_1 \cdot \mathbf{B}_1) \, dV$$

$$+ \int_{V_1} \left(\int_0^{\mathbf{B}} \mathbf{H} \cdot d\mathbf{B} - \frac{1}{2} \mathbf{H}_1 \cdot \mathbf{B}_1 \right) dV \quad (36.18)$$

Induction within volume V_2 is $\mathbf{B} = \mu_1\mathbf{H}$. The first term reduces therefore to $\frac{1}{2}\int\limits_{V_2}(\mathbf{H}-\mathbf{H}_1)(\mathbf{B}+\mathbf{B}_1)\,dV$. By virtue of the identical initial and final current distributions, curl $(\mathbf{H}-\mathbf{H}_1) = 0$. In addition, div $(\mathbf{B}+\mathbf{B}_1) = 0$.

Now we make use of the boundary conditions at the boundary of volume V_1:

$$(\mathbf{B}+\mathbf{B}_1)_n^{(2)} = (\mathbf{B}+\mathbf{B}_1)_n^{(1)}, \quad (\mathbf{H}-\mathbf{H}_1)_t^{(2)} = (\mathbf{H}-\mathbf{H}_1)_t^{(1)} \tag{36.19}$$

In the second of these relations it is assumed that there are no surface currents on the boundary. As curl is zero, we can set $\mathbf{H}-\mathbf{H}_1 = -\text{grad}\,\psi$. We shall also denote $\mathbf{B}+\mathbf{B}_1 \equiv \mathbf{Q}$. The following relation holds for integration both over volume V_1 and over volume V_2:

$$\int\limits_{V_1,\,V_2}(\text{grad}\,\psi\cdot\mathbf{Q})\,dV = \int \text{div}\,(\psi\mathbf{Q})\,dV - \int \psi\,\text{div}\,\mathbf{Q}\,dV = \oint\limits_{\sigma}\psi Q_n\,d\sigma$$

Hence,

$$\int\limits_{V_1+V_2}(\text{grad}\,\psi\cdot\mathbf{Q})\,dV = \int\limits_{\sigma}\psi Q_n^{(2)}\,d\sigma - \int\limits_{\sigma}\psi Q_n^{(1)}\,d\sigma = 0 \tag{36.20}$$

Equality to zero in (36.20) follows from the first relation in (36.19). Obviously here we assume the surface integral over the external boundary of volume V vanishing, for example, by removing this boundary to infinity. We have thus proved that

$$\int\limits_{V_1+V_2}(\mathbf{H}-\mathbf{H}_1)\cdot(\mathbf{B}+\mathbf{B}_1)\,dV = \int\limits_{V_1}\ldots + \int\limits_{V_2}\ldots = 0$$

that is

$$\frac{1}{2}\int\limits_{V_2}(\mathbf{H}-\mathbf{H}_1)\cdot(\mathbf{B}+\mathbf{B}_1)\,dV = -\frac{1}{2}\int\limits_{V_1}(\mathbf{H}-\mathbf{H}_1)\cdot(\mathbf{B}+\mathbf{B}_1)\,dV \tag{36.21}$$

The significance of this formula consists in that integration in (36.18) is now reduced to integration only over the volume of the body, V_1. Namely,

$$W = \frac{1}{2}\int\limits_{V_1}\left(\mathbf{H}_1\cdot\mathbf{B} - \mathbf{H}\cdot\mathbf{B}_1 - \mathbf{H}\cdot\mathbf{B} + 2\int\limits_0^{\mathbf{B}}\mathbf{H}\cdot d\mathbf{B}\right)dV \tag{36.22}$$

This is the expression for magnetization energy within volume V_1.

In a particular case of the body being linear in its magnetic properties as well as the surrounding medium, namely, $\mathbf{B} = \mu_2\mathbf{H}$ formula (36.22) reduces to

$$W = \frac{1}{2}\int\limits_{V_1}(\mathbf{H}_1\cdot\mathbf{B} - \mathbf{H}\cdot\mathbf{B}_1)\,dV = \frac{1}{2}\int\limits_{V_1}(\mu_2-\mu_1)\,\mathbf{H}\cdot\mathbf{H}_1\,dV$$

If magnetization induced by the external field is introduced by an ordinary formula $\mathbf{B} = \mu_0 (\mathbf{H} + \mathbf{M})$, then $\mathbf{M} = (\mu_2/\mu_0 - 1)\,\mathbf{H} = (1/\mu_0 - 1/\mu_2)\,\mathbf{B}$. And finally, if the ambient medium is the vacuum, so that $\mu_1 = \mu_0$, then

$$W = \frac{1}{2} \int_{V_1} \mathbf{M} \cdot \mathbf{B}_1 \, dV \tag{36.23}$$

This formula can be compared to relation (31.23) in electrostatics; note, however, that the signs are opposite.

If the body occupying volume V_1 has nonzero magnetization, the calculation of its energy in a magnetic field becomes too complicated and is not considered here[12]. However, the calculation is significantly simplified if we assume that constant magnetization of each element of ferromagnetic is not effected by the field generated by other sources. The expression $\mathbf{M}\, dV = d\mathbf{m}$ must be interpreted as the dipole moment of element dV of the magnet. Similarly to formula (11.21) for electric dipole, we can determine potential energy $dU = -\mathbf{B}\, d\mathbf{m}$ of a magnetic dipole in field $\mathbf{B}$. This field can be written in the form $\mathbf{B} = \mathbf{B}_1 + \mathbf{B}_2$, where $\mathbf{B}_1$ is the field of external sources, and $\mathbf{B}_2$ is generated by all other elements of the same magnet. These elements are kept together by nonmagnetic forces, and their total potential energy relative to one another is $-\int_V \mathbf{M}\mathbf{B}_2 \, dV$ (integration should be extended over the whole volume of the magnet). Their potential energy in the external field is $-\int \mathbf{M}\mathbf{B}_1 \, dV$. Hence, the total potential energy is

$$U = -\int \mathbf{M}\mathbf{B} \, dV \tag{36.24}$$

The difference between (36.23) and (36.24) originates with the fact that the former takes account of the work necessary to produce magnetization $\mathbf{M}$, while the latter assumes that magnetization is fixed beforehand and is not affected.

§ 37. Phenomenological description of superconductivity

37.1. The phenomenon of superconductivity (discovered by Kamerlingh Onnes in 1911 and observed at temperatures in the vicinity of the absolute zero in a large number of materials) can be understood only by using the modern methods of quantum field theory. However, some substantial characteristics of this phenomenon can

[12] See, for example, J. A. Stratton, *Electromagnetic Theory*, McGraw Hill, New York, 1941.

be described phenomenologically if the Maxwell equations are supplemented with additional conditions. It should be emphasized that these conditions (Londons' equations) were an important leading idea for the development of the modern theory[13].

It is clear from the term that the effect consists in the electric resistance vanishing at sufficiently low temperature (resistance becomes equal to zero within the accuracy of the existing, highly accurate, measurement techniques). The transition to the superconducting state is a phase transition of the matter. The behaviour of superconductors in a magnetic field is very peculiar. It is customary to divide superconductors into two classes according to their magnetic properties: superconductors of the first kind (usually, these are chemical elements) and of the second kind (usually, these are alloys). Superconductors of the first kind are spectacular in that they are ideal diamagnetics. Namely, the magnetic flux is completely zero inside such a superconductor placed in an external magnetic field (the *Meissner effect*: magnetic field is squeezed out of the superconductor). The external magnetic field, however, must not exceed a certain critical value, otherwise the transition to superconducting state is impossible (a sufficiently strong magnetic field as if destroys the superconducting state). Properties of superconductors of the second kind are somewhat different, and in what follows we shall discuss only superconductors of the first kind, and only sufficiently large specimens (important changes would be necessary for very small specimens).

The abovementioned ideal diamagnetism distinguishes the substance in the superconducting state from the "ideal conductor" (cf. § 35), with electric resistivity tending to zero but magnetic lines of force "frozen in". This aspect needs special analysis.

Consider first an ideal conductor. The condition of freezing (35.6′) at $\mathbf{v} = 0$ takes the form $\partial \mathbf{B}/\partial t = 0$. In other words, the distribution of the magnetic field inside an ideal conductor cannot change. Assume that the material of interest becomes an ideal conductor at temperatures T lower than a certain critical temperature T_{crit}. At $T > T_{\text{crit}}$ electric conductivity is finite. Compare now two different experiments involving external magnetic field. In the first experiment, field $\mathbf{B}$ is initially absent at $T > T_{\text{crit}}$. The specimen is then cooled to temperature $T < T_{\text{crit}}$, after which magnetic field is switched on. As follows from the freezing condition, this will not change the field inside the ideal conductor: $\mathbf{B} = 0$. In the second situation field $\mathbf{B}$ is switched on at $T > T_{\text{crit}}$. Usually the material above temperature T_{crit} is a linear magnetic; furthermore, it can often be assumed that $\mu \simeq \mu_0$. Therefore the distribution of the

[13] The exposition of this theory can be found, for example, in: A. C. Rose-Innes and E. H. Roderick, *Introduction to Superconductivity*, Pergamon Press, Oxford, 1969.

magnetic field that sets up within the specimen is completely determined by external sources. Now we cool the specimen to a temperature below the phase transition point. If the material becomes an ideal conductor, the distribution of the magnetic field inside the specimen will not change. Moreover, if the magnetic field is now switched off, then by virtue of the condition $\partial\mathbf{B}/\partial t = 0$ the field inside the ideal conductor will remain the same. This behaviour of the ideal conductor in an external magnetic field is caused by the possibility of generating induced surface currents whose magnetic field completely balances out the effects of changes in the external field.

In the first of the above cases superconductors behave as ideal conductors. However, in the second experiment their behaviour is completely reversed; prior to the phase transition the field in the medium was nonzero, but after the transition to superconducting state took place, the field inside the specimen vanishes. This is the phenomenon known as the Meissner effect mentioned above. It can be said that surface currents are generated on the surface of a superconducting specimen and screen its inner regions from the external magnetic field. The same phenomenon can be treated in terms of negative magnetization $\mathbf{M} = -\mathbf{H}_{\text{ext}}$, where $\mathbf{H}_{\text{ext}}$ is the external magnetic field (cf. § 36).

37.2. In the phenomenological description of superconductivity, electrons in the superconductor are divided into "normal" electrons (which behave exactly as in ordinary metals: are scattered and undergo resistance) and "superconducting" electrons which are transferred in the metal without resistance ("two-liquid" model). At constant applied voltage the current at temperatures below the transition point is transferred only by "superconducting" electrons. Naturally, the current intensity remains finite despite the zero resistance (it is limited by the internal resistance of the power supply source, for example, a battery). However, if temperature is not exactly 0 K, a fraction of electrons remain in the normal state. This effect is revealed in alternating fields: in this case part of the current is transferred by normal electrons. Above the transition point all electrons are in the normal state. Note that in the case of direct current no electric field can exist inside the superconductor, otherwise the superconducting electrons would be accelerated and the current would grow infinitely. But if electric field is zero, the current of normal electrons which survived in the superconductor is also absent as nothing can drive their motion.

We can thus express the current density as the sum

$$\mathbf{j} = \mathbf{j}_n + \mathbf{j}_s, \tag{37.1}$$

Here $\mathbf{j}_n$ is the current density of normal electrons, with $\mathbf{j}_n = \sigma'\mathbf{E}$, where σ' is the electric conductivity due to these electrons. At this juncture we are mostly interested in the superconducting component of current density, $\mathbf{j}_s$, and so shall not consider $\mathbf{j}_n$ at all. As the superconducting electrons meet no resistance, each of them is uniformly accelerated by electric field $\mathbf{E}$:

$$m\dot{\mathbf{v}}_s = e\mathbf{E} \tag{37.2}$$

On the other hand, $\mathbf{j}_s = n_s e \mathbf{v}_s$, where n_s is the number of superconducting electrons per unit volume. We see from (37.2) that

$$\frac{\partial}{\partial t}\mathbf{j}_s + \frac{n_s e^2}{m}\mathbf{E} \tag{37.3}$$

Assume that the superconductor possesses no ferromagnetic properties, and that the field varies so slowly that we can ignore the displacement current. The Maxwell equations can now be written in the form

$$\frac{\partial \mathbf{B}}{\partial t} = -\operatorname{curl}\mathbf{E}, \quad \operatorname{curl}\mathbf{B} = \mu_0 \mathbf{j}_s \tag{37.4}$$

From (37.3) and (37.4) we find

$$\frac{\partial \mathbf{B}}{\partial t} = -\frac{m}{n_s e^2}\operatorname{curl}\frac{\partial \mathbf{j}_s}{\partial t} \tag{37.5}$$

and, consequently,

$$\frac{\partial \mathbf{B}}{\partial t} = -a\operatorname{curl}\operatorname{curl}\dot{\mathbf{B}}$$

where $a \equiv m/(\mu_0 n_s e^2)$. Finally, transforming curl curl as usual in Cartesian coordinates and taking into account that $\operatorname{div}\mathbf{B} = 0$, we obtain

$$\Delta\dot{\mathbf{B}} = \frac{1}{a}\dot{\mathbf{B}} \tag{37.6}$$

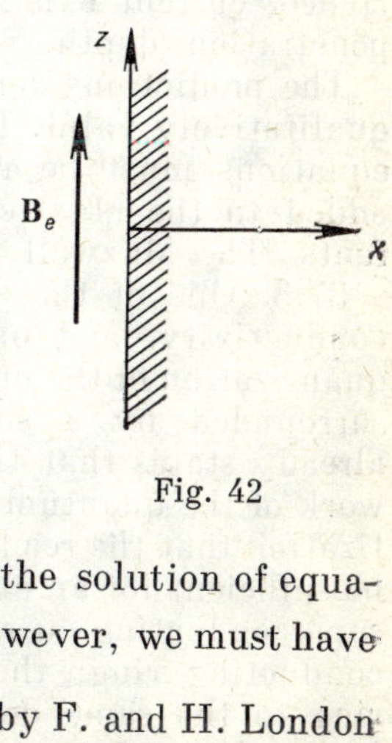

Fig. 42

This equation shows that $\dot{\mathbf{B}}$ drops off exponentially with increasing depth into the superconductor.

We have mentioned above that superconductors are essentially diamagnetic, that is, contain no magnetic flux. But from (37.6) we only find that $\dot{\mathbf{B}} = 0$ at sufficiently large distances from the surface of the superconductor (this is readily found if we analyze a configuration shown in Fig. 42, recalling that the solution of equation (37.6) has the form $\dot{\mathbf{B}}(x) = \dot{\mathbf{B}}_e e^{-x/\sqrt{a}}$). However, we must have $\mathbf{B} = 0$.

The main assumption of the theory suggested by F. and H. London (1935) states that the field in a superconductor satisfies not only equation (37.6) but also the equation

$$\Delta \mathbf{B} = \frac{1}{a} \mathbf{B} \tag{37.7}$$

For the situation shown in Fig. 42 it means

$$\mathbf{B}(x) = \mathbf{B}_e e^{-x/\sqrt{a}} \tag{37.8}$$

Formula (37.7) can be derived exactly as (37.6) if $\dot{\mathbf{B}}$ is everywhere replaced by **B**. In particular, (37.5) must be replaced by the relation

$$\mathbf{B} = -\mu_0 a \operatorname{curl} \mathbf{j}_s \tag{37.9}$$

Equations (37.9) and (37.3) are called the *London equations*. They can be used to calculate distribution of a magnetic field in a superconducting specimen, with (37.9) describing the property of ideal diamagnetics.

At depth $x = \sqrt{a}$ into the superconductor the field diminishes by a factor of e. This characteristic is called the London penetration depth of magnetic field, λ_L, that is

$$\lambda_L = \sqrt{a} = \left(\frac{m}{\mu_0 n_s e^2}\right)^{1/2} \tag{37.10}$$

The current distribution in a superconductor can be derived from the second equation in (37.4). In the case presented in Fig. 42 we have $-\partial B/\partial x = \mu_0 j_y$, and (37.8) yields

$$j_y = j_e e^{-x/\lambda_L}, \quad j_e \equiv \frac{B_e}{\mu_0 \lambda_L} \tag{37.11}$$

Hence, current exists only within a surface layer bounded by the penetration depth.

The predictions derived on the basis of the London equations are qualitatively valid. It is clear from the derivation itself that these equations must be regarded as complementary *ad hoc* conditions added to the Maxwell equations to describe superconducting currents. The Maxwell equations remain valid in this case as well.

37.3. One of the spectacular corollaries of the theory of superconductivity, and one that was confirmed experimentally, is the quantization of the magnetic field in the nonsuperconducting medium surrounded by a superconductor ring. The term "quantization" already states that the effect can be interpreted only in the framework of the quantum theory. However, the Bohr-Sommerfeld quantization that the reader has learnt in the atomic physics course will be sufficient for an elementary description. We have seen above that superconducting currents flow in a λ_L-deep surface layer of a superconducting ring; the external magnetic field penetrates into the ring to the same depth. Consider a closed contour drawn in this layer of the ring and enclosing the hole. The charges forming the current are driven by the external magnetic field. This motion can be described in terms of the nonrelativistic Hamiltonian introduced at the end of § 8. Apply the quantization condition to the charge momentum integrated along the mentioned contour (this integration is used similarly in the atomic theory to an electron moving along a closed path). Namely,

$$\oint \boldsymbol{\pi} \cdot d\mathbf{s} = nh \tag{37.12}$$

where n is an arbitrary integer, and h is the Planck constant. We have shown in § 8 that $\boldsymbol{\pi} = m\mathbf{v} + q\mathbf{A}$ (we are using the SI system of units). Substitute this into (37.12) and take into account the definition of the superconducting current which immediately follows equation (37.2) (for reasons that will be clear later, we substitute symbol q for e; besides, $\mathbf{v} = \mathbf{v}_s$). Another equation we shall use is

$$\oint \mathbf{A} \cdot d\mathbf{s} = \int \mathbf{B} \cdot \mathbf{n}\, d\sigma$$

obtained in § 33. The quantization condition (37.12) then takes the form

$$\frac{m}{n_s q^2} \oint \mathbf{j} \cdot d\mathbf{s} + \Phi = \frac{nh}{q} \tag{37.13}$$

Quantity Φ in the left-hand side of this equation is the familiar magnetic flux through the contour. On the whole, the left-hand side of (37.13) determines a variable quantized by the Bohr-Sommerfeld condition and called the *fluxoid*. If the integration contour is now moved sufficiently deep into the ring, then the part of the fluxoid which is determined by current $\mathbf{j}$ vanishes and equation (37.13) becomes the quantization condition for magnetic flux Φ through the hole in the ring: $\Phi = n\Phi_0$. The magnetic flux quantum $\Phi_0 = h/q$ is called the *fluxon*. Quantization of magnetic flux in the described conditions was predicted by F. London already in 1950. Its experimental confirmation was obtained in 1961 (see the monographs cited above). Measurements of the fluxon charge show that $q = 2e$ (here e is the electron charge), that is, $\Phi_0 = h/2e = 2.07 \times 10^{-15}$ Wb. The superconducting current is thus associated with the motion of particles possessing a charge twice that of the electron. Correspondingly, mass m in (37.13) must be set equal to twice the electron mass.

By the time the fluxon was experimentally discovered, another quantum theory of superconductivity, much more profound than the London theory, was already in existence; its ideas were developed almost simultaneously (at the end of the 50s) by Bardeen, Cooper and Schrieffer in the USA and by Bogoliubov *et al.* in the

USSR.[14] This theory associates the superconductivity effect with the properties of a collective of electron pairs formed in a superconductor. Neither the interactions between electrons binding them into pairs, nor the reasons for which electric resistance vanishes in superconductors can be explained by classical physics.

Let us return to the elementary theory of superconductivity, namely, to our derivation of equation $\Phi = n\Phi_0$. The physical meaning of this equation may seem not quite clear because it was obtained with a contour drawn through a region in the superconductor in which both the currents and the magnetic field were zero. But the magnetic flux was obtained by transforming integral $\oint \mathbf{A} \cdot d\mathbf{s}$ taken along this contour. We have to assume, therefore, that function $\mathbf{A}(\mathbf{r})$ is not identically zero at the points of the contour, while field $\mathbf{B}$ on the same contour is zero. Relation $\mathbf{B} = \text{curl}\,\mathbf{A}$ between the field and the potential shows that such a situation is possible if the potential is a gradient function, $\mathbf{A} = \text{grad}\,\zeta$, where ζ is, in the general case, an arbitrary scalar function which is not necessarily a single-valued function along the contour. Denote $\chi = \frac{2\pi}{\Phi_0}\zeta$. Then the equation $\Phi = n\Phi_0$ demands that each complete circumvention of the closed contour in the superconductor must change function χ by $2\pi n$. Function χ is associated with the phase of a wave function describing the charge carriers in superconductivity from the standpoint of the quantum theory. Thus we return to the Bohr-Sommerfeld quantization; wave mechanics proves that this quantization is valid precisely when the length of a closed orbit is equal to an integral number of the wavelengths of probability describing the particle.

---

[14] An elementary exposition of this theory can be found in the monographs cited on p. 288. For details see: J. R. Schrieffer, *Theory of Superconductivity*, W. A. Benjamin, New York, 1964. An outstanding role in the development of the theory was played by the monograph: N. N. Bogoliubov, V. V. Tolmachev and D. V. Shirkov, *A New Method in the Theory of Superconductivity*, Plenum Press Consultants bureau, New York, 1959.

CHAPTER 9

ALTERNATING ELECTROMAGNETIC FIELD IN CONTINUOUS MEDIA

§ 38. Electromagnetic waves in conductors. Waveguide and cavity

38.1. The propagation of electromagnetic field in homogeneous conducting media will be analyzed in the approximation of linear medium, that is, assuming the validity of relations $\mathbf{B} = \mu\mathbf{H}$, $\mathbf{D} = \varepsilon\mathbf{E}$, $\mathbf{j} = \sigma\mathbf{E}$, so that the Maxwell equations take the form

$$\operatorname{curl}\mathbf{H} - \varepsilon\frac{\partial\mathbf{E}}{\partial t} = \sigma\mathbf{E}, \quad \operatorname{curl}\mathbf{E} + \mu\frac{\partial\mathbf{H}}{\partial t} = 0, \quad \operatorname{div}\mathbf{E} = 0, \quad \operatorname{div}\mathbf{H} = 0 \tag{38.1}$$

For field $\mathbf{E}$ this yields

$$\Delta\mathbf{E} = \mu\varepsilon\frac{\partial^2\mathbf{E}}{\partial t^2} + \mu\sigma\frac{\partial\mathbf{E}}{\partial t} \tag{38.2}$$

A similar equation is obtained for $\mathbf{H}$.

We shall seek a solution of this equation in the form $\mathbf{E} = \mathbf{E}_0 \exp(-i\omega t)$, so that the time derivative in (38.2) can be replaced by multiplying by $-i\omega$. Equation (38.2) is then rewritten as the Helmholtz equation:

$$(\Delta + \hat{k}^2)\,\mathbf{E} = 0 \tag{38.3}$$

where

$$\hat{k}^2 \equiv \omega^2\mu\left(\varepsilon + i\frac{\sigma}{\omega}\right) \tag{38.4}$$

By introducing a "complex dielectric permittivity"

$$\hat{\varepsilon} \equiv \varepsilon + i\frac{\sigma}{\omega} \tag{38.5}$$

we achieve a formal analogy of the problem of electromagnetic field propagation in dielectrics. It is therefore natural to introduce complex quantities

$$\hat{v} \equiv \frac{1}{\sqrt{\mu\hat{\varepsilon}}}, \quad \hat{n} \equiv \frac{c}{\hat{v}} = c\sqrt{\mu\hat{\varepsilon}} = \frac{c}{\omega}\hat{k} \tag{38.6}$$

where $\hat{n}$ is the complex refractive index. We rewrite it in the form

$$\hat{n} = n(1 + i\varkappa) \tag{38.7}$$

in which $\varkappa$ is called the *extinction coefficient*. On the one hand, then $\hat{n}^2 = n^2 (1 + 2i\varkappa - \varkappa^2)$, and on the other hand, we obtain from (38.6) $\hat{n}^2 = \mu\hat{\varepsilon}c^2 = \mu c^2 (\varepsilon + i\sigma/\omega)$, whence $n^2 (1 - \varkappa^2) = \mu c^2\varepsilon$ and $n^2\varkappa = c^2\mu\sigma/2\omega$. By eliminating $\varkappa$, we obtain

$$n^2 = \frac{\mu c^2}{2} \left(\sqrt{\varepsilon^2 + \frac{\sigma^2}{\omega^2}} + \varepsilon \right) \tag{38.8}$$

It is important to remember that quantities ε, σ, μ in the preceding formulas are functions of frequency ω and differ from the corresponding static quantities characterizing the properties of media in constant fields. These functions and their physical consequences are studied in the theory of dispersion which is discussed in § 39.

The results obtained above show that equation (38.3) has for a solution a plane wave

$$\mathbf{E} = \mathbf{E}_0 \exp [i (\hat{k}\mathbf{r} \cdot \mathbf{s} - \omega t)] \tag{38.9}$$

where $\mathbf{s}$ is a unit vector of direction, that is, according to (38.6) and (38 7),

$$\mathbf{E} = \mathbf{E}_0 \exp \left(-\frac{\omega}{c} n\varkappa \mathbf{r} \cdot \mathbf{s} \right) \exp \left[i\omega \left(\frac{n}{c} \mathbf{r} \cdot \mathbf{s} - t \right) \right] \tag{38.10}$$

Hence, coefficient $\varkappa$ indeed describes exponential damping of electromagnetic waves in conductors.

It will be of interest to find and analyze energy density W of the wave. This requires that $\mathbf{E}^2$ be time-averaged. The result is

$$W = W_0 \exp (-\chi \mathbf{r} \cdot \mathbf{s}) \tag{38.11}$$

Here χ is the absorption coefficient (χ = const):

$$\chi = \frac{2\omega}{c} n\varkappa = \frac{4\pi}{\lambda} \varkappa \tag{38.12}$$

and λ is the wavelength in the medium. Therefore at a depth $d = \chi^{-1} = \lambda/4\pi\varkappa$ energy density is reduced by a factor of e. We have obtained an effect similar to that described at the beginning of § 34 (introduced there as the *skin effect*).

We have already discussed the dependence of material parameter on the frequency of radiation; it is thus easy to understand now that the same material may be considered a good conductor in one frequency range and a poor conductor in another. Formula (38.4) can be rewritten in the form

$$\hat{k} = \alpha + i\beta \tag{38.13}$$

Then

$$\left. \begin{matrix} \alpha \\ \beta \end{matrix} \right\} = \left(\frac{\mu\varepsilon}{2} \right)^{1/2} \omega \left[\sqrt{1 + \left(\frac{\sigma}{\omega\varepsilon} \right)^2} \pm 1 \right]^{1/2} \tag{38.14}$$

It is natural to classify a conductor as poor if $\frac{\sigma}{\omega\varepsilon} \ll 1$. To within the terms of the first order in this small parameter, we then obtain

$$\hat{k} \simeq \sqrt{\mu\varepsilon}\,\omega + \frac{i}{2}\sqrt{\frac{\mu}{\varepsilon}}\,\sigma \tag{38.15}$$

In these calculations σ is assumed real. If σ is independent of ω, then the damping of the wave is also independent of ω (of course, this is valid for the frequency range in which the above inequality holds sufficiently well). On the other hand, a conductor is good if $\frac{\sigma}{\omega\varepsilon} \gg 1$, so that the small parameter is $\omega\varepsilon/\sigma$. In this case

$$\hat{k} \simeq (1+i)\,(\sigma\omega\mu/2)^{1/2} \tag{38.16}$$

For transverse fields in conducting media we obtain from the second curl equation (38.1)

$$\mathbf{H} = \frac{1}{\mu\omega}\,\hat{\mathbf{k}} \times \mathbf{F} \tag{38.17}$$

if both fields vary according to formula (38.9). If $\mathbf{H}_0$ denotes the amplitude of the magnetic field, we have

$$\mathbf{H}_0 = \frac{1}{\mu\omega}\,(\alpha + i\beta)\,\mathbf{s} \times \mathbf{E}_0$$

Consequently, phases of fields $\mathbf{H}$ and $\mathbf{E}$ in conducting media are different. If, as usual, we define the modulus $|\hat{k}| = (\alpha^2 + \beta^2)^{1/2}$ and the phase $\varphi = \arctan \frac{\beta}{\alpha}$, we immediately obtain

$$\mathbf{H}_0 = \sqrt{\frac{\varepsilon}{\mu}}\left[1 + \left(\frac{\sigma}{\omega\varepsilon}\right)^2\right]^{1/4} e^{i\varphi}\,\mathbf{s} \times \mathbf{E}_0$$

If the medium is a good conductor, the energy is mostly concentrated in the magnetic field, with the phase of the magnetic field lagging behind that of the electric field by almost 45°.

38.2. Let us turn now to the properties of radiation enclosed in a volume bounded by well-conducting walls. Only elementary information concerning these properties will be given here[1], and we shall assume that the walls are made of the ideal conductor, that is $\sigma \to \infty$. Formula (38.5) shows that formally this means $\varepsilon \to i\infty$. The Maxwell equations take the following form in the case of monochromatic radiation:

$$\operatorname{curl} \mathbf{H} = -ik\varepsilon\mathbf{E}, \quad \operatorname{curl} \mathbf{E} = ik\mu\mathbf{H} \tag{38.18}$$

Consequently, calculation of the limit $\varepsilon \to i\infty$ is possible only under an assumption $\mathbf{E} \equiv 0$. But we then obtain from the second

[1] See the monographs given in p. 131. In the present section the formulas are given in the Gaussian units.

equation in (38.18) that it entails $\mathbf{H} \equiv 0$. Hence, electromagnetic field is identically zero inside an ideal conductor. Note that currents on the surface of such a conductor can exist; furthermore, surface currents are possible only if the conductor is ideal. This immediately follows from Ohm's law, $\mathbf{j} = \sigma\mathbf{E}$: current density per unit length cannot vanish unless $\sigma \to \infty$. Hence, $E_t = 0$, $H_t = 0$ immediately below the surface of the ideal conductor (inside the conductor), and it follows from boundary conditions (4.14) that immediately above its surface

$$E_t = 0, \quad \mathbf{n} \times \mathbf{H} = \frac{1}{c}\mathbf{i} \tag{38.19}$$

Here $\mathbf{n} \times \mathbf{H}$ represents the tangential component of vector $\mathbf{H}$, $\mathbf{n}$ is the outward normal to the conductor surface, and $\mathbf{i}$ is the surface current density.

Obviously, real conducting walls always have properties different from those of the ideal conductor; the differences are caused by inevitable energy losses in real conductors (the Joule heat within the skin layer). These losses are the larger the smaller is the thickness of this layer.

The cavity bounded by the walls on all sides is called the *resonator* (or endovibrator). And if the cavity is a cylinder of arbitrary cross section, infinite or with open bases, it is called the *waveguide*. We shall assume the vacuum in the cavity, and begin the analysis with waveguides.

A characteristic feature of waveguides is the existence of travelling waves with longitudinal components of vectors $\mathbf{E}$ and $\mathbf{H}$. If the magnetic field of the wave is completely transverse, and vector $\mathbf{E}$ has a longitudinal component, this wave is called the electric or E-wave (another frequently used term is the TM-type wave). If, on the opposite, electrical field is completely transverse and vector $\mathbf{H}$ has a longitudinal component, the wave is called the magnetic or H-wave (or the TE-type wave).

Let us prove that waves of these two types can indeed exist in waveguides (i.e., in particular, satisfy the abovementioned boundary conditions).

Let axis z of Cartesian coordinates be directed along the waveguide axis. Then for E-waves we have $E_z \neq 0$ and $H_z = 0$, while for H-waves the situation is reversed, $E_z = 0$ and $H_z \neq 0$.

A wave of electric type may be described by the Hertz electric vector $\mathbf{\Pi}$ (see the end of § 2). For the region with no sources (for monochromatic field, with $\alpha = 1$ and $\mu = \varepsilon = 1$ in formulas of § 2) we can then write $\mathbf{A} = -ik\mathbf{\Pi}$, $\varphi = -\operatorname{div}\mathbf{\Pi}$, so that

$$\mathbf{E} = \operatorname{grad}\operatorname{div}\mathbf{\Pi} + k^2\mathbf{\Pi}, \quad \mathbf{H} = -ik\operatorname{curl}\mathbf{\Pi} \tag{38.20}$$

and

$$(\Delta + k^2)\,\mathbf{\Pi} = 0 \tag{38.21}$$

We choose the Hertz vector in the form

$$\Pi_x = 0, \quad \Pi_y = 0, \quad \Pi_z = \Pi(x, y)\, e^{ik_\parallel z} \tag{38.22}$$

where $\Pi(x, y)$ is a function not yet defined. The z-component of equation (38.21) is a two-dimensional Helmholtz equation:

$$\left(\frac{\partial^2}{\partial x^2} + \frac{\partial^2}{\partial y^2} + k_\perp^2\right)\Pi = 0, \quad k_\perp^2 = k^2 - k_\parallel^2 \tag{38.23}$$

From formulas (38.20) we obtain

$$E_x = ik_\parallel \frac{\partial \Pi}{\partial x} e^{ik_\parallel z}, \quad E_y = ik_\parallel \frac{\partial \Pi}{\partial y} e^{ik_\parallel z}, \quad E_z = k_\perp^2 \Pi e^{ik_\parallel z} \tag{38.24}$$

$$H_x = -ik \frac{\partial \Pi}{\partial y} e^{ik_\parallel z}, \quad H_y = ik \frac{\partial \Pi}{\partial x} e^{ik_\parallel z}, \quad H_z = 0 \tag{38.24'}$$

We have thus indeed obtained the electric-type wave. Now we have to show that boundary condition $E_t = 0$ at the waveguide wall can also be satisfied. Denote by C the contour formed by the intersection of a plane orthogonal to the waveguide axis with its walls. Let us require that $\Pi = 0$ on contour C. The third formula in (38.24) immediately shows that in this case $E_z = 0$ on the waveguide wall. But if s is the direction of the tangent to this contour, then from $E_s = E_x \frac{\partial x}{\partial s} + E_y \frac{\partial y}{\partial s} = ik_\parallel \frac{\partial \Pi}{\partial s} e^{ik_\parallel z}$ we obtain $E_s = 0$.

As $H_z = 0$, magnetic lines of force are in this case plane curves. Mathematical analysis of equation (38.23) shows that it has a solution only at specific positive values of parameter $k_\perp^2$ which can be arranged into an ordered sequence: $k_{\perp 1}^2 \leqslant k_{\perp 2}^2 \leqslant \ldots$ (a discrete spectrum of eigenvalues). Each such $k_{\perp i}^2$ corresponds to two possible values of $k_\parallel$ and to a function $\Pi(x, y)$ (eigenfunction).[2] The values of $k_\parallel^2$ may be both positive and negative. In the first case $k_\parallel = \pm\sqrt{k^2 - k_\perp^2}$ is a real number; the wave propagates along the waveguide without damping. In the second case $k_\parallel = \pm i\sqrt{k_\perp^2 - k^2}$ and the wave is damped. This case is analogous to quasistationary fields in the short-range zone of the emitter (cf. § 17); damping of these waves is not associated with energy dissipation.

The magnetic-type waves can be derived from formulas of § 2, which were used to introduce the Hertz magnetic vector $\mathbf{\Pi}^*$. With this vector, electromagnetic field of a specified frequency is calculated by the formulas

$$\mathbf{E} = ik \operatorname{curl} \mathbf{\Pi}^*, \quad \mathbf{H} = \operatorname{grad} \operatorname{div} \mathbf{\Pi}^* + k^2 \mathbf{\Pi}^* \tag{38.25}$$

In addition,

$$(\Delta + k^2)\, \mathbf{\Pi}^* = 0 \tag{38.26}$$

[2] The spectrum of $k_{\perp i}^2$ depends on the waveguide geometry or more precisely on the shape of waveguide cross section by planes $z = \text{const}$.

If the Hertz vector is chosen in the form

$$\Pi_x^* = 0, \quad \Pi_y^* = 0, \quad \Pi_z^* = \Pi^*(x, y)\, e^{ik_\parallel z} \tag{38.27}$$

then relations (38.25) yield

$$E_x = ik \frac{\partial \Pi^*}{\partial y} e^{ik_\parallel z}, \quad E_y = -ik \frac{\partial \Pi^*}{\partial x} e^{ik_\parallel z}, \quad E_z = 0 \tag{38.28}$$

$$H_x = ik_\parallel \frac{\partial \Pi^*}{\partial x} e^{ik_\parallel z}, \quad H_y = ik_\parallel \frac{\partial \Pi^*}{\partial y} e^{ik_\parallel z}, \quad H_z = k_\perp^2 \Pi^* e^{ik_\parallel z} \tag{38.28'}$$

Equation (38.26) again transforms to the two-dimensional form:

$$\left(\frac{\partial^2}{\partial x^2} + \frac{\partial^2}{\partial y^2} + k_\perp^2\right) \Pi^* = 0 \tag{38.29}$$

It can be shown that boundary condition $E_t = 0$ will be satisfied if $\partial \Pi^* / \partial n = 0$ on contour C defined above.

The strict theory of waveguides proves that a set of E-type and H-type waves forms a complete system for a waveguide, that is, each wave can be represented by a linear combination of a number (possibly, infinite) of the waves of these types.

38.3. Let us turn to the case of a cavity (resonator) with ideally conducting walls; only elementary results will be given. As before, the boundary condition for all bounding surfaces of the resonator is $E_t = 0$.

First we choose a cubic cavity with edge length L, and Cartesian coordinates in which the cube is defined by inequalities $0 < x < L$, $0 < y < L$, $0 < z < L$. The boundary conditions then take the following form:

$$\begin{aligned} &E_y = E_z = 0 \quad \text{for } x = 0 \text{ and } x = L \\ &E_z = E_x = 0 \quad \text{for } y = 0 \text{ and } y = L \\ &E_x = E_y = 0 \quad \text{for } z = 0 \text{ and } z = L \end{aligned} \tag{38.30}$$

Inside the cavity field $\mathbf{E}$ must satisfy the equation

$$(\Delta + k^2)\, \mathbf{E} = 0 \tag{38.31}$$

This wave equation is solved by separating the variables and taking into account both the boundary conditions and the necessity to satisfy the condition $\operatorname{div} \mathbf{E} = 0$ for no sources inside the cube. It can be verified that vector $\mathbf{E}$ with components

$$\begin{aligned} E_x &= A \cos \frac{n_1 \pi x}{L} \sin \frac{n_2 \pi y}{L} \sin \frac{n_3 \pi z}{L} \\ E_y &= B \sin \frac{n_1 \pi x}{L} \cos \frac{n_2 \pi y}{L} \sin \frac{n_3 \pi z}{L} \\ E_z &= C \sin \frac{n_1 \pi x}{L} \sin \frac{n_2 \pi y}{L} \cos \frac{n_3 \pi z}{L} \end{aligned} \tag{38.32}$$

satisfies all these requirements. Here A, B, C are constants and n_1, n_2, n_3 are non-negative integers. Condition div $\mathbf{E} = 0$ is then satisfied if

$$n_1A + n_2B + n_3C = 0 \tag{38.33}$$

The following expressions for wave number and frequency are obtained from (38.32) and (38.31):

$$k = \frac{\pi}{L}\sqrt{n_1^2 + n_2^2 + n_3^2}, \quad \nu = \frac{c}{2L}\sqrt{n_1^2 + n_2^2 + n_3^2} \tag{38.34}$$

These expressions show that the number of vibrations in the interval from ν to $\nu + d\nu$ can be calculated exactly as it has been done at the end of § 22.

The magnetic field is found by substituting solution (38.32) into the second equation of (38.18).

A more general case of cylindrical resonator is obtained by partitioning the waveguide analyzed above by conducting plane walls (for example, at points $z = 0$ and $z = L$). The boundary condition on these walls is $E_x = E_y = 0$ at $z = 0$ and $z = L$. The problem of the waves propagating in such a resonator is solved similarly to the waveguide problem, but now the boundary conditions make it necessary to consider not travelling but standing waves. Electric-type waves are found by using the Hertz electric vector $\Pi_x = 0$, $\Pi_y = 0$, $\Pi_z = \Pi(x, y) \cos k_{\|}z$. Vectors $\mathbf{E}$ and $\mathbf{H}$ are then calculated easily. In particular, components E_x and E_y are proportional to $\sin k_{\|}z$; hence, the boundary condition for $z = 0$ is satisfied automatically. But the boundary condition for $z = L$ is satisfied if $\sin k_{\|}L = 0$, that is $k_{\|} = p\pi/L$, where p is an arbitrary integer. Consequently, $k = \sqrt{k_\perp^2 + (p\pi/L)^2}$ ($p = 0, 1, 2, \ldots$). Similarly, magnetic-type standing waves are determined by the Hertz magnetic vector $\Pi_z^* = 0$, $\Pi_y^* = 0$, $\Pi_z^* = \Pi^*(x, y) \sin k_{\|} z$.

§ 39. Dispersion of electromagnetic field in the medium. Waves in anisotropic media

39.1. An alternating electromagnetic field penetrates into the material and interacts with its particles (electrons, nuclei), causing their vibrations. Vibrating particles generate secondary radiation. The ensemble of these microscopic processes results in the frequency-dependent absorption of the field in the medium. In the framework of classical physics the material is modelled by a set of oscillators each of which interacts with the electromagnetic field via a mechanism described in § 25. This picture is evidently a rough approximation (even if we ignore the fact that a correct description of the effects in question can be provided only by the quantum theory). However, this approach is adequate for interpreting the qualitative side of the processes.

Let us assume for simplicity that all the oscillators are identical and have the same natural frequency ω_0. With damping taken into account, forced vibrations of each oscillator in an electromagnetic field are described by equation (25.21). For example, we can specify that oscillators are electrons bound by a quasielastic force to their equilibrium positions. A deviation from the equilibrium position results in a dipole moment $\mathbf{p}(t) = q\mathbf{r}(t)$. As usual, time-averaged energy consumed to produce this dipole moment is given by squared modulus of amplitude $\mathbf{p}_0$ if we write $\mathbf{p}(t) = \mathbf{p}_0 \exp(-i\omega t)$. Equation (25.21) shows that $\mathbf{p} = \alpha\mathbf{E}$, where

$$\alpha \equiv \frac{q^2}{m}[(\omega_0^2 - \omega^2) - i\omega\tilde{\Gamma}]^{-1}$$

is the complex polarizability coefficient. The macroscopic polarization of a chosen volume of matter produced by electromagnetic field can be expressed via α. For example, consider a rarefied medium in which we can assume, first, that the field on each oscillator is equal to the field of the radiation transmitted through the medium, and second, that $\mu = \mu_0$. Polarization of a unit volume is $\mathbf{P} = N\mathbf{p} = N\alpha\mathbf{E}$, where N is the number of oscillators per unit volume. Assuming $\mathbf{D} = \varepsilon\mathbf{E}$ in the relation $\mathbf{D} = \varepsilon_0\mathbf{E} + \mathbf{P}$ and defining $\hat{n}^2 = \varepsilon/\varepsilon_0$, we obtain

$$\hat{n}^2 = 1 + (N/\varepsilon_0)\alpha$$

where $\hat{n}$ is the complex refractive index phenomenologically similar to the quantity introduced for metals in formula (38.7). In dense media it would be necessary to take into account that the field acting on the oscillators differs from the external field. A comparison of the given formulas already shows with sufficient clarity that the complex nature of refractive index (and with it, as in § 38, of ε and μ) is caused by damping of vibrations of each elementary oscillator.

Obviously, any real substance possesses an extremely large number of different resonance frequencies ω_0. When electromagnetic radiation passes through a medium, the properties observed in the process depend on what these resonance frequencies are.

A phenomenological investigation of the matter-field interaction must be based on an equation of the type of (4.2), which relates the field and the induction (of course, a similar equation can be written for magnetic quantities). This equation already incorporates the assumption on the linearity of the above relation. In many cases, including the case of sufficiently strong fields, this assumption proves inadequate (see § 41). Nevertheless, this assumption covers (a) a possible anisotropy of the material (tensor nature of ε_{ik}); (b) the fact that the "response" of particles to the external field is not instantaneous (*frequency dispersion* corresponding to the dependence of ε_{ik}

on $t - t'$); and (c) the fact that in the general case the field-matter interaction is never absolutely local (*spatial dispersion* revealed as the dependence of ε_{ik} on $\mathbf{r} - \mathbf{r}'$). In other words, polarization of the medium at a given point is determined by the values of the field not only at this point but in a certain region surrounding it. Naturally, these three effects may be pronounced to a different degree (for example, one of them may be dominating) in different materials (and in the same material) for different frequencies of radiation.[3] Thus, frequency dispersion is the main factor in optics. Spatial dispersion proved very important in relation with comparatively new problems of physics, namely, the study of properties of plasma and the phenomenological description of excitations in crystals. This approach successfully gives a consistent description of some well-known effects, such as gyrotropy (optical activity), that is, rotation of the plane of polarization of linearly polarized waves transmitted through a medium (see the cited above monograph by Agranovich and Ginzburg).

39.2. It will be convenient to describe dispersion not via relation (4.2) but by its Fourier transform which, as we have seen in § 4, gives the formula

$$D_i(\omega, \mathbf{k}) = \varepsilon_{ij}(\omega, \mathbf{k}) E_j(\omega, \mathbf{k}) \tag{39.1}$$

We can assume here, in accordance with the standard expression $\mathbf{D} = \varepsilon_0\mathbf{E} + \mathbf{P}$, that

$$\varepsilon_{ij}(\omega, \mathbf{k}) = \varepsilon_0\delta_{ij} + \chi_{ij}(\omega, \mathbf{k}) \tag{39.2}$$

Here χ_{ij} is the dielectric permittivity multiplied by ε_0.

In a similar manner one can derive for magnetic parameters a formula analogous to (39.1). However, with an exception of some aspects appearing in the optics of ferromagnetics, we can assume that $\mu_{ij} = \mu_0\delta_{ij}$ provided we neglect the region of very low frequencies of electromagnetic fields (much lower than the optical frequencies). The corresponding estimates can be found in the monograph of Landau and Lifshitz cited above (§ 60). Therefore, in what follows we shall operate only with tensor $\varepsilon_{ij}(\omega, \mathbf{k})$. This tensor makes it possible to take into account the relation between vectors **B** and **D**. Indeed, the Fourier transform of equation curl $\mathbf{E} = -\partial\mathbf{B}/\partial t$ yields $\omega\mathbf{B}(\omega, \mathbf{k}) = \mathbf{k} \times \mathbf{E}(\omega, \mathbf{k})$, so that the effect of **B** on **D** can be considered known if the relation between **D** and **E** is given. In fact,

[3] The theory of dispersion is analyzed in detail in L. D. Landau and E. M. Lifshitz, *Electrodynamics of Continuous Media* (§§ 58-64 and 76-80), Course of Theoretical Physics, vol. 8, Pergamon Press, Oxford, 1960, and in: V. M. Agranovich and V. L. Ginzburg, *Spatial Dispersion in Crystal Optics and the Theory of Excitons*, Interscience, New York, 1966. These are the books recommended for study, but we considered it necessary to give in the present text at least a schematic introduction to the field, with a view to facilitate further progress.

field $\mathbf{B}$ affects polarization currents

$$\mathbf{j} = \frac{\partial \mathbf{P}}{\partial t} = \frac{\partial}{\partial t} (\mathbf{D} - \varepsilon_0 \mathbf{E}).$$

We shall neglect spatial dispersion and thus assume

$$\lim_{k \to 0} \varepsilon_{ij} (\omega, \mathbf{k}) = \varepsilon_{ij} (\omega) \tag{39.3}$$

This is the case (most frequent in practical situations) that we are going to analyze in more detail. For the sake of simplicity we assume first that the medium is isotropic, and shall analyze the properties of function $\varepsilon (\omega)$. This means that the kernel of the integrand $\varepsilon (t - t')$ in formula (4.2) depends only on time difference and there is no integration over the spatial variable (indeed, we have seen above that ε becomes a function of $\mathbf{k}$ only as a result of the Fourier transformation with respect to $\mathbf{r}$).

Of course, only real values of argument ω have clear physical meaning. But we have found that function $\varepsilon (\omega)$ may assume complex values. We can write therefore

$$\varepsilon (\omega) = \varepsilon_1 (\omega) + i\varepsilon_2 (\omega) \tag{39.4}$$

where ε_1 and ε_2 are real functions. Formula (39.2) takes the form

$$\varepsilon (\omega) = \varepsilon_0 + \chi (\omega)$$

where

$$\chi (\omega) = \int_0^\infty \chi (\tau) e^{i\omega\tau} d\tau, \quad \tau = t - t' \tag{39.5}$$

This immediately shows that

$$\varepsilon (-\omega) = \varepsilon^* (\omega) \tag{39.6}$$

From (39.4) and (39.6) we obtain

$$\varepsilon_1 (-\omega) = \varepsilon_1 (\omega), \quad \varepsilon_2 (-\omega) = -\varepsilon_2 (\omega) \tag{39.7}$$

For sufficiently low frequencies we can retain only the first terms of expansions of these functions into the Taylor series. It is natural to assume that in dielectrics $\lim\limits_{\omega \to 0} \varepsilon_1 (\omega) = \varepsilon_1 (0)$, where $\varepsilon_1 (0)$ is the dielectric permittivity in the static field. As function ε_2 is odd, the series representing ε_2 begins, in the general case, with a term proportional to frequency ω.

Function $\varepsilon (\omega)$ for metals can be found by using the condition that displacement current $\partial \mathbf{D}/\partial t$ be formally equal to conduction current $\sigma \mathbf{E}$ for $\omega \to 0$. In the monochromatic field of sufficiently low frequency we obtain $-i\omega\varepsilon\mathbf{E} = \sigma\mathbf{E}$; hence, for $\omega \to 0$ we must have $\varepsilon = i\sigma/\omega$, that is, $\varepsilon (\omega)$ has a simple pole at $\omega = 0$. This relation can be compared to formula (38.5) valid both for good and for poor conductors.

At sufficiently high frequencies ω polarization in the medium is too slow to develop, so that we can assume that $\varepsilon(\omega)$ tends to electric constant ε_0 for $\omega \to \infty$ (obviously, this condition means that in the limit under discussion $\mathbf{P} = 0$).

Let us show that the imaginary part ε_2 of dielectric permittivity represents the absorption of the field energy in the medium, and let us use macroscopic arguments. Electric energy changes by $\mathbf{E}\,\partial\mathbf{D}/\partial t$, where $\mathbf{E}$ and $\mathbf{D}$ are, of course, real. In the complex notation $\mathbf{E}$ must be replaced by $1/2\,(\mathbf{E} + \mathbf{E}^*)$, and $\mathbf{D}$ must be replaced by $1/2\,(\varepsilon\mathbf{E} + \varepsilon^*\mathbf{E}^*)$. In the case of monochromatic fields (the dependence on time is given by $e^{-i\omega t}$) $\partial\mathbf{D}/\partial t$ is equal to $1/2(-i\omega\varepsilon\mathbf{E}+i\omega\varepsilon^*\mathbf{E}^*)$. Here we are interested only in the mean variation of energy during a sufficiently long time interval and, exactly as we had in § 17, $\langle\mathbf{E}^2\rangle_t \simeq 0$, $\langle\mathbf{E}^{*2}\rangle_t \simeq 0$. Finally, therefore,

$$\left\langle \mathbf{E}\frac{\partial\mathbf{D}}{\partial t}\right\rangle_t \simeq \frac{1}{4}\, i\omega\,(\varepsilon^*-\varepsilon)\,|\mathbf{E}|^2 = \frac{\omega}{2}\,\varepsilon_2|\mathbf{E}|^2 \qquad (39.8)$$

Evidently, the calculation of the magnetic energy will lead to a similar result but, as we have mentioned above, we assume $\mu_2 = 0$. The calculated loss of energy is observed as a release of an equivalent quantity of heat ΔQ in the medium. The fact that $\Delta Q > 0$ follows from the irreversibility of the dissipation process $\left(\Delta S' = \frac{\Delta Q}{T} > 0\right)$. This shows that $\varepsilon_2 > 0$. If ε_2 is very small at a certain frequency ω, then absorption drops sharply, that is, the medium is transparent for the field of this frequency.

39.3. An analysis of function $\varepsilon(\omega)$ in complex plain $\omega = \omega_1 + i\omega_2$ reveals a number of important properties of th s function (its behaviour on the real axis has already been described). First of all, equation (39.5) shows that in the upper halfplane (i.e. for $\omega_2 > 0$) the integrand contains factor $\exp(-\omega_2\tau)$ because $\tau > 0$. And function $\chi(\tau)$ can be considered nonzero only in a finite range of values of τ. Indeed, the processes of polarization buildup in an external electromagnetic field are characterized by a relaxation time which determines this range of τ. Consequently, function $\varepsilon(\omega)$ has no singularities in the upper halfplane. It is very important to emphasize that this property follows from the causality requirement (expressed simply by integration over the values $\tau \geqslant 0$, i.e. $t \geqslant t'$). In addition, we have just seen that in the case of dielectrics $\varepsilon(\omega)$ has no singularity on the real axis and can have a simple pole at point $\omega = 0$ in the case of metals. On the imaginary axis function $\varepsilon(\omega)$ is real. This follows from (39.5). Indeed, $\varepsilon(-\omega^*) = \varepsilon^*(\omega)$ for any complex ω, and therefore $\varepsilon_2(\omega) = 0$ when $\omega = i\omega_2$. It is also important to note that (39.5) also yields that $\varepsilon \to \varepsilon_0$ for ω

tending to infinity along any path in the upper halfplane (and not only along the real axis as we mentioned above). This is the property of the integrand $e^{i\omega\tau} = e^{i\omega_1\tau}e^{-\omega_2\tau}$.

The analyticity of function $\varepsilon(\omega)$ in the upper halfplane of complex variable ω is a corollary of causality requirement, and enables us to derive an important relation between its real and imaginary parts. Let $\omega_0 > 0$ be a point on the real axis. Consider an integral over a closed contour C shown in Fig. 43:

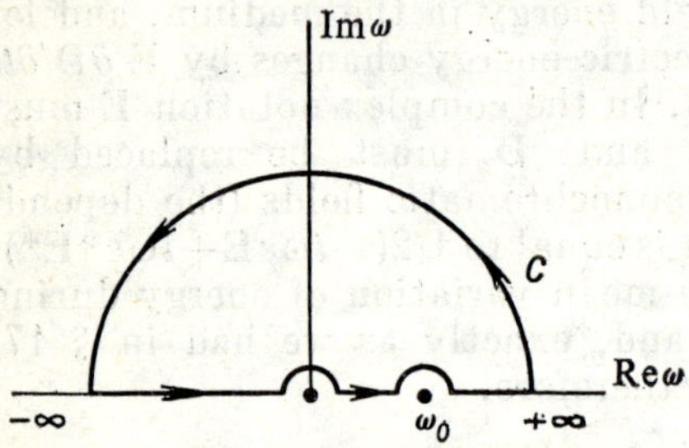

Fig. 43

$$I = \int_C \frac{\varepsilon(\omega) - \varepsilon_0}{\omega - \omega_0} d\omega \tag{39.9}$$

A pole at point $\omega = \omega_0$ and a pole (in the case of metals) at point $\omega = 0$ are cut out by semicircles of infinitely small radius $\rho \to 0$. The radius of the semicircle closing the contour tends to infinity. We have seen already that at infinity $\varepsilon(\omega) \to \varepsilon_0$, so that the integral over the closing semicircle tends to zero. And since the integrand has no singularities within the contour, we have $I = 0$. As a result of circumventing point ω_0 clockwise we obtain, at the limit $\rho \to 0$, $-i\pi[\varepsilon(\omega_0) - \varepsilon_0]$. This gives for a dielectric, with no pole at point $\omega = 0$,

$$\text{P.V.} \int_{-\infty}^{+\infty} \frac{\varepsilon(\omega) - \varepsilon_0}{\omega - \omega_0} d\omega - i\pi[\varepsilon(\omega_0) - \varepsilon_0] = 0 \tag{39.10}$$

The integral here is meant as the principal value on the complete real axis, that is, by definition

$$\text{P.V.} \int_{-\infty}^{+\infty} \frac{\varepsilon - \varepsilon_0}{\omega - \omega_0} d\omega = \lim_{\rho \to 0} \left[\int_{-\infty}^{-\rho+\omega_0} \frac{\varepsilon - \varepsilon_0}{\omega - \omega_0} d\omega + \int_{\rho+\omega_0}^{\infty} \frac{\varepsilon - \varepsilon_0}{\omega - \omega_0} d\omega \right]$$

By separating the real and imaginary parts in formula (39.10), we obtain

$$\varepsilon_1(\omega_0) - \varepsilon_0 = \frac{1}{\pi} \text{P.V.} \int_{-\infty}^{+\infty} \frac{\varepsilon_2(\omega)}{\omega - \omega_0} d\omega \tag{39.11}$$

$$\varepsilon_2(\omega_0) = -\frac{1}{\pi} \text{P.V.} \int_{-\infty}^{+\infty} \frac{\varepsilon_1(\omega) - \varepsilon_0}{\omega - \omega_0} d\omega \tag{39.12}$$

These formulas are called the *Kramers-Kronig dispersion relations.* When the pole in point $\omega = 0$ in the case of metals is taken into account, an additional term σ/ω_0 appears in the right-hand side of (39.12).

Function $\varepsilon_2(\omega)$ being odd, relation (39.11) can be rewritten in the form

$$\varepsilon_1(\omega_0)-\varepsilon_0=-\frac{4\pi q^2}{m}\,\text{P.V.}\int\limits_0^\infty \frac{f(\omega)\,d\omega}{\omega_0^2-\omega^2}$$

Here we have introduced a frequently used notation

$$f(\omega)\equiv\frac{m}{2\pi^2q^2}\,\omega\varepsilon_2(\omega) \tag{39.13}$$

Quantity $f(\omega)\,d\omega$ is called the *oscillator strength* in the range $d\omega$. The origin of this term becomes clear if we compare the last two formulas with the expression for the polarization coefficient α given at the beginning of the present section (assuming $\tilde{\Gamma}=0$). Indeed, $f(\omega)$ can be interpreted as the density of distribution of oscillating dipole moments emitting radiation in the indicated frequency range.

39.4. Let us return now to the general case (39.1), taking into account both anisotropy and spatial dispersion. For real values of ω and $\mathbf{k}$ the tensor $\varepsilon_{ij}(\omega, \mathbf{k})$ takes on, in the general case, complex values; in addition, it is not necessarily symmetric (the medium is gyrotropic if $\varepsilon_{ij}\neq\varepsilon_{ji}$). The arguments given above for the case of frequency dispersion clearly show that the behaviour of this tensor in the complex region of arguments ω, $\mathbf{k}$ is of considerable interest. Note that $\mathbf{k}=\mathbf{k}'+i\mathbf{k}''$, where $\mathbf{k}'$ and $\mathbf{k}''$ are real vectors. A wave is called uniform if these vectors are parallel. Then $\mathbf{k}=(k'+ik'')\,\mathbf{s}$, where $\mathbf{s}$ is a unit real vector, and the extension of vector $\mathbf{k}$ to the complex region reduces to operating with a single complex variable $k'+ik''$ (and not three variables as in the general case). Now it is important to discuss the extent to which variables ω and $\mathbf{k}$ are independent. Assume that distribution of field sources, that is, of currents $\mathbf{j}_{\text{ext}}$ and charges ρ_{ext}, is fixed and is independent of the transmitted waves. The Maxwell equations make it possible to express $\mathbf{E}(\omega, \mathbf{k})$ in terms of $\mathbf{j}_{\text{ext}}(\omega, \mathbf{k})$ and $\rho_{\text{ext}}(\omega, \mathbf{k})$. In the general case it is therefore always possible to choose the sources in such a manner that field $\mathbf{E}$ could be found for any values of mutually independent parameters ω and $\mathbf{k}$. Contrary to this, let us assume that no independent field sources are present in the medium through which an electromagnetic wave propagates. We have already seen, for example, in § 38, that the mode of its propagation is determined by the wave frequency and, in the general case, by the direction of its propagation. The refractive index is a function of frequency with $\mathbf{k}=\frac{\omega}{c}\hat{n}\mathbf{s}$ and $\hat{n}=\hat{n}(\omega, \mathbf{s})$, and hence $\mathbf{k}=\mathbf{k}(\omega)$.

By taking into account causality in an analysis of function $\varepsilon_{ij}(\omega, \mathbf{k})$ in the complex plane of variable ω (with $\mathbf{k}$ regarded as a parameter) one can derive dispersion relations of the type (39.11) and (39.12). In these equations one only has to replace $\varepsilon_1(\omega_0)$ and $\varepsilon_2(\omega_0)$ by, respectively, real $\varepsilon_{(1)ij}(\omega, \mathbf{k})$ and imaginary $\varepsilon_{(2)ij}(\omega, \mathbf{k})$ parts of tensor $\varepsilon_{ij}(\omega, \mathbf{k})$ and ε_0 by $\varepsilon_0\delta_{ij}$. The derivation of these relations is completely analogous to that given for $\varepsilon(\omega, \mathbf{k})$ (see the monograph by Agranovich and Ginzburg, cited earlier).

The field propagation through a medium under conditions $j_{\text{ext}} = 0$, $\rho_{\text{ext}} = 0$, and $\mu = \mu_0$ is described by the Maxwell equations in the form

$$\operatorname{curl} \mathbf{H} = \frac{1}{c}\frac{\partial \mathbf{D}}{\partial t}, \quad \operatorname{curl} \mathbf{E} = -\frac{1}{c}\frac{\partial \mathbf{H}}{\partial t}, \quad \operatorname{div} \mathbf{D} = 0, \quad \operatorname{div} \mathbf{H} = 0 \tag{39.14}$$

As an exception, we are using here Gaussian units which give clearer form to the derived formulas. By substituting

$$\mathbf{E} = \mathbf{E}_0 e^{i(\mathbf{k}\cdot\mathbf{r}-\omega t)}, \quad \mathbf{D} = \mathbf{D}_0 e^{i(\mathbf{k}\cdot\mathbf{r}-\omega t)}, \quad \mathbf{H} = \mathbf{H}_0 e^{i(\mathbf{k}\cdot\mathbf{r}-\omega t)} \tag{39.15}$$

where $\mathbf{E}_0$, $\mathbf{D}_0$, and $\mathbf{H}_0$ are constant amplitudes, we obtain

$$\mathbf{D} = -\frac{c}{\omega}\mathbf{k}\times\mathbf{H}, \quad \mathbf{H} = \frac{c}{\omega}\mathbf{k}\times\mathbf{E}, \quad \mathbf{k}\cdot\mathbf{D} = 0, \quad \mathbf{k}\cdot\mathbf{H} = 0 \tag{39.16}$$

Here $\mathbf{H}$, $\mathbf{D}$ and $\mathbf{E}$ can be considered functions of ω and $\mathbf{k}$. By eliminating $\mathbf{H}$ from the first two equations in (39.16)

$$\frac{\omega^2}{c^2}\mathbf{D} = -[\mathbf{k}\times(\mathbf{k}\times\mathbf{E})] = k^2\mathbf{E} - \mathbf{k}(\mathbf{k}\cdot\mathbf{E}) \tag{39.17}$$

and using (39.1), we obtain

$$\left[\frac{\omega^2}{c^2}\varepsilon_{ij}(\omega, \mathbf{k}) - k^2\delta_{ij} + k_i k_j\right] E_j(\omega, \mathbf{k}) = 0 \tag{39.18}$$

This is a homogeneous system of algebraic equations. If

$$\det\left\{\frac{\omega^2}{c^2}\varepsilon_{ij}(\omega, \mathbf{k}) - k^2\delta_{ij} + k_i k_j\right\} = 0 \tag{39.19}$$

this system has a solution $E_j(\omega, \mathbf{k})$ not equal identically to zero. If function ε_{ij} is known, equation (39.19) makes it possible to express, in principle, ω as a function of $\mathbf{k}$ (and vice versa), its solutions of the type

$$\omega_m = \omega_m(\mathbf{k}) \quad (m = 1, 2, \ldots) \tag{39.20}$$

describe all types of electromagnetic waves possible in the medium.

39.5. Assume now that components ε_{ij} are independent both of ω and of $\mathbf{k}$. In this particular case we are interested only in the effect of anisotropy of the medium on wave propagation. This problem is encountered in the crystal optics (of course, crystals are considered macroscopically continuous). Substitute $\mathbf{k} = \frac{\omega}{c}\mathbf{n}$ into equation

(39.19) (the magnitude of vector $\mathbf{n}$ defined in this way is, in the general case, a function of its direction). If, in addition, we choose for Cartesian coordinate axes x, y, z the principal axes of tensor ε_{ij} and denote by $\varepsilon_{(x)}$, $\varepsilon_{(y)}$, $\varepsilon_{(z)}$ the corresponding principal values, equation (39.19) can be rewritten in the form

$$n^2(\varepsilon_{(x)}n_x^2+\varepsilon_{(y)}n_y^2+\varepsilon_{(z)}n_z^2) - \{n_x^2\varepsilon_{(x)}(\varepsilon_{(y)}+\varepsilon_{(z)})+n_y^2\varepsilon_{(y)}(\varepsilon_{(x)}+\varepsilon_{(z)})+n_z^2\varepsilon_{(z)}(\varepsilon_{(x)}+\varepsilon_{(y)})\} + \varepsilon_{(x)}\varepsilon_{(y)}\varepsilon_{(z)}=0 \quad (39.21)$$

This is the *Fresnel equation.*

In our case of constant $\varepsilon_{(x)}$, $\varepsilon_{(y)}$, $\varepsilon_{(z)}$ the Fresnel equation gives the magnitude of the wave vector in a given direction. If the angles between vector $\mathbf{n}$ and coordinate axes are fixed, then equation (39.21) is a quadratic equation for n^2. This means that in the general case each direction of vector $\mathbf{n}$ corresponds to two possible magnitudes of the vector. And in the space with coordinates n_x, n_y, n_z the Fresnel equation defines in the general case a fourth order surface (the wave vector surface or the optical indicatrix).

The specific form of the optical indicatrix is completely determined by the properties of dielectric tensor ε_{ij} characterizing a specific crystal. In this respect we distinguish cubic, uniaxial, and biaxial crystals (see Landau and Lifshitz, §§ 78 and 79). In crystals with cubic symmetry all principal values of tensor ε_{ij} are equal, so that optically these crystals behave as isotropic media. Crystals are called uniaxial if coordinate axes can be chosen so that $\varepsilon_{(x)} = \varepsilon_{(y)} \equiv \varepsilon_{\perp}$, with the value of $\varepsilon_z \equiv \varepsilon_{\parallel}$ being not equal to the preceding two components. In this case equation (39.21) separates into the following two quadratic equations:

$$n^2=\varepsilon_{\perp}, \quad \frac{n_z^2}{\varepsilon_{\perp}}+\frac{n_x^2+n_y^2}{\varepsilon_{\parallel}}=1 \quad (39.21')$$

The wave vector surface thus separates into two surfaces: a sphere and an ellipsoid of revolution. If $\varepsilon_{\perp} > \varepsilon_{\parallel}$, the sphere is tangent to the ellipsoid from outside, and if $\varepsilon_{\perp} < \varepsilon_{\parallel}$, it touches the ellipsoid from inside (Fig. 44). The spherical surface corresponds to a wave vector independent of direction. It describes the so-called *ordinary* waves; with respect to these waves the crystal behaves like an isotropic body. On the opposite, the waves corresponding to the ellipsoidal indicatrix, called *extraordinary* waves, are characterized by wave vector whose magnitude depends on the angle between the vector and the optical axis z of the crystal.

In biaxial crystals all three principal values of tensor ε_{ij} are different. An analysis of their optical properties requires the general Fresnel equation (39.21); this goes beyond the scope of this book.

Let us return to the first two equations in (39.16). They show that vectors **D** and **H** are orthogonal to each other and to vector **k**. Moreover, vector **H** is orthogonal to three vectors **D**, **E**, **k**, and thus these three vectors lie in a plane. It is important to note that anisotropy makes vectors **E** and **D** nonparallel, and since energy flux **S** is determined by vector product **E** × **H**, wave vector **k** (or **n**) is not

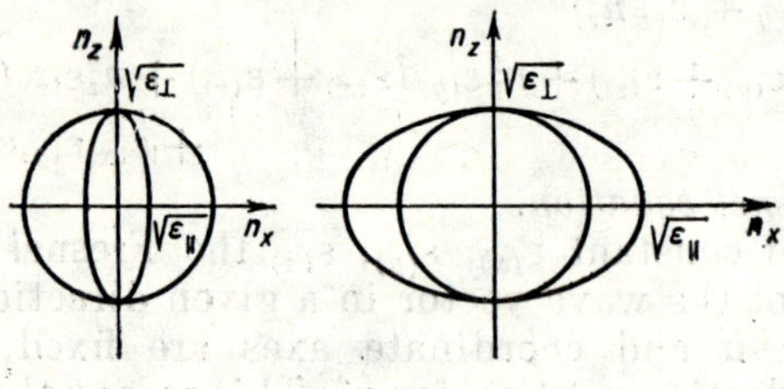

Fig. 44

parallel to the direction of energy flux propagation (Fig. 45). This last direction is called the direction of light ray propagation in a crystal.

Consequently, in isotropic media the direction of ray propagation coincides with that of the normal to the wavefront of the light wave

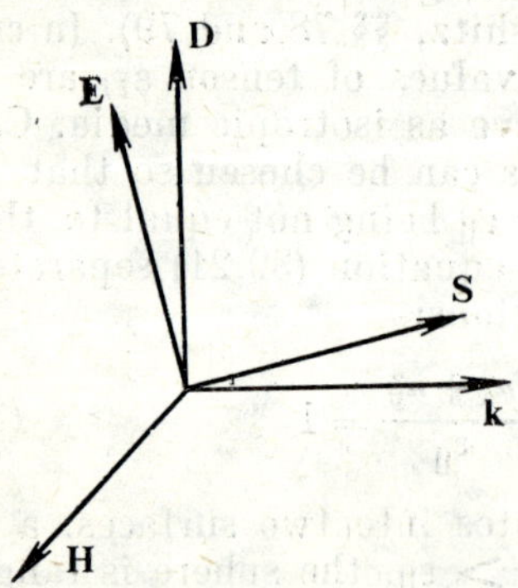

Fig. 45

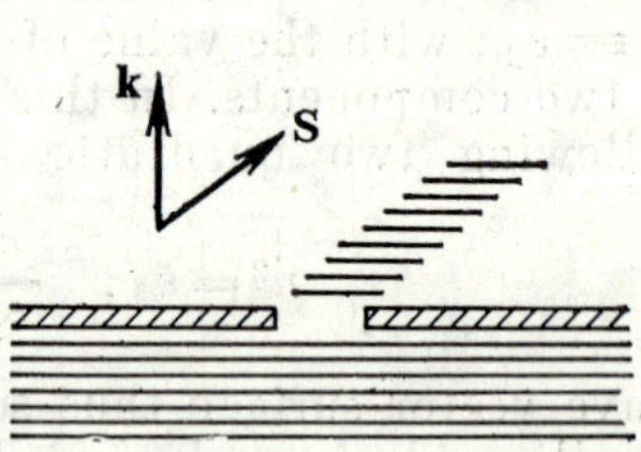

Fig. 46

(we have seen it, for example, in § 21), and these directions are different in anisotropic media. The wave vector is determined by the pattern of propagation of the constant-phase surfaces (such are, by definition, light wavefronts), and so the velocity of propagation associated with the wave vector is called the *phase velocity*. In the general case energy flux propagates at a different velocity (different in magnitude and direction), called the *group velocity*. This phenomenon can be illustrated by imagining a flux of electromagnetic waves entering an anisotropic medium through a narrow aperture as shown in Fig. 46 (here we neglect diffraction). This aperture defines the light ray. The reason why vectors **k** and **S** are not parallel is made quite obvious by the figure.

39.6. Let us describe light rays in anisotropic media in more detail. We introduce a "ray vector" **s** whose direction coincides with that of the Poynting vector **S**, and whose magnitude is conveniently specified by the condition

$$\mathbf{n}\cdot\mathbf{s} = 1 \tag{39.22}$$

In principle, we could consider vector **s** also as a unit vector. Similarly to **S**, vector **s** satisfies orthogonality relations:

$$\mathbf{s}\cdot\mathbf{H} = 0, \quad \mathbf{s}\cdot\mathbf{E} = 0 \tag{39.23}$$

If **n** is substituted for **k**, formulas (39.16) take the form

$$\mathbf{H} = \mathbf{n}\times\mathbf{E}, \quad \mathbf{D} = -\mathbf{n}\times\mathbf{H}, \quad D_i = \varepsilon_{ij}E_j \tag{39.24}$$

Here again we give the relation between **D** and **E** via tensor ε_{ij}. From (39.24) and (39.23) together with (39.22) we obtain

$$\mathbf{H} = \mathbf{s}\times\mathbf{D}, \quad \mathbf{E} = -\mathbf{s}\times\mathbf{H}, \quad E_i = \varepsilon_{ij}^{-1}D_j \tag{39.25}$$

Here we have added the expression for components of **E** in terms of components of **D**, and the matrix composed of the components of tensor ε_{ij}^{-1} is an inverse of matrix $\|\varepsilon_{ij}\|$. A comparison of formulas (39.24) and (39.25) shows that the duality principle holds, namely, if there is a relation for waves, then the corresponding relation for rays is obtained by replacing **E** by **D**, **n** by **s**, ε_{ij} by ε_{ij}^{-1}, and vice versa. Thus, the Fresnel equation (39.21) for the optical indicatrix is transformed into the equation of light ray surface:

$$s^2\left(\varepsilon_{(y)}\varepsilon_{(z)}s_x^2 + \varepsilon_{(x)}\varepsilon_{(z)}s_y^2 + \varepsilon_{(x)}\varepsilon_{(y)}s_z^2\right) - \left[s_x^2\left(\varepsilon_{(y)}+\varepsilon_{(z)}\right) + s_y^2\left(\varepsilon_{(x)}+\varepsilon_{(z)}\right) + s_z^2\left(\varepsilon_{(x)}+\varepsilon_{(y)}\right)\right] + 1 = 0 \tag{39.26}$$

This equation for uniaxial crystals yields results quite analogous to those obtained for the indicatrix. It must be kept in mind here that according to the duality principle, the substitution of **s** for **n** in equation (39.21′) must be accompanied by the substitution of $1/\varepsilon_\perp$ for $\varepsilon_\perp$ and $1/\varepsilon_\parallel$ for $\varepsilon_\parallel$.

As we have seen above, the refraction of a light wave incident on the surface of a uniaxial crystal generates two waves: an ordinary wave and an extraordinary wave. This phenomenon is called the *birefringence* in crystals. Experimentally we observe not the waves but the rays corresponding to them. Note that the abovementioned interrelation between waves and rays operates only within the crystal. On the surface of the crystal the reflection and refraction of light rays differ from the reflection and refraction of light waves. This is caused by the difference in boundary conditions for vector fields **D** and **E**. As a result, for example, an extraordinary ray leaves the incidence plane while the wave vector lies within this plane.

39.7. The difference between the phase and group velocities considered above for anisotropic media also takes place in isotropic media with dispersion, when $\omega = \omega(k)$ (and $\omega(k) = \omega(-k)$ because dispersion must be independent of the direction of transmitted radiation). Obviously, these velocities coincide for monochromatic radiation. Dispersion affects propagation of "wave packets" through the medium, that is, transmission of radiation formed by a superposition of monochromatic waves with unequal frequencies. In isotropic media the directions of the phase and group velocities coincide but their magnitudes differ. An analysis of the propagation of wave packets through a medium makes it possible to find the properties of dispersive media when they contain no sources of electromagnetic field, independent of the transmitted radiation (see above, p. 307). The term "dispersion" is often referred to the radiation penetrating into the medium and thus indicates that different frequencies correspond to different phase velocities ("dispersion of the wave packet").

Consequently, we need to consider how to determine the group velocity of the radiation in isotropic media with dispersion. The general solution of a homogeneous wave equation, representing an arbitrary wave propagating in the medium along axis x, is a superposition of monochromatic waves:

$$f(x, t) = \int_{-\infty}^{+\infty} A(k)\, e^{i(kx-\omega(k)t)}\, dk \qquad (39\ 27)$$

One monochromatic wave with wave vector k_0 corresponds to $A(k) = \delta(k-k_0)A_0$. The wave packet is usually defined as a superposition in which function $A(k)$ falls off rapidly on both sides of a certain value of k_0. As can be seen from (39.27), this property means that the "main part" of the wave packet is confined to a finite region in space. By using an expansion

$$\omega(k) = \omega_0 + \left.\frac{d\omega}{dk}\right|_0 (k-k_0) + \ldots, \quad \left.\frac{d\omega}{dk}\right|_0 \equiv \left.\frac{d\omega}{dk}\right|_{k=k_0}$$

we can transform formula (39.27) to

$$f(x, t) \simeq \exp\left\{i\left[k_0 \left.\frac{d\omega}{dk}\right|_0 - \omega_0\right]t\right\} \int_{-\infty}^{+\infty} A(k) \exp\left\{i\left[x - \left.\frac{d\omega}{dk}\right|_0 t\right]k\right\} dk \qquad (39.28)$$

Simultaneously, the inversion of Fourier transform (39.27) at $t = 0$ shows that

$$A(k) = \frac{1}{2\pi} \int_{-\infty}^{+\infty} f(x, 0)\, e^{-ikx}\, dx$$

Therefore, the integral in (39.28) is equal to $f\left(x - \left.\frac{\partial\omega}{\partial k}\right|_0 t,\ 0\right)$. Hence

$$f(x, t) \simeq f\left(x - \left.\frac{d\omega}{dk}\right|_0 t, 0\right) \exp\left\{i\left[k_0 \left.\frac{d\omega}{dk}\right|_0 - \omega_0\right] t\right\}$$

This shows that amplitude of the wave packet moves at a velocity

$$v_{\mathrm{gr}} = \left.\frac{d\omega}{dk}\right|_0 \tag{39.29}$$

Energy density is determined by squared modulus of the amplitude. The velocity of energy propagation in the wave is therefore precisely the group velocity (39.39). Phase velocity equal to $v_{\mathrm{ph}} = \omega(k)/k = c/n(k)$ may, in particular cases, be larger than c (if $n(k) < 1$). We must emphasize that the arguments of the theory of relativity concerning the light velocity in vacuo being the maximum possible velocity (with respect to light propagation velocity in any material) deal with the propagation of light signals, that is, precisely with the group velocity. As $n = ck/\omega$, we can also write $k(\omega) = \omega n(\omega)/c$ if the refractive index is assumed to be a function of frequency. Hence,

$$v_{\mathrm{gr}} = \frac{1}{dk/d\omega} = \frac{c}{n(\omega) + \omega\, dn/d\omega} \tag{39.30}$$

Usually $dn/d\omega > 1$ (normal dispersion). From $n > 1$ follows $v_{\mathrm{gr}} < c$. However, there are also regions of anomalous dispersion in which $dn/d\omega < 0$. If the magnitude of $dn/d\omega$ in such ranges of frequency is sufficiently high, a formal application of definition (39.30) may give $v_{\mathrm{gr}} > c$. However, this definition is inapplicable in these conditions, since the derivation was based at the very beginning on the assumption of sufficiently slow variation of function $\omega(k)$ or, which is the same, $n(\omega)$, while in the case of anomalous absorption the variation must be rapid (derivative $dn/d\omega$ must be large!).

In anisotropic media, where we ignored frequency dispersion, the difference in behaviour of rays and waves still can be related to the concepts of group and phase velocities; for this we have to assume that the former of them characterizes propagation of quantities quadratic with respect to the wave amplitude.

§ 40. Waves in magnetohydrodynamics

40.1. Peculiar wave phenomena are produced in conducting continuous media interacting with magnetic fields. The basic equations describing such processes were discussed in § 35. Thus, formulas

(35.8) and (35.9) have the form

$$\frac{\partial \varkappa}{\partial t}+\operatorname{div}(\varkappa \mathbf{v})=0 \tag{40.1}$$

$$\varkappa\left(\frac{\partial \mathbf{v}}{\partial t}+(\mathbf{v}\cdot\operatorname{grad})\,\mathbf{v}\right)=-\mu_0\mathbf{H}\times\operatorname{curl}\mathbf{H}-\operatorname{grad}p \tag{40.2}$$

Here $\varkappa$ is the density of the conducting liquid, equation (40.2) uses definition (35.11) of the derivative with respect to time.

In magnetic fields

$$\operatorname{div}\mathbf{H}=0 \tag{40.3}$$

if $\mathbf{B}=\mu_0\mathbf{H}$. And finally, let us recall condition (35.6) which is a corollary of our assumption of infinite electric conductivity. By applying operation curl to (35.6) and using the law of induction $\operatorname{curl}\mathbf{E}=-\mu_0\,\partial\mathbf{H}/\partial t$, we can transform this condition to

$$\frac{\partial \mathbf{H}}{\partial t}=\operatorname{curl}(\mathbf{v}\times\mathbf{H}) \tag{40.4}$$

In what follows we assume that this condition always holds.

First we consider the following particular formulation of the problem. Let density $\varkappa$ be constant, and $\mathbf{H}=\mathbf{H}_0+\mathbf{h}$. In addition, $\mathbf{H}_0$ is assumed constant, and $\mathbf{h}$ is a function of coordinates and time. Field $\mathbf{h}$ plays the role of a fluctuation superposed on the initial field $\mathbf{H}_0$. We direct axis x of the Cartesian coordinate system along $\mathbf{H}_0$, so that $\mathbf{H}_0=(H_0,\ 0,\ 0)$. Recall the equation

$$\frac{\partial \mathbf{v}}{\partial t}=-(\mathbf{v}\cdot\operatorname{grad})\,\mathbf{v}-\frac{1}{\varkappa}\operatorname{grad}\left(p+\frac{\mu_0H^2}{2}\right)+\frac{1}{\varkappa}\mu_0(\mathbf{H}\cdot\operatorname{grad})\,\mathbf{H} \tag{40.5}$$

which was obtained at the end of § 35 by using the transformation of equation (35.9), that is (40.2) (see (35.16)). The last term in the right-hand side of (40.5) is transformed to $\frac{\mu_0}{\varkappa}\left(\mathbf{H}_0\frac{\partial \mathbf{h}}{\partial x}+(\mathbf{h}\cdot\operatorname{grad})\,\mathbf{h}\right)$. The continuity equation for an incompressible liquid is $\operatorname{div}\mathbf{v}=0$. And of course, $\operatorname{div}\mathbf{h}=0$.

Let us show that

$$\mathbf{v}=\pm\mathbf{h}\,(\mu_0/\varkappa)^{1/2} \tag{40.6}$$

is a possible solution of equation (40.5). If (40.6) is substituted into (40.5), then terms $-(\mathbf{v}\cdot\operatorname{grad})\,\mathbf{v}$ and $\frac{\mu_0}{\varkappa}(\mathbf{h}\cdot\operatorname{grad})\,\mathbf{h}$ in (40.5) cancel out. Therefore, we obtain

$$\frac{\partial \mathbf{v}}{\partial t}=-\frac{1}{\varkappa}\operatorname{grad}\left(p+\frac{\mu_0}{2}(\mathbf{H}_0+\mathbf{h})^2\right)+\frac{\mu_0}{\varkappa}H_0\frac{\partial \mathbf{h}}{\partial x} \tag{40.7}$$

Apply operation div to both sides of (40.7). Taking into account the above arguments, we obtain the necessary condition for the solution to have the suggested form (40.6), namely,

$$\Delta\left[p+\frac{\mu_0}{2}(\mathbf{H}_0+\mathbf{h})^2\right]=0$$

Assuming that fluctuation is restricted to a certain region in space outside which $\mathbf{h}=0$ and pressure $p=p_0$, we obtain

$$p+\frac{\mu_0}{2}(\mathbf{H}_0+\mathbf{h})^2=p_0+\frac{\mu_0 H_0^2}{2}=\text{const} \tag{40.8}$$

In other words, the hydrostatic pressure is balanced by the magnetic pressure everywhere in the region of fluctuation (see § 35) and

$$\text{grad}\left(p+\frac{\mu_0}{2}(\mathbf{H}_0+\mathbf{h})^2\right)=0 \tag{40.9}$$

Equation (40.7) takes the form

$$\frac{\partial \mathbf{v}}{\partial t}=\frac{\mu_0}{\varkappa}H_0\frac{\partial \mathbf{h}}{\partial x} \tag{40.10}$$

Now turn to formula (40.4) in which $\mathbf{H}$ must be replaced by $\mathbf{H}_0+\mathbf{h}$. According to (B.20) and taking into account that $\text{div}\,\mathbf{v}=0$ and $\text{div}\,\mathbf{h}=0$, we have

$$\text{curl}\,(\mathbf{v}\times\mathbf{H})=(\mathbf{H}\cdot\text{grad})\,\mathbf{v}-(\mathbf{v}\cdot\text{grad})\,\mathbf{H}$$
$$=H_0\frac{\partial \mathbf{v}}{\partial x}+(\mathbf{h}\cdot\text{grad})\,\mathbf{v}-(\mathbf{v}\cdot\text{grad})\,\mathbf{h}$$

But as follows from (40.6), the last two terms cancel out. Consequently, in our case equation (40.4) takes the form

$$\frac{\partial \mathbf{h}}{\partial t}=H_0\frac{\partial \mathbf{v}}{\partial x} \tag{40.11}$$

Equations (40.10) and (40.11) must be solved simultaneously. They show that

$$\frac{\partial^2 \mathbf{v}}{\partial t^2}=\frac{\mu_0}{\varkappa}H_0^2\frac{\partial^2 \mathbf{v}}{\partial x^2} \tag{40.12}$$

For $\mathbf{h}$ we obtain an absolutely identical equation. Hence, vectors $\mathbf{v}$ and $\mathbf{h}$ satisfy one-dimensional wave equations. If an arbitrary initial distribution of velocities $\mathbf{v}$ is given and satisfies (40.6), then (40.12) yields that this distribution propagates along axis x (i.e. along the initial constant magnetic field $\mathbf{H}_0$) at a velocity

$$V=\pm H_0\sqrt{\mu_0/\varkappa} \tag{40.13}$$

Such waves in a conducting liquid are called magnetohydrodynamic (MHD) waves. The solution of the type of (40.6) of the magnetohydrodynamic equations was first found by Alfvén in 1942.

An expression for V^2 is $2p_0^{(m)}/\varkappa$ where $p_0^{(m)}$ is the magnetic pressure. Relation (40.13) can therefore be compared to the relation for sound velocity $v = (\gamma p_0/\varkappa)^{1/2}$, which is obtained if the adiabatic relation $p = k\varkappa^\gamma$ is valid (γ is the ratio of specific heats). A more detailed analysis would show that MHD waves can be represented as transverse oscillations of magnetic lines of force, which are to a high degree similar to vibrations of a string. Transversality of these vibrations distinguishes them from sound waves but, like sound waves, they are associated with an unusual magnetic pressure generated in a conducting liquid by the magnetic field.

40.2. So far we avoided assuming that fluctuations, that is $\mathbf{v}$ and $\mathbf{h}$, are small compared to some characteristic quantities. A much more general statement of the problem can be analyzed if such an assumption is made. As before, formulas (40.1)-(40.4) can be considered the basic equations of magnetohydrodynamics (in the case of infinite electric conductivity). If the motion of the medium is assumed adiabatic, then the continuity equation must hold for entropy S of unit volume (the law of entropy conservation):

$$\frac{\partial S}{\partial t} + (\mathbf{v}\cdot\text{grad})\, S = 0 \tag{40.14}$$

which must be added to the mentioned basic formulas. Let the initial state of the liquid be characterized by a constant density $\varkappa$; in a constant magnetic field $\mathbf{H}_0$ the liquid flows at a constant velocity $\mathbf{v}_0$. Let a perturbation factor cause small fluctuations in these parameters: $\varkappa \to \varkappa + \varkappa_1$, $\mathbf{H}_0 \to \mathbf{H}_0 + \mathbf{H}_1$, $\mathbf{v}_0 \to \mathbf{v}_0 + \mathbf{v}_1$. Simultaneously $S \to S + S_1$. Substitute all these perturbed variables into equations (40.1)-(40.4) and (40.14) and take into account the smallness of the fluctuations, that is, ignore their products. By introducing, similarly to (40.13), notations $\mathbf{V} = \mathbf{H}_0\, (\mu_0/\varkappa)^{1/2}$ and $\mathbf{V}' = \mathbf{H}_1\, (\mu_0/\varkappa)^{1/2}$, we obtain the following system of equations:

$$\frac{\partial \mathbf{V}'}{\partial t} + (\mathbf{v}_0\cdot\text{grad})\, \mathbf{V}' = (\mathbf{V}\cdot\text{grad})\, \mathbf{v}_1 - \mathbf{V}\,\text{div}\,\mathbf{v}_1$$

$$\frac{\partial \mathbf{v}_1}{\partial t} + (\mathbf{v}_0\cdot\text{grad})\, \mathbf{v}_1 = -\frac{1}{\varkappa}\,\text{grad}\,[(p' + \varkappa\mathbf{V})\cdot\mathbf{V}'] + (\mathbf{V}\cdot\text{grad})\, \mathbf{V}'$$

$$\frac{\partial \varkappa_1}{\partial t} + (\mathbf{v}_0\cdot\text{grad})\, \varkappa_1 = -\varkappa\,\text{div}\,\mathbf{v}_1$$

$$\frac{\partial S_1}{\partial t} + (\mathbf{v}_0\cdot\text{grad})\, S_1 = 0, \quad \text{div}\,\mathbf{V}' = 0 \tag{40.15}$$

If the equation of state of the liquid has the form $p = p\,(\varkappa, S)$, then pressure in the perturbed state will be given by the formula

$$p' = p + dp = p + \left(\frac{\partial p}{\partial \varkappa}\right)_S d\varkappa + \left(\frac{\partial p}{\partial S}\right)_\varkappa dS \tag{40.16}$$

Here $d\varkappa = \varkappa_1$, $dS = S_1$; if additionally we denoted $dp \equiv p_1$, $b \equiv (\partial p/\partial S)_\varkappa$ and take into account that in the general case the sound velocity is $w = \sqrt{(\partial p/\partial \varkappa)_S}$ (this is proved in thermodynamics), then

$$p_1 = w^2 \varkappa_1 + bS_1 \tag{40.17}$$

Let us look for a solution of system (40.15) in the form of plane waves $\exp[i(\mathbf{k}\cdot\mathbf{r} - \omega t)]$ with constant amplitudes, denoting $\omega_0 = \omega - \mathbf{k}\cdot\mathbf{v}_0$. Then system (40.15) yields the relations

$$\begin{gathered}\omega_0 \mathbf{V}' + (\mathbf{k}\cdot\mathbf{V})\,\mathbf{v}_1 - \mathbf{V}(\mathbf{k}\cdot\mathbf{v}_1) = 0 \\ \omega_0 \mathbf{v}_1 + (\mathbf{k}\cdot\mathbf{V})\,\mathbf{V}' - \varkappa^{-1}[p_1 + \varkappa(\mathbf{V}\cdot\mathbf{V}')]\,\mathbf{k} = 0 \\ \omega_0 \varkappa_1 - \varkappa(\mathbf{k}\cdot\mathbf{v}_1) = 0, \quad \mathbf{k}\cdot\mathbf{V}' = 0, \quad \omega_0 S_1 = 0\end{gathered} \tag{40.18}$$

We also have to keep in mind the equation of state (40.17).

For system (40.18) to have a nonzero solution, its determinant must equal zero. After rather cumbersome manipulations this condition can be written in the form

$$\omega_0^2\,[\omega_0^2 - (\mathbf{k}\cdot\mathbf{V})^2]\,[\omega_0^4 - k^2(w^2 + V^2)\,\omega_0^2 + k^2 w^2 (\mathbf{k}\cdot\mathbf{V})^2] = 0 \tag{40.19}$$

This equation makes it possible to classify the possible types of waves[4]. The relation between ω and ω_0 simply takes into account the Doppler effect in changing to the frame of reference co-moving with the liquid at a velocity v_0.

First of all, there is a solution

$$\omega_0 = \pm\mathbf{k}\cdot\mathbf{V} \tag{40.20}$$

Waves of this type are called the MHD waves (or the Alfvén waves) and correspond to the phenomenon discussed at the beginning of this section. Equations (40.20) and (40.18) yield $p_1 = 0$, $\varkappa = 0$, $S_1 = 0$, and also

$$\mathbf{v}_1 = \pm\mathbf{V}', \quad \mathbf{k}\cdot\mathbf{V}' = 0, \quad \mathbf{V}\cdot\mathbf{V}' = 0 \tag{40.21}$$

Vector $\mathbf{v}_1$ characterizing the direction of vibrations is thus orthogonal to vector $\mathbf{V}$, that is, to vector $\mathbf{H}_0$. At the same time, we find from (40.20) that the propagation velocity is $\pm(\mu_0/\varkappa)^{1/2} H_0 \cos\vartheta$, where ϑ is the angle between vectors $\mathbf{k}$ and $\mathbf{H}_0$.

Another solution corresponds to

$$\omega_0 = 0 \tag{40.22}$$

Once generated, a perturbation propagates together with the medium. Used in equations (40.18), this solution yields

$$\varkappa_1 = -\frac{b}{w^2} S_1, \quad \mathbf{v}_1 = 0, \quad \mathbf{V}' = 0, \quad p_1 = 0$$

---

[4] Our exposition of the theory of waves in magnetohydrodynamics is based on the article: S. I. Syrovatsky, Magnetohydrodynamics, *Soviet Physics, Uspekhi*, 52, 247-303 (1957). See also: H. Alfvén and C.-G. Fälthammer, *Cosmical Electrodynamics* (2nd ed.), Clarendon Press, Oxford, 1963.

Hence, it describes the interrelated fluctuations of density and entropy (entropy waves).

And finally, equation (40.19) also allows a relation

$$\omega_0^4 - k^2 (V^2 + w^2)\, \omega_0^2 + k^2 w^2 (\mathbf{k} \cdot \mathbf{V})^2 = 0 \tag{40.23}$$

It corresponds to two waves propagating at velocities

$$V_\pm^2 = \omega_0^2/k^2 = (1/2)\left(w^2 + V^2 \pm \sqrt{(w^2 + V^2)^2 - 4w^2 V^2 \cos^2 \vartheta}\right)$$

Angle ϑ is defined as earlier in this subsection. Such waves are called the *magnetoacoustical waves.*[5]

As can be seen from the above formulas, there is no frequency dispersion in the case under consideration because the wave propagation velocity is independent of ω. If in contrast to our assumptions the electric conductivity of the medium is finite, it becomes necessary to take account of the magnetic viscosity of the medium (see § 35), and also of its hydrodynamic viscosity if it is appreciably large. Introduction of these terms into the initial equations results in dissipative phenomena, that is, absorption of energy and damping of vibrations. These effects lead to frequency dispersion.

One also has to pay attention to the fact that formulas of the type of (40.20) define the dispersion of the phase velocity of the wave as a function of the direction of its propagation. Contrary to this, the group velocity is equal, in accordance with (39.29), to $\partial\omega/\partial\mathbf{k} = \mathbf{V}$, and is the same in any direction.

§ 41. Fundamentals of nonlinear optics

41.1. Nonlinearity of **D** as a function of **E** or, which is the same, of polarization **P** as a function of **E** (and of **B** or **M** as functions of **H**) is in many cases a significant factor. We have already mentioned the nonlinear properties of ferromagnetic and ferroelectric crystals. Another example of deviations from linearity is given by saturation effects. For example, the magnetic moment of unit volume of a paramagnetic substance is approximately linear in a certain range of variation of the external magnetic field but in sufficiently strong fields further increase of the magnetic moment becomes impossible and magnetization approaches a constant level. In this section we are interested in the nonlinear properties of the media, which are mostly significant in alternating electromagnetic fields if energy density in these fields is very high; experimentally these nonlinearities are observed as a number of optical effects. A study of these properties was stimulated during the last fifteen years by the appearance of sources of high-power coherent radiation (lasers, etc.).

---

[5] An analysis of such waves was given, for example, in the article by Syrovatsky cited above.

Similarly to our approach to the theory of dispersion (§ 39) we shall give here a rough but illustrative model of microscopic properties of the medium resulting in nonlinear effects. The model simply states that for sufficiently strong external perturbation factors the "elementary oscillators" of the medium cannot be assumed harmonic any more. Restoring force **F** applied to the oscillator stops being quasielastic at sufficiently large deviations from the equilibrium position. Instead we can write $F = kx + k'x^2 + \ldots$ (it will be sufficient to consider the one-dimensional case). The equation of motion of the oscillator driven by this force and by an external electric field $E(t)$ takes the form

$$m\frac{d^2x}{dt^2} = F(x) + qE(t) - \widetilde{\Gamma}\frac{dx}{dt} \tag{41.1}$$

As usual, the last term in the right-hand side serves to take account of damping (see § 25). A deviation x from the equilibrium position generates a dipole moment $p = qx$. The macroscopic polarization of the medium can be written in the form $P = \gamma qx$, where factor γ is a function of the density of the dipole moments distribution. Consequently, equation (41.1) also serves to determine $P(t)$. A solution of equation (41.1) will not be considered here. What is important to us here is only that in many cases this equation can be solved by the method of successive approximations, with polarization represented by

$$\mathbf{P}(t) = \sum_{n=1}^{\infty} \mathbf{P}^{(n)}(t) \tag{41.2}$$

where the nth term includes an n-fold product of field strengths.[6]

Each individual term in formula (41.2) can be written in a more extended form:

$$P_i^{(n)}(t) = \int_0^{\infty} d\tau_1 \ldots \int_0^{\infty} d\tau_n \chi_{ij_1 \ldots j_n}^{(n)}(\tau_1, \ldots, \tau_n) \times E_{j_1}(t-\tau_1) \ldots E_{j_n}(t-\tau_n) \tag{41.3}$$

Similarly to § 39, here we take into account possible frequency dispersion in the medium. Integration from zero (and not from $-\infty$) is a corollary of causality, as the value of polarization at any given moment is determined exclusively by the values of electric field prior to this moment. Practically one has to take into account the terms in the polarization expression up to the third order inclusively.

[6] See G. C. Baldwin, *Introduction to Nonlinear Optics*, Plenum Press, New York, 1969 and a review article by S. A. Akhmanov and R. V. Khokhlov, *Soviet Physics, Uspekhi* 88, 439 (1966) and 95, 231 (1968). Our exposition was meant to demonstrate how an elementary description of nonlinear effects stems from an extension of material relations added to the Maxwell equations.

41.2. A sufficiently complete picture of the nonlinear optical effects will be obtained if we limit the analysis to the case of

$$\mathbf{E}=\mathbf{E}_0+\mathbf{E}^{\omega}\cos\omega t \tag{41.4}$$

where $\mathbf{E}_0$ and $\mathbf{E}^{\omega}$ are constant amplitudes. Assume also

$$P_i=\chi_{ij}^{(1)}E_j+\chi_{ijk}^{(2)}E_jE_k+\chi_{ijkl}^{(3)}E_jE_kE_l \tag{41.5}$$

Coefficients $\chi^{(2)}$ and $\chi^{(3)}$ must be symmetric in all their indices except the first one. By substituting (41.4) into (41.5) and recalling the abovementioned symmetry and elementary relations $\cos^2\omega t = 1/2\,(\cos 2\omega t+1)$ and $\cos^3\omega t=(1/4)(\cos 3\omega t+3\cos\omega t)$, we can rewrite relation (41.5) in the form

$$\mathbf{P}=\mathbf{P}^0+\mathbf{P}^{\omega}\cos\omega t+\mathbf{P}^{2\omega}\cos 2\omega t+\mathbf{P}^{3\omega}\cos 3\omega t \tag{41.6}$$

Here $\mathbf{P}_0$ is a constant component of polarization. We shall write these terms and briefly discuss their meaning. Note that equation (41.5) ignores dispersion. If, however, it is taken into account, that is, if we use expressions of the type of (41.3), we can find that after substituting (41.4) the coefficients are represented by their Fourier transforms and are functions of ω. But in the further analysis dispersion will be ignored although then the reader must remember that in fact it exists.

Let us successively analyze the terms of polarization given in (41.6). It will be convenient to begin with $\mathbf{P}^{2\omega}$. The corresponding component of polarization oscillates at a frequency twice that of the frequency of the field generating these oscillations. Such oscillations generate radiation at this doubled frequency, and this radiation is observed when light is transmitted through the matter (the second harmonics generation). It can be shown that

$$P_i^{2\omega}=(1/2)\,\chi_{ijk}^{(2)}E_j^{\omega}E_k^{\omega}+3\chi_{ijkl}^{(3)}E_j^{0}E_k^{\omega}E_l^{\omega} \tag{41.7}$$

We see that the effect of the second harmonic generation is made up of two components: a quadratic component determined only by the alternating component of the field (41.4), and a cubic component appearing only if the field has a constant component.

A similar effect is the third harmonic generation, that is, emission at a triple frequency represented by the term with coefficient

$$P_i^{3\omega}=\frac{1}{4}\,\chi_{ijkl}^{(3)}E_j^{\omega}E_k^{\omega}E_l^{\omega}$$

Let us analyze the remaining two terms in (41.6). The constant component is

$$P_i^{(0)}=\chi_{ij}^{(1)}E_j^{0}+\chi_{ijk}^{(2)}E_j^{0}E_k^{0}+\chi_{ijkl}^{(3)}E_j^{0}E_k^{0}E_l^{0} + (1/2)\,\chi_{ijk}^{(2)}E_j^{\omega}E_k^{\omega}+(3/2)\,\chi_{ijkl}^{(3)}E_j^{0}E_k^{\omega}E_l^{\omega} \tag{41.8}$$

The first three terms represent the effect of static electric field. Under certain conditions the quadratic and cubic terms are observed in ferroelectrics. The fourth term describes constant polarization appearing when the second harmonic is excited. And the last term represents an additional change in static polarization in the case when the second harmonic is excited in the presence of the constant electric field.

And finally the component of polarization, oscillating at the same frequency as the exciting wave, is

$$P_i^{\omega} = \chi_{ij}^{(1)} E_j^{\omega} + 2\chi_{ijk}^{(2)} E_j^0 E_k^{\omega} + 3\chi_{ijkl}^{(3)} E_j^0 E_k^0 E_l^{\omega} + (3/4)\, \chi_{ijkl}^{(3)} E_j^{\omega} E_k^{\omega} E_l^{\omega} \quad (41.9)$$

The properties of this component can be expressed in terms of the refractive index. The first term here is already familiar. The other two terms represent effects which were known long before the other nonlinear optical effects were discovered. Indeed, it is clear that their observation only requires a strong constant electric field. The first of them corresponds to *Pockel's effect* (discovered in 1893): a sufficiently strong constant electric field affects the direction of propagation of light (affects the refractive index of the medium). Symmetry-based arguments (which unfortunately cannot be dealt with here) show that Pockel's effect, being a second-order effect, can be observed only in crystals without the center of inversion among their symmetry elements. However, if the center of inversion is one of the symmetry elements of the crystal, the corresponding effect may have only the third order. It is then described by the third term in (41.9) and is called the *Kerr effect*. Experimentally it is observed as birefringence of light waves in constant electric fields even if in zero field the medium is isotropic. And finally, the last term represents a change in the refractive index of the medium produced by the wave transmitted through the medium.

All the physical effects listed above were confirmed experimentally. Their theoretical description requires more detailed information on the structure of the media in which they are observed. The description given above is elementary and schematic. There are also a number of very interesting problems in nonlinear optics (such as self-focusing of optical beams in nonlinear media) which unfortunately have no place in this book. However, even a brief outline as given above must convincingly demonstrate that the possibilities of description of electromagnetic phenomena are greatly expanded when the assumptions concerning the properties of media are generalized (we again refer the reader to the monographs cited on p. 319). This may be regarded as an illustration of universal nature of the method whose foundation was laid down by Maxwell.

APPENDIX

A. Basic formulas of tensor analysis

We assume that the reader is familiar with the axioms of vector space. Specifically, we assume that in space V_N there are N linearly independent vectors but that any $(N+1)$ vectors are linearly dependent (axiom of dimensions). If $\mathbf{e}_i$ $(1 \leqslant i \leqslant N)$ is a set of linearly independent vectors, then an arbitrary vector $\mathbf{x}$ can be presented in the form

$$\mathbf{x} = x^i \mathbf{e}_i \tag{A.1}$$

Here, as throughout the book, we use Einstein's summation notation: if an index in a formula appears twice, it means that summation is carried out over all values of this index (i.e. from 1 to N). The set of vectors $\mathbf{e}_i$ is called the *basis* of vector space.

Let $\| A_{i'}^{i} \| \equiv A$ be a square matrix, such that $\det (A_{i'}^{i}) \neq 0$ (here the superscript enumerates the columns and subscript the rows of matrix A). With this matrix we can go to a new basis formed by vectors $\mathbf{e}_{i'}$:

$$\mathbf{e}_{i'} = A_{i'}^{i} \mathbf{e}_i \tag{A.2}$$

so that

$$\mathbf{x} = x^{i'} \mathbf{e}_{i'} = (A_{i'}^{i} x^{i'})\, \mathbf{e}_i = x^i \mathbf{e}_i \tag{A.3}$$

that is

$$x^i = A_{i'}^{i} x^{i'} \tag{A.4}$$

Note that transformation (A.4) is realized by matrix A^T transposed with respect to matrix A because summation is carried out over subscript i'. By resolving equation (A.4) for variables $x^{i'}$ (this is possible since $\det A^T = \det A \neq 0$), we find new components of vector $\mathbf{x}$ in terms of x^i:

$$x^{i'} = A_i^{i'} x^i \tag{A.5}$$

Elements $A_i^{i'}$ make up a matrix inverse with respect to A^T. According to the definition of the inverse matrix,

$$A_{i'}^{i} A_k^{k'} = \delta_{i'}^{k'}, \quad A_{i'}^{i} A_k^{i'} = \delta_k^i \tag{A.6}$$

A set of linear transformations A with nonzero determinants forms a group with standard multiplication of matrices (multiplication is associative; for each matrix there exists an inverse; there exists a unit element, namely, the matrix of identical transformation). This group constitutes a group of affine transformations in vector space.

Any physical interpretation of vector space must be based on the fundamental fact that, as follows from (A.3), the concept of vector is invariant. In contrast to this, the numerical description of a vector by components x^i is meaningful only with respect to a given basis; a change in components, given by (A.5) for a changed basis, states that the vector itself, $\mathbf{x}$, is unaltered. However, the vector is only one example among geometrical objects of vector space, which are independent of the choice of basis. Another example is a linear numerical (scalar) function $\varphi(\mathbf{x})$ defined in space V_N.[1] In this case we denote $\varphi_i = \varphi(\mathbf{e}_i)$ and $\varphi_{i'} = \varphi(\mathbf{e}_{i'})$, so that

$$\varphi(\mathbf{x}) = \varphi(x^i \mathbf{e}_i) = x^i \varphi_i = x^{i'} \varphi_{i'} = x^{i'} (A^i_{i'} \varphi_i)$$

that is

$$\varphi_{i'} = A^i_{i'} \varphi_i \tag{A.7}$$

Here matrix $\| A^i_{i'} \|$ is the same as in (A.1). Quantities φ_i are regarded as components of a geometric object referred to as covariant tensor of rank 1 (or *covariant vector*). Components x^i given earlier define a contravariant tensor of rank 1 (*contravariant vector*).

In a more general case a geometric object can be defined by giving for each basis a matrix of components $T^{i_1 i_2 \ldots i_k}_{j_1 j_2 \ldots j_l}$ with components corresponding to different bases related by the following transformation formula:

$$T^{i'_1 i'_2 \ldots i'_k}_{j'_1 j'_2 \ldots j'_l} = A^{i'_1}_{i_1} A^{i'_2}_{i_2} \ldots A^{i'_k}_{i_k} A^{j_1}_{j'_1} A^{j_2}_{j'_2} \ldots A^{j_l}_{j'_l} T^{i_1 i_2 \ldots i_k}_{j_1 j_2 \ldots j_l} \tag{A.8}$$

This geometric object is called the tensor of rank $k + l$, k times contravariant and l times covariant. Obviously, in the general case a permutation of indices may modify the matrix. Therefore, we should additionally mark the place occupied by covariant subscripts with respect to contravariant superscripts (for instance, $T_{k\cdot m}^{\cdot l}$, since in the general case $T^l_{\cdot km} \neq T_{k\cdot m}^{\cdot l}$). This was omitted in (A.8) in order not to encumber the formula.

The simplest tensor of rank 2 is the direct product of two vectors whose components are given by a table of products of the type $x^k y^l$, $x_k y_l$, or $x^k y_l$ (these are three different tensors, each with its own law of transformation of components).

[1] The linearity condition signifies that $\varphi(\alpha\mathbf{x} + \beta\mathbf{y}) = \alpha\varphi(\mathbf{x}) + \beta\varphi(\mathbf{y})$, where α and β are scalar constants.

We shall enumerate algebraic operations over tensors. The operations will be sufficiently well illustrated by formulas for tensors of low ranks.

(1) *Addition.* A tensor obtained by adding the components of two tensors with identical law of transformation has the same law of transformation. For example, $T_1^{ij} + T_2^{ij} = T_3^{ij}$.

(2) *Multiplication by a scalar.* Multiplication of all components of a tensor by the same quantity (for example, $\alpha T^{ik}{}_{..l}$) does not alter the law of transformation for this tensor.

(3) *Direct product of tensors.* For a tensor T_1 of rank n_1, k_1 times contravariant and l_1 times covariant (so that $k_1 + l_1 = n_1$) and a second arbitrary tensor T_2 of rank n_2, k_2 times contravariant and l_2 times covariant ($k_2 + l_2 = n_2$), the matrix of all possible products of components of tensor T_1 by components of tensor T_2 is transformed as a tensor of rank $n_1 + n_2$, which is $k_1 + k_2$ times contravariant and $l_1 + l_2$ times covariant.

For example, from tensor components $u_k^{\cdot l}$ and v_{mn} we can form components $u_k^{\cdot l} v_{mn}$ of the new tensor.

(4) *Contraction of a tensor.* If a tensor has k contravariant and l covariant indices, a table of new components can be composed of its components in the following manner. Let us choose those components in which the values of one contravariant index are equal to those of one covariant index, and form the sums of such components over these equal indices. The set of these sums forms a new tensor which is $k - 1$ times contravariant and $l - 1$ times covariant.

For instance, if $T^{kl}{}_{..m}$ is a tensor, then $T^{km}{}_{..m}$ and $T^{ml}{}_{..m}$ are also tensors (in this case each of them retains only one index and is transformed as a vector).

Operations of multiplication and contraction are often combined, for instance, $u^{ml} v_l \equiv w^m$.

In the general case, a tensor of rank n has N^n components in N-dimensional space. If its table of components has certain symmetry properties, the components are algebraically related and the number of linearly independent components diminishes. Thus, by definition, a tensor is symmetric if its components are unchanged under any permutation of indices. A completely antisymmetric tensor is the one in which the components change sign under permutation of any pair of indices (hence, under any odd permutation of indices). Formula (A.8) shows that symmetry properties of a tensor are conserved regardless of the choice of a basis.

No antisymmetric nonzero tensor of rank above N is possible in the N-dimensional space. An antisymmetric tensor $A_{i_1 i_2 \cdots i_N}$ (this tensor is also called pseudoscalar) of the highest possible rank has components $A_{12 \cdots N}$ distinct from zero; all other nonzero components of the tensor differ from this component by a permutation of indices 1, 2, . . ., N. If the permutation is even, the corresponding component is equal to $A_{12\ldots N}$, and if the permutation is odd, the component has the sign opposite to that of $A_{12\ldots N}$. If $A_{12\ldots N}=1$, we denote the tensor by a symbol $\varepsilon_{i_1 i_2 \ldots i_N}$. Application of formula (A.8) yields

$$\varepsilon_{1'2'\ldots N'}=\det(A^{i}_{i'})\cdot\varepsilon_{12\ldots N} \qquad \text{(A.9)}$$

In the simplest case of a rank-2 tensor T_{ik} having no symmetry, we can construct from it a symmetric and an antisymmetric tensor. Indeed, expressions

$$T^{(s)}_{ik}=\frac{1}{2}(T_{ik}+T_{ki}),\quad T^{(a)}_{ik}=\frac{1}{2}(T_{ik}-T_{ki}),\quad T_{ik}=T^{(s)}_{ik}+T^{(a)}_{ik} \qquad \text{(A.10)}$$

define tensors, with $T^{(s)}_{ik}=T^{(s)}_{ki}$ and $T^{(a)}_{ik}=-T^{(a)}_{ki}$. It must be kept in mind that formulas (A.10) are applicable only to tensors of rank 2. In more general cases, the general definition of operations of symmetrization and alternation, not encountered in this book, have to be used. In particular, a symmetrized product $T^{(s)}_{ik}$ and the so-called bivector $T^{(a)}_{ik}$ can be constructed from the direct product $T_{ik}\equiv\equiv x_i y_k$:

$$T^{(s)}_{ik}=\frac{1}{2}(x_i y_k+x_k y_i),\quad T^{(a)}_{ik}=\frac{1}{2}(x_i y_k-x_k y_i) \qquad \text{(A.10')}$$

Euclidean N-dimensional space E_N is defined by a bilinear scalar function in vector space V_N; thus function assumes real values and is denoted by $\varphi(\mathbf{x},\mathbf{y})\equiv(\mathbf{x},\mathbf{y})$. This function is called the scalar product of vectors $\mathbf{x}$ and $\mathbf{y}$, or the metric of space E_N. It is assumed that $(\mathbf{x},\mathbf{y})=(\mathbf{y},\mathbf{x})$. Squared length of a vector is defined as $(\mathbf{x},\mathbf{x})\equiv\equiv\mathbf{x}^2$, and orthogonality of two vectors as $(\mathbf{x},\mathbf{y})=0$. Function $(\mathbf{x},\mathbf{y})$ need not be of fixed sign, so that three situations are possible for various $\mathbf{x}$: $\mathbf{x}^2>0$, $\mathbf{x}^2<0$, or $\mathbf{x}^2=0$. If $\mathbf{x}^2\geqslant 0$ for any $\mathbf{x}$, the space is called the *properly Euclidean space*, otherwise it is called the *pseudo-Euclidean space*.

Denote $(\mathbf{e}_i,\mathbf{e}_j)\equiv g_{ij}$ if $\mathbf{e}_i$ is a basis in V_N. Then formula (A.2) yields the following relations:

$$g_{i'j'}=(\mathbf{e}_{i'},\mathbf{e}_{j'})=A^{i}_{i'}A^{j}_{j'}(\mathbf{e}_i,\mathbf{e}_j)=A^{i}_{i'}A^{j}_{j'}g_{ij} \qquad \text{(A.11)}$$

The set of quantities g_{ij} is therefore the matrix of components of a tensor which is called *metric*. Metric tensors are symmetric and determine completely the metric of the space. In what follows we assume that

$$\det(g_{ij})\neq 0 \qquad \text{(A.12)}$$

This means that E_N does not contain nonzero vectors orthogonal to all vectors of the space.

An analysis of real quadratic forms (see, for example, P. K. Rashevsky, *Rimanian Geometry and Tensor Analysis*, Gostekhizdat, 1953,

§ 42) shows that among all possible bases in space E_N there is a class of such bases in which quadratic form $\mathbf{x}^2$ reduces to a sum of squares. Such bases are called *orthonormalized*; the unit vectors $\mathbf{e}_i$ of the basis satisfy the relations

$$\mathbf{e}_i^2 = 1 \quad (1 \leqslant i \leqslant k), \quad \mathbf{e}_i^2 = -1 \; (k+1 \leqslant i \leqslant N)$$
$$(\mathbf{e}_i, \mathbf{e}_j) = 0 \quad (i \neq j) \tag{A.13}$$

The vectors of the basis normalized to (+1) or (—1) are called unit vectors or imaginary unit vectors, respectively, with the number in each subset conserved in each orthonormal basis (the law of inertia of quadratic forms). Thus, the following formulas hold in any orthonormal basis:

$$\mathbf{x}^2 = (x^1)^2 + \ldots + (x^k)^2 - (x^{k+1})^2 - \ldots - (x^N)^2$$
$$(\mathbf{x}, \mathbf{y}) = x^1y^1 + \ldots + x^ky^k - x^{k+1}y^{k+1} - \ldots - x^Ny^N \tag{A.14}$$

In accordance with the tensor terminology, components x^i of any vector $\mathbf{x}$, defined by the axiom of dimensions (A.1), are called contravariant. A vector can also be defined by fixing N numbers with respect to any basis:

$$x_i = (\mathbf{x}, \mathbf{e}_i) = x^j (\mathbf{e}_j, \mathbf{e}_i) = g_{ij}x^j \tag{A.15}$$

These quantities are transformed under a change of a basis by formulas (A.7) and are called covariant components of vector $\mathbf{x}$. In orthonormal bases formula (A.15) takes the form

$$x_i = g_{ii}x^i \text{ (without summation!)} \tag{A.16}$$

It can be proved that if equations (A.15) are resolved in x^j and the solution is written in the form

$$x^i = g^{ij}x_j \tag{A.17}$$

then g^{ij} are components of a twice contravariant tensor. This tensor defines the metric just as well as the one introduced earlier:

$$(\mathbf{x}, \mathbf{y}) = g_{ij}x^iy^j = g^{ij}x_iy_j \tag{A.18}$$

In orthonormal bases $g^{ij} = g_{ij} = g_{ii}\delta_{ij}$.

Formulas (A.15) and (A.17) can be called the rules of raising and lowering of tensor indices; these rules can be extended to tensors of arbitrary rank by writting, for example,

$$T^{i}_{\cdot k} = g^{ij}T_{jk}, \quad T_{jk} = g_{jl}T^{l}_{\cdot k} \tag{A.19}$$

and so on. The corresponding formulas for orthonormalized bases are simplified in accordance with (A.16).

In order to describe transitions from one orthonormalized basis to another, we have to select among affine transformations all those which transform relations (A.13) for components g_{ij} in the former

basis into similar relations in the new basis; in other words, $(\mathbf{e}_{i'}, \mathbf{e}_{j'}) = 0$ $(i' \neq j')$, and normalization of vectors $\mathbf{e}_{i'}$ is conserved. In order to find these transformations, we compare formulas $x_{i'} = A^i_{i'} x_i$ and $x^{i'} = A^{i'}_i x^i$ with relation (A.16). For example, the transformation law for contravariant components can be rewritten by applying (A.16) in the form $g^{i'i'} x_{i'} = g^{ii} A^{i'}_i x_i$, that is $x_{i'} = g_{i'i'} g^{ii} A^{i'}_i x_i$, so that

$$A^i_{i'} = g_{i'i'} g^{ii} A^{i'}_i \quad \text{(without summation!)} \tag{A.20}$$

Transformations satisfying condition (A.20) are called pseudo-orthogonal. It is readily verified that they form a group. In the case of the properly Euclidean space we obtain, recalling the arguments given on p. 321 ($A^{-1} = A^T$), that the condition is $A^i_{i'} = A^{i'}_i$. Such transformations are called *orthogonal*.

In going from one orthonormalized basis to another by pseudo-orthogonal transformations, components g_{ij} conserve their values, matrix $\| g_{ij} \|$ remains diagonal, and its determinant $\det(g_{ij})$ is always equal to $+1$ or -1 (this depends only on the number of imaginary unit vectors characterizing a given space). Formula (A.11) then yields for such transformations $[\det(A^i_{i'})]^2 = 1$, that is

$$\det(A^i_{i'}) = \det(A^{i'}_i) = \pm 1 \tag{A.21}$$

The first of equalities (A.21) is readily proved by using (A.20). Pseudo-orthogonal transformations with determinant equal to $+1$ are called proper transformations, or rotations, and the transformations with determinant equal to -1 are called improper, or reflection-containing transformations.

One of the important concepts in physics is that of a *tensor field*, that is a tensor whose components are position functions in some space, for example, $F^{kl}(\mathbf{x})$. Usually it is assumed that these functions are continuous and continuously differentiable a sufficient number of times in a certain range of variation of arguments. In the general case the ranges of continuity may be separated by surfaces on which the components of the tensor may have finite discontinuities. In the simplest case this is a scalar function $\varphi(\mathbf{x})$ defined in a given range of vector space. Derivatives $\partial\varphi/\partial x^i$ can be found with respect to any basis in this space. When the basis is changed, these derivatives are transformed as follows:

$$\frac{\partial\varphi}{\partial x^{i'}} = \frac{\partial x^i}{\partial x^{i'}} \frac{\partial\varphi}{\partial x^i} = A^i_{i'} \frac{\partial\varphi}{\partial x^i} \tag{A.22}$$

Differentiation operators $\partial/\partial x^i$ thus form covariant components of vector operator $\boldsymbol{\nabla}$ (*gradient operator*, denoted more often in the text by symbol grad), for which $(\boldsymbol{\nabla} \cdot \mathbf{e}_i) = \partial/\partial x^i$. Differentiation operators $\partial/\partial x_i$ with respect to covariant components of the argument form contravariant components of vector ∇. If we take in space in unit vector s, the derivative in the direction of s is found from the formula

$$(\nabla \cdot \mathrm{s})\,\varphi \equiv \frac{\partial \varphi}{\partial s} \tag{A.23}$$

A number of other operators, with specific properties of transformation, can be formed using operator ∇. For example,

$$\nabla^2 \equiv \Delta = g^{ij}\frac{\partial^2}{\partial x^i\,\partial x^j} \tag{A.24}$$

is the so-called *Laplace operator*. Scalar *operation of divergence* of a vector field is defined by the formula

$$\operatorname{div}\mathbf{A} \equiv (\nabla \cdot \mathbf{A}) = \frac{\partial A^i}{\partial x^i} \tag{A.25}$$

The gradient operation applied to a tensor field of arbitrary rank produces a new tensor field with a rank higher by unity than that of the initial field, for instance, $\partial F^{kl}/\partial x^m$. An analog of the divergence operation in the general case is the contraction of a tensor field with gradient operator, lowering the rank of a derivative of a tensor by two, for example, $\partial F^{km}/\partial x^m$.

Evidently, a scalar operation of the second order (A.24) gives a tensor of the same rank as that of the differentiated one.

B. Vector analysis in three-dimensional Euclidean space

Formulas of tensor calculus given in Appendix A can be easily applied to the case of the three-dimensional proper Euclidean space E_3. Tensor indices assuming values 1, 2, 3 will be denoted, as in the main text, by Greek letters.

Consider a completely antisymmetric tensor of the highest rank $\varepsilon_{\alpha\beta\gamma}$ transformed by formula (A.9). Property (A.21) of orthogonal transformations ensures that this tensor is completely characterized by fixing a single of its components, for example, ε_{123}, which is conserved (behaves as a scalar) with respect to rotations, and changes sign under reflections. It is thus natural to call tensor $\varepsilon_{\alpha\beta\gamma}$ a *pseudoscalar*. Let us assume $\varepsilon_{\alpha\beta\gamma}$ to be a unit pseudoscalar, that is $\varepsilon_{123} = 1$.

The following relations are easily verified:

$$\varepsilon^{\alpha\beta\gamma}\varepsilon_{\alpha\varkappa\lambda} = \delta^{\beta}_{\varkappa}\delta^{\gamma}_{\lambda} - \delta^{\gamma}_{\varkappa}\delta^{\beta}_{\lambda}$$
$$\varepsilon^{\alpha\beta\gamma}\varepsilon_{\alpha\beta\lambda} = 2\delta^{\gamma}_{\lambda}, \quad \varepsilon^{\alpha\beta\gamma}\varepsilon_{\alpha\beta\gamma} = 3! \tag{B.1}$$

If A^α is a 3-dimensional vector, and $T^{\alpha\beta}$ is an antisymmetric tensor, then contraction yields the following quantities:

$$A^{\alpha}\varepsilon_{\alpha\beta\gamma} = \overset{*}{T}_{\beta\gamma}, \quad \frac{1}{2}\,T^{\alpha\beta}\varepsilon_{\alpha\beta\gamma} = \overset{*}{A}_{\gamma} \tag{B.2}$$

Note that according to (A.16) there is no difference between covariant and contravariant indices in space E_3, so that any one of the indices can be raised or lowered without changing the meaning of the formulas. Taking into account formulas (A.8) and (A.9), we conclude that $\overset{*}{A}_\gamma$ and $\overset{*}{T}_{\beta\gamma}$ are transformed under rotations as the components of a vector and of an antisymmetric tensor of rank 2, respectively, and are multiplied by -1 under reflections. Consequently, $\overset{*}{A}_\gamma$ is known as a pseudovector, and $\overset{*}{T}_{\beta\gamma}$ as a pseudotensor. In the three-dimensional vector analysis, pseudovectors are often referred to as *axial vectors*; correspondingly, "ordinary" vectors are sometimes called *polar vectors*.

Elementary geometric arguments show that doubled bivector (A.10′) constructed of components of vectors **x** and **y** has components $2T^{(a)}_{\alpha\beta}$ numerically equal to the projections of the area of a parallelogram, composed of these vectors, onto coordinate planes (α, β). Formula (B.2) enables us to put this bivector in correspondence with a pseudovector:

$$\overset{*}{A}{}^{\alpha}=\varepsilon^{\alpha\beta\gamma}T^{(a)}_{\beta\gamma}=\frac{1}{2}\,\varepsilon^{\alpha\beta\gamma}(x_\beta y_\gamma-x_\gamma y_\beta)=\varepsilon^{\alpha\beta\gamma}x_\beta y_\gamma \tag{B.3}$$

In this case the notation $\overset{*}{A}{}^{\alpha}=(\mathbf{x}\times\mathbf{y})^\alpha$ serves as a definition of components of the vector product $\mathbf{x}\times\mathbf{y}$ of vectors **x** and **y**.

Definition (B.3) easily yields the well-known properties of the product of three vectors. Thus, for example,

$$\mathbf{a}\cdot(\mathbf{b}\times\mathbf{c})=a_\alpha(\mathbf{b}\times\mathbf{c})^\alpha=\varepsilon^{\alpha\beta\gamma}a_\alpha b_\beta c_\gamma \tag{B.4}$$

From $\varepsilon^{\alpha\beta\gamma}a_\alpha b_\beta c_\gamma=\varepsilon^{\beta\alpha\gamma}a_\beta b_\alpha c_\gamma=-\varepsilon^{\alpha\beta\gamma}b_\alpha a_\beta c_\gamma$ (here the first relation is obtained by permutation of summation indices, and the second follows from antisymmetry of tensor $\varepsilon^{\alpha\beta\gamma}$) and, similarly, from $\varepsilon^{\alpha\beta\gamma}a_\alpha b_\beta c_\gamma=\varepsilon^{\gamma\beta\alpha}a_\gamma b_\beta c_\alpha=-\varepsilon^{\alpha\beta\gamma}c_\alpha b_\beta a_\gamma$, we obtain

$$\mathbf{a}\cdot(\mathbf{b}\times\mathbf{c})=\mathbf{b}\cdot(\mathbf{c}\times\mathbf{a})=\mathbf{c}\cdot(\mathbf{a}\times\mathbf{b}) \tag{B.5}$$

Similarly, we can derive

$$\mathbf{a}\times(\mathbf{b}\times\mathbf{c})=\mathbf{b}\,(\mathbf{a}\cdot\mathbf{c})-\mathbf{c}\,(\mathbf{a}\cdot\mathbf{b}) \tag{B.6}$$

The operation of antisymmetrization over tensor indices will be denoted by bracketing these indices. Namely,

$$x^1_{[i_1}\dots x^m_{i_m]}=\frac{1}{m!}\sum_{(P)}\pm Px^1_{i_1}\dots x^m_{i_m}$$

Here summation is carried out over all possible $m!$ permutations P of indices $i_1, \dots, i_m$; in each summand, obtained by an even permutation from some arrangement of indices chosen to be initial, the plus sign is taken, and the minus sign is taken in the case of an odd permutation. If $\mathbf{a} = d\mathbf{x}$ and $\mathbf{b} = d\mathbf{y}$ are infinitesimal displacements originating at a point on a two-dimensional surface σ and lying in a tangent plane to this surface, formula (B.3) defines the area of an oriented infinitesimal element of this surface:

$$d\sigma^{\alpha} = \varepsilon^{\alpha\beta\gamma}\, dx_{[\beta}\, dy_{\gamma]} \tag{B.7}$$

Similarly, formula (B.4) gives the oriented 3-dimensional volume for noncoplanar displacements $\mathbf{a} = d\mathbf{x}$, $\mathbf{b} = d\mathbf{y}$, $\mathbf{c} = d\mathbf{z}$:

$$d\overset{*}{V} = \varepsilon^{\alpha\beta\gamma}\, dx_{[\alpha}\, dy_{\beta}\, dz_{\gamma]} \tag{B.8}$$

We specify the right-handed orientation to be positive throughout the exposition. Absolute values of area $(\sum_{\alpha} \mid d\sigma^{\alpha} \mid^2)^{1/2}$ and volume $\mid dV^* \mid$ are independent of the choice of orientation.

Consider now vector fields in space E_3 and differential operations applied to these fields.

The results for operations grad, div, and div grad given in formulas (A.22)-(A.25) are extended in a straightforward manner to the particular case E_3. In Cartesian coordinates the Laplacian is

$$\Delta \equiv \operatorname{div}\operatorname{grad} = \frac{\partial^2}{\partial x_1^2} + \frac{\partial^2}{\partial x_2^2} + \frac{\partial^2}{\partial x_3^2} \tag{B.9}$$

One specific feature of the three-dimensional space is the possibility of defining a pseudovectorial operation curl similarly to the given above definition of the vector product of two vectors. By combining components $\partial/\partial x^{\alpha}$ of operator grad and components b^{β} of an arbitrary differentiable vector field $\mathbf{b}$ in Cartesian coordinates, we can form quantities

$$\frac{\partial b^{\beta}}{\partial x^{\alpha}} - \frac{\partial b^{\alpha}}{\partial x^{\beta}}$$

which constitute components of an antisymmetric tensor of rank 2. By using the general rule (B.2) we define a field

$$\overset{*}{A}{}^{\gamma} = \frac{1}{2}\, \varepsilon^{\gamma\alpha\beta} \left(\frac{\partial b^{\beta}}{\partial x^{\alpha}} - \frac{\partial b^{\alpha}}{\partial x^{\beta}} \right) = \varepsilon^{\gamma\alpha\beta}\, \frac{\partial b^{\beta}}{\partial x^{\alpha}} \tag{B.10}$$

This field $\overset{*}{A}{}^{\gamma}$ transforms as a pseudovector and is called the *curl* (or vortex) of vector field $\mathbf{b}$. The corresponding notation is

$$\overset{*}{A}{}^{\gamma} = (\operatorname{curl} \mathbf{b})^{\gamma} \tag{B.11}$$

The tensor notation of differential operations, which we have chosen, makes it possible to derive readily and in a single-valued manner all standard formulas of vector analysis, with the character of transformation of the introduced quantities being clear at all stages of transformation.

For example, if $\mathbf{b} = \operatorname{grad} \varphi$, then

$$\overset{*}{A}{}^{\gamma} = \varepsilon^{\gamma\alpha\beta} \frac{\partial^2 \varphi}{\partial x^\alpha \, \partial x^\beta} \equiv 0$$

(because $\partial^2/\partial x^\alpha \, \partial x^\beta$ is symmetric in indices α and β, and $\varepsilon^{\gamma\alpha\beta}$ is antisymmetric), that is

$$\operatorname{curl} \operatorname{grad} \varphi \equiv 0 \tag{B.12}$$

Further,

$$\frac{\partial \overset{*}{A}{}^{\gamma}}{\partial x^\gamma} = \varepsilon^{\gamma\alpha\beta} \frac{\partial^2 b^\beta}{\partial x^\alpha \, \partial x^\gamma} \equiv 0$$

that is

$$\operatorname{div} \operatorname{curl} \mathbf{b} = 0 \tag{B.13}$$

If φ and ψ are scalar functions, then elementary application of the formula for differentiation of a product and of the basic definitions given above yields

$$\operatorname{grad} (\varphi\psi) = \varphi \operatorname{grad} \psi + \psi \operatorname{grad} \varphi \tag{B.14$_1$}$$

$$\operatorname{div} (\varphi \mathbf{a}) = \varphi \operatorname{div} \mathbf{a} + (\operatorname{grad} \varphi \cdot \mathbf{a}) \tag{B.14$_2$}$$

$$\operatorname{curl} (\varphi \mathbf{a}) = \varphi \operatorname{curl} \mathbf{a} + \operatorname{grad} \varphi \cdot \mathbf{a} \tag{B.14$_3$}$$

Note that as follows from formulas (B.1), (B.10), and (B.11),

$$\varepsilon_{\gamma\alpha\beta} (\operatorname{curl} \mathbf{b})^\gamma = \varepsilon_{\gamma\alpha\beta} \varepsilon^{\gamma\mu\nu} \frac{\partial b^\nu}{\partial x^\mu} = \frac{\partial b^\beta}{\partial x^\alpha} - \frac{\partial b^\alpha}{\partial x^\beta} \tag{B.15}$$

This relation can be used to derive the expression for $\operatorname{grad} (\mathbf{a} \cdot \mathbf{b})$. First, we calculate $\operatorname{grad} (\mathbf{a}^2) = \operatorname{grad} (a^\alpha a_\alpha)$. The expression for components is

$$\frac{1}{2} \frac{\partial (a^\alpha a_\alpha)}{\partial x^\beta} = a_\alpha \frac{\partial a^\alpha}{\partial x^\beta} = a_\alpha \left(\frac{\partial a^\alpha}{\partial x^\beta} - \frac{\partial a^\beta}{\partial x^\alpha} \right) + a_\alpha \frac{\partial a^\beta}{\partial x^\alpha}$$

$$= a^\alpha \varepsilon_{\gamma\beta\alpha} (\operatorname{curl} \mathbf{a})^\gamma + a_\alpha \frac{\partial a^\beta}{\partial x^\alpha}$$

that is, with the frequently used notation

$$a_\alpha \frac{\partial}{\partial x^\alpha} \equiv (\mathbf{a} \cdot \operatorname{grad}) \tag{B.16}$$

and formula (B.3), we obtain

$$\frac{1}{2} \operatorname{grad} (\mathbf{a}^2) = \mathbf{a} \times \operatorname{curl} \mathbf{a} + (\mathbf{a} \cdot \operatorname{grad}) \, \mathbf{a} \tag{B.17}$$

By applying formula (B.17) to $\operatorname{grad} [(\mathbf{a} + \mathbf{b})^2]$ and using linearity of operation grad, we readily obtain

$$\operatorname{grad} (\mathbf{a} \cdot \mathbf{b}) = \mathbf{a} \times \operatorname{curl} \mathbf{b} + \mathbf{b} \times \operatorname{curl} \mathbf{a}$$

$$+ (\mathbf{a} \cdot \operatorname{grad}) \, \mathbf{b} + (\mathbf{b} \cdot \operatorname{grad}) \, \mathbf{a} \tag{B.18}$$

The following formulas of vector analysis are often used:

$$\operatorname{div}(\mathbf{a}\times\mathbf{b}) = \mathbf{b}\cdot\operatorname{curl}\mathbf{a} - \mathbf{a}\cdot\operatorname{curl}\mathbf{b} \tag{B.19}$$

$$\operatorname{curl}(\mathbf{a}\times\mathbf{b}) = (\mathbf{b}\cdot\operatorname{grad})\,\mathbf{a} - (\mathbf{a}\cdot\operatorname{grad})\,\mathbf{b} + \mathbf{a}\operatorname{div}\mathbf{b} - \mathbf{b}\operatorname{div}\mathbf{a} \tag{B.20}$$

The derivation of the last relation is as follows:

$$\operatorname{curl}_\nu(\mathbf{a}\times\mathbf{b}) = \varepsilon_{\nu\lambda\varkappa}\frac{\partial}{\partial x^\lambda}(\mathbf{a}\times\mathbf{b})^\varkappa = \varepsilon_{\nu\lambda\varkappa}\varepsilon^{\varkappa\mu\rho}\frac{\partial}{\partial x^\lambda}(a_\mu b_\rho)$$

$$= a_\nu\frac{\partial b_\rho}{\partial x^\rho} + b_\rho\frac{\partial a_\nu}{\partial x^\rho} - b_\nu\frac{\partial a_\nu}{\partial x^\mu} - a_\mu\frac{\partial b_\nu}{\partial x^\mu} \tag{B.20$'$}$$

Here we have used the first of formulas (B.1).

Formula (B.19) is proved in a similar manner. We can also prove a relation valid only in Cartesian coordinates:

$$\operatorname{curl}\operatorname{curl}\mathbf{a} = \operatorname{grad}\operatorname{div}\mathbf{a} - \operatorname{div}\operatorname{grad}\mathbf{a} \tag{B.21}$$

Formula (B.20′) shows that a pseudovectorial operation curl applied to pseudovector $\mathbf{a}\times\mathbf{b}$ yields a polar vector written in the right-hand side of (B.20′). Of course, a similar result will be obtained for curl of any pseudovector, as well as for the "vector product" of a vector by a pseudovector. Indeed, if $\overset{*}{B}_\lambda = (1/2)\,\varepsilon_{\lambda\iota\varkappa}b_{\iota\varkappa}$, where $b_{\iota\varkappa} = -b_{\varkappa\iota}$, then the first of equations in (B.1) readily gives, for example, $\operatorname{curl}_\mu \overset{*}{\mathbf{B}} = \partial b_{\mu\varkappa}/\partial x^\varkappa$. In all such cases it is sufficient to find whether spatial reflection reverses the sign of the expression of interest. For example, if the right-hand side of equality $\operatorname{curl}\mathbf{a} = \mathbf{b}$ is known to be a polar vector, we conclude that $\mathbf{a}$ is a pseudovector.

The main role in the theory of vector fields is played by the integral theorems formulated below.

Consider first a three-dimensional region V bounded by a closed two-dimensional surface σ, and specify the positive direction of a unit normal $\mathbf{n}$ at each point of this surface as that toward the region in space external with respect to the volume under consideration.

The *flux* of vector field $\mathbf{a}$ across surface σ is defined by the expression

$$\oint \mathbf{a}\cdot\mathbf{n}\,d\sigma \equiv \oint a_n\,d\sigma$$

A definition of the divergence operation $\operatorname{div}\mathbf{a}$ independent of the preliminary choice of coordinate system in space E_3 can be given in terms of the flux of the vector field. Let us surround an arbitrary point x in space by a closed surface σ; the volume within this surface will be denoted by V. If surface σ is contracted to point $\mathbf{x}$ so that $V\to 0$, the definition will take the form

$$\lim_{V\to 0}\frac{1}{V}\oint a_n\,d\sigma = \operatorname{div}\mathbf{a}(\mathbf{x}) \tag{B.22}$$

It is proved in the courses of vector analysis that in Cartesian coordinates definition (B.22) leads to formula (A.25) used earlier, that is div $\mathbf{a} = \partial a^\alpha / \partial x^\alpha$ (the necessary condition is that field $\mathbf{a}$ should have continuous partial derivatives with respect to all coordinates in the region where its divergence is considered).

The Gauss theorem. If field $\mathbf{a}(\mathbf{x})$ has continuous (or piecewise continuous) divergence div $\mathbf{a}(\mathbf{x})$, then in any region V with boundary σ

$$\oint \mathbf{a} \cdot \mathbf{n}\, d\sigma = \int_V \operatorname{div} \mathbf{a}\, dV \tag{B.23}$$

In complete analogy to the conventional proof of this theorem, we can derive a formula for tensor field $T^{\alpha\beta}(\mathbf{x})$:

$$\oint T^{\alpha\beta} n_\beta\, d\sigma = \int_V \frac{\partial T^{\alpha\beta}}{\partial x^\beta}\, dV \tag{B.23$'$}$$

Let us take now a closed linear contour s in space E_3, with a fixed direction of circling the contour, that is with a positive direction of tangent vectors, and an arbitrary two-dimensional surface σ bounded by this contour. One of the two possible directions of the normal to surface σ at any point on this surface can be chosen as positive. The normal will be considered positive if its direction is related to the positive direction of circling the contour by the right-handed screw rule.

The *circulation* of vector field $\mathbf{a}$ along a closed contour $\mathbf{s}$ is defined by

$$\oint_s \mathbf{a} \cdot \mathbf{s}\, ds \equiv \oint a_s\, ds \tag{B.24}$$

The concept of circulation makes it possible to define operation curl $\mathbf{a}$ without resorting to Cartesian coordinates. Namely, if contour $\mathbf{s}$ is contracted to point $\mathbf{x}$ lying on the mentioned surface σ, then by definition

$$\operatorname{curl}_n \mathbf{a} = \lim_{S \to 0} \frac{1}{S} \oint \mathbf{a} \cdot \mathbf{s}\, ds \tag{B.25}$$

The left-hand side is the projection of curl $\mathbf{a}$ on normal $\mathbf{n}$ to surface σ at point $\mathbf{x}$. Surface σ passing through point $\mathbf{x}$ can be chosen arbitrary, so that formula (B.25) determines at the same time the projection of curl $\mathbf{a}$ on any direction at point $\mathbf{x}$, that is completely defines vector curl $\mathbf{a}$. We have fixed the positive orientation in E_3 in advance, so that curl $\mathbf{a}$ is also defined as a pseudovector. It can be shown that in Cartesian coordinates relation (B.25) yields the definition of curl used earlier and given in formulas (B.10) and (B.11).

The Stokes theorem. If region S on surface σ is bounded by a contour s, then

$$\oint_s \mathbf{a} \cdot \mathbf{s}\, ds = \int_S \operatorname{curl} \mathbf{a} \cdot \mathbf{n}\, d\sigma \tag{B.26}$$

where the unit normal $\mathbf{n}$ to the surface is in the positive direction of the unit vector of the tangent to contour s.

Green's formula. Let vector $\mathbf{a}$ in formula (B.23) have the form $\mathbf{a} = \psi \operatorname{grad} \varphi$. Using for div $\mathbf{a}$ formula (B.14$_2$), we obtain

$$\int \{\psi \Delta\varphi + (\operatorname{grad} \psi \cdot \operatorname{grad} \varphi)\}\, dV = \oint \psi \frac{\partial \varphi}{\partial n}\, d\sigma \tag{B.27}$$

where derivative $\partial\varphi/\partial n$ is defined by (A.23). Permute symbols φ and ψ in formula (B.27) and subtract the thus obtained equation from (B.27) term by term. As a result, we obtain

$$\int (\psi \Delta\varphi - \varphi \Delta\psi)\, dV = \oint \left(\psi \frac{\partial \varphi}{\partial n} - \varphi \frac{\partial \psi}{\partial n}\right) d\sigma \tag{B.28}$$

This relation is called Green's formula; it has important applications in the theory of integration of equations in partial derivatives.

C. Basic formulas for delta function and its derivatives

The exposition of the properties of the delta function as given below is not mathematically rigorous. The readers wishing to find rigorous proofs of these properties must turn to special treatises[2].

We begin with one-dimensional argument x. Formally delta function $\delta(x - \xi)$ can be defined by the following equation:

$$\int \delta(x-\xi) f(x)\, dx = f(\xi) \tag{C.1}$$

It can be shown, however, that there is no such a position function $\delta(x - \xi)$ which would satisfy equation (C.1). But there exist infinite sequences of functions $\{\varphi_n(x)\}$, such that equation (C.1) is satisfied in the following sense:

$$\lim_{n\to\infty} \int \varphi_n(x-\xi) f(x)\, dx = f(\xi) \tag{C.2}$$

Nevertheless, property (C.1) is often formulated in terms of an "improper" function $\delta(x - \xi)$ equal to zero at $x \neq \xi$ and tending to infinity at $x = \xi$, so that

$$\int \delta(x-\xi)\, dx = 1$$

Although relation (C.1) is mathematically incorrect, it can nevertheless be used for symbolic derivation of further properties of delta function, which can be rigorously substantiated (see references cited above).

[2] See, for example, I. M. Gelfand and G. E. Shilov, *Generalized Functions*, vol. 1: Properties and Operators, Academic Press, New York, 1964, and M. J. Lighthill, *Introduction to Fourier Analysis and Generalized Functions*, Cambridge Univ. Press, Cambridge, 1958.

Integrating (C.2) by parts, we obtain

$$\int_a^b f(x)\frac{\partial\delta(x-\xi)}{\partial x}\,dx = f(x)\,\delta(x-\xi)\Big|_a^b - \int_a^b \delta(x-\xi)\frac{\partial f}{\partial x}\,dx$$

$$= -\frac{\partial f}{\partial x}\Big|_{x=\xi} \qquad \text{(C.3)}$$

Here we assume $a<\xi<b$. In fact, (C.3) must be regarded as a definition of the derivative of delta function $\partial\delta(x-\xi)/\partial x$.

Similarly,

$$\int_a^b f(x)\,\delta^{(n)}(x-\xi)\,dx = (-1)^n f^{(n)}(\xi) \quad (a<\xi<b) \qquad \text{(C.4)}$$

Standard rules of change of the integration variable yield

$$\delta(-x) = \delta(x), \quad \delta'(-x) = -\delta'(x) \qquad \text{(C.5)}$$

In addition,

$$x\delta(x) = 0 \qquad \text{(C.6)}$$

Besides,

$$\delta(ax) = \frac{\delta(x)}{|a|} \qquad \text{(C.7)}$$

Let interval $a<x<b$ contain simple roots x_s of equation $\varphi(x)=0$. In this case

$$\delta(\varphi(x)) = \sum_s \frac{\delta(x-x_s)}{|\varphi'(x)|} = \sum_s \frac{\delta(x-x_s)}{|\varphi'(x_s)|} \qquad \text{(C.8)}$$

In particular,

$$\delta(x^2-a^2) = \frac{\delta(x-a)+\delta(x+a)}{2|a|} \qquad \text{(C.9)}$$

Relation (C.8) is not valid in the case of multiple roots.

For a symbolic derivation of formula (C.8) consider the case of a single root $x=\xi$ of equation $\varphi(x)=0$ in the indicated interval. By introducing a new variable φ instead of x, we obtain

$$\int_a^b f(x)\,\delta[\varphi(x)]\,dx = \int \frac{\tilde{f}(\varphi)}{|\varphi'(x)|}\,\delta[\varphi]\,d\varphi = \frac{\tilde{f}(\varphi)}{|\varphi'(x)|}\Big|_{\varphi=0}$$

$$= \frac{f(x)}{|\varphi'(x)|}\Big|_{x=\xi}$$

In the n-dimensional case formula (C.1) transforms to

$$\int dx_1 \ldots \int dx_n\,\delta(x_1-\xi_1,\ \ldots,\ x_n-\xi_n)\,f(x_1,\ \ldots,\ x_n) = f(\xi_1,\ \ldots,\ \xi_n)$$

We can set

$$\delta(x_1-\xi_1, \ldots, x_n-\xi_n)=\prod_{k=1}^{n} \delta(x_k-\xi_k) \tag{C.10}$$

Denote the Fourier transform of function f by $(\mathcal{F}f)(\lambda)$ (here λ is the argument of function $\mathcal{F}f$). The definition of the Fourier transform $(\mathcal{F}\delta)(x)$ of delta function follows from a formal equality

$$\int \delta(\lambda-\xi)(\mathcal{F}f)(\lambda)\, d^n\lambda=\int f(x)(\mathcal{F}\delta)(x)\, d^nx \tag{C.11}$$

By (λ, x) we denote the scalar product in the argument space. As

$$(\mathcal{F}f)(\lambda)=\int e^{i(\lambda, x)} f(x)\, d^nx$$

(C.11) can be rewritten in the form

$$\int\int e^{i(\lambda, x)}\delta(\lambda-\xi) f(x)\, d^nx\, d^n\lambda$$

$$=\int d^nx f(x)\left[\int d^n\lambda e^{i(\lambda, x)}\delta(\lambda-\xi)\right] \tag{C.12}$$

A comparison of (C.11) and (C.12) yields

$$(\mathcal{F}\delta_\xi)(x)=e^{i(\xi, x)} \tag{C.13}$$

where $\delta_\xi \equiv \delta(\lambda-\xi)$. In particular, $\mathcal{F}\delta_0=1$. Therefore, in n-dimensional space the inverse Fourier transform $\mathcal{F}^{-1}$ with the normalizing factor is written in the form

$$\delta(\xi-\lambda)=\delta_\xi(\lambda)=\frac{1}{(2\pi)^n}\int e^{i(\xi-\lambda, x)}\, d^nx \tag{C.14}$$

Note that the concept of delta function is related to the description of a pointlike charge, and the concept of derivatives of delta function—to the description of pointlike multipoles (cf. § 11).

D. Integration over hypersurfaces in the Minkowski space

1. Let us define function $\rho(\vec{x})$ as in § 23. Consider a hypersurface Σ from a family of timelike hypersurfaces in the Minkowski space, defined by the equations of the type $\rho(\vec{x})=$ const (so that gradient $\partial\rho/\partial x^r$, being normal to the hypersurface, is a spacelike vector). Denote by $-\vec{p}$ a unit vector of the normal to hypersurface Σ, and by $\vec{v}=\vec{u}/c$ a unit timelike tangent to this hypersurface at one of its points (Fig. 47).

Let us assume that an infinitesimal element $d\Sigma_l$ of hypersurface Σ at point $\vec{x}$ is oriented in direction $-\vec{p}$ (this condition will prove convenient). We can write then $d\Sigma_r = -p_r\, d\Sigma$. Let $\vec{dx}$ be an infinitesimal displacement along $\vec{p}$, so that $dx^r = p^r\, dl$. Consider a 4-dimensional volume $d\Omega$ of a cylinder with generatrix $\vec{dx}$; the base of the cylinder is a 3-dimensional element of hypersurface $d\Sigma_r$. Clearly,

$$d\Omega = d\Sigma\, dl \tag{D.1}$$

The same volume $d\Omega$ can be calculated somewhat differently (Fig. 48). Consider a spacelike hypersurface Π passing through point $\vec{x}$ and

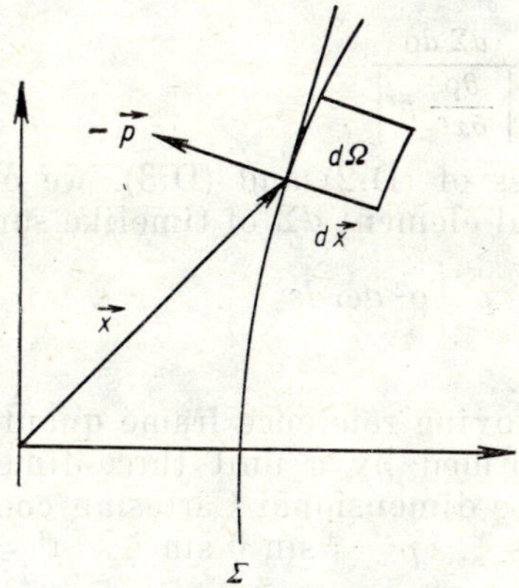

Fig. 47

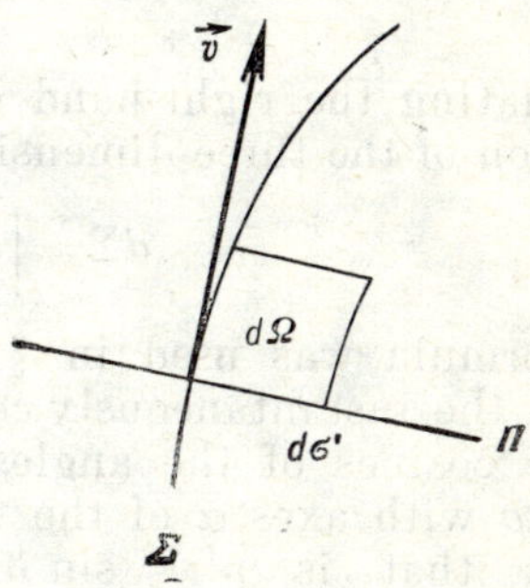

Fig. 48

defined by the equation $(\vec{dx}, \vec{\nu}) = 0$. Vector $\vec{p}$ lies in hyperplane Π. Volume $d\Omega$ can now be sliced into hyperplanes parallel to Π, having a common timelike normal $\vec{\nu}$. Element $d\sigma'$ of hyperplane Π can be found as follows. Since $\vec{R} = \rho\,(\vec{\nu} + \vec{p})$, the projection of this vector onto hyperplane Π is $\vec{R}_\Pi = \rho\vec{p}$. Let us change to the rest frame at point $\vec{x}$. In this frame element $d\sigma'$ is found simply as an element of 3-dimensional volume in hyperplane Π. By introducing 3-dimensional Cartesian coordinates in this hyperplane, we can transform to spherical coordinates, with radius vector coinciding with $\vec{R}_\Pi$ and angles ϑ and φ defined in the standard manner. In this spherical coordinate system $d\sigma' = \rho^2\, d\rho\, d\omega$, where $d\omega = \sin\vartheta\, d\vartheta\, d\varphi$, and an element of 4-volume takes the form

$$d\Omega = \rho^2\, d\rho\, d\omega\, ds \tag{D.2}$$

where ds is an infinitely small displacement in the direction of vector $\vec{\nu}$ orthogonal to Π. As the four-dimensional volume is invariant under the Lorentz transformations, it is numerically equal to the right-hand side of (D.2) in any reference frame. Furthermore, it is clear that to within infinitesimals of higher orders expressions (D.1) and (D.2) must coincide. The first of them is obtained by slicing the volume by a family of timelike hypersurfaces $\rho = \text{const}$, while the second results from slicing the same volume into a family of three-dimensional spacelike hyperplanes parallel to hyperplane Π.

On the other hand, if we return to formula (D.1), we can write

$$d\rho = \left| \frac{\partial \rho}{\partial x^r} dx^r \right| = dl \left| \frac{\partial \rho}{\partial x^r} p^r \right|$$

that is

$$d\Omega = \frac{d\Sigma \, d\rho}{\left| \frac{\partial \rho}{\partial x^r} p^r \right|} \tag{D.3}$$

By equating the right-hand sides of (D.2) and (D.3), we obtain a definition of the three-dimensional element $d\Sigma$ of timelike surface Σ:

$$d\Sigma = \left| \frac{\partial \rho}{\partial x^r} p^r \right| \rho^2 \, d\omega \, ds \tag{D.4}$$

This formula was used in § 23.

2. In the instantaneously co-moving reference frame quantities p^α are the cosines of the angles formed by a unit three-dimensional vector p with axes α of the three-dimensional Cartesian coordinate system, that is $p^1 = \sin\vartheta \cos\zeta$, $p^2 = \sin\vartheta \sin\zeta$, $p^3 = \cos\vartheta$. In addition,

$$\int p^\alpha p_\beta \, d\omega = -\frac{4\pi}{3} \delta^\alpha_\beta \tag{D.5}$$

where, according to the rule of lowering spacial indices in space-time, we have taken into account that $p_\beta = -p^\beta$. As everywhere in the book,

$$d\omega = \sin\vartheta \, d\vartheta \, d\zeta, \quad \int d\omega = 4\pi \tag{D.6}$$

Any integral of a product of an odd number of components p^α equals zero. Denote spatial components of vector $\vec{p}$ in an arbitrary reference frame by p'^α. As $p^0 = 0$, transformation formulas (5.12) take the form $p'^\alpha = p^\alpha + (\gamma - 1) \frac{\mathbf{p} \cdot \mathbf{v}}{v^2} v^\alpha$. By taking into account that $\mathbf{p} \cdot \mathbf{v} = \sum p^\gamma v^\gamma = -\sum p_\gamma v^\gamma$, we therefore obtain

$$p'^\alpha p'_\beta = p^\alpha p_\beta + (\gamma - 1)^2 \frac{v^\alpha v_\beta}{v^4} v_\gamma v^\delta p^\gamma p_\delta$$
$$- (\gamma - 1) \frac{v^\alpha v_\gamma}{v^2} p^\gamma p_\beta - (\gamma - 1) \frac{v_\beta v^\gamma}{v^2} p^\alpha p_\gamma$$

Integrate both parts of this equation over $d\omega$ (solid angle in the rest frame of reference) and use (D.5). The last three terms give $\frac{4\pi}{3}\frac{v^\alpha v_\beta}{v_2}$ multiplied by $(\gamma-1)^2+2(\gamma-1)=\gamma^2-1=\gamma^2\beta^2$, that is $\frac{4\pi}{3}\frac{u^\alpha u_\beta}{c^2}$. In this calculation we have to take into account that $v^2=\sum(v^\gamma)^2=-\sum v_\gamma v^\gamma$. Hence (dropping primes in the left-hand side),

$$I^\alpha_\beta=\int p^\alpha p_\beta\, d\omega=-\frac{4\pi}{3}\left(\delta^\alpha_\beta-\frac{u^\alpha u_\beta}{c^2}\right)$$

As $\vec{u}\vec{p}=0$, we obtain $p_0=-\frac{u^\beta}{u^0}p_\beta$, whence

$$\int p_0 p^\alpha\, d\omega=-\frac{u^\beta}{u^0}I^\alpha_\beta \quad \text{and} \quad \int p_0^2\, d\omega=\frac{1}{(u^0)^2}u_\alpha u^\beta I^\alpha_\beta$$

By using these formulas we easily find that

$$\int p_i p^k\, d\omega=-\frac{4\pi}{3}\left(\delta^k_i-\frac{u_i u^k}{c^2}\right) \tag{D.7}$$

Integrals of products of an odd number of components of 4-vector p^i vanish; this is obvious from the given above formula for p_0 and from the derivation of the formula for I^α_β.

The relativistic invariance of (D.7) is explained as follows. While the components of vector $\vec{p}$ are taken in an arbitrary reference frame, integration is always carried out over an element of solid angle taken in the instantaneous co-moving frame of reference. In §§ 15 and 23 we have used precisely this way of calculating the integrals.

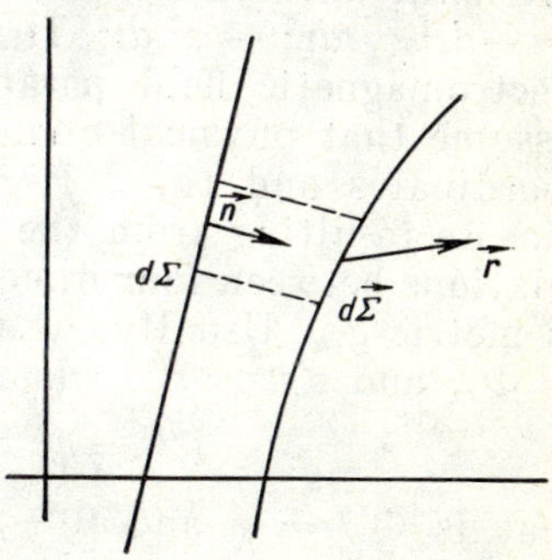

Fig. 49

3. A volume element of light cone is defined as follows. Consider first a hypersphere with a spacelike radius given by the equation $\vec{R}^2=-\lambda^2$. An element of volume of this hypersphere is $d\bar{\Sigma}^m = r^m\, d\bar{\Sigma}=\frac{R^m}{\lambda}d\bar{\Sigma}$ (here r^m is a unit vector of the normal). Consider now an arbitrary hyperplane with spacelike unit normal $\vec{n}$. An element of volume of this hyperplane is $d\Sigma^m=n^m\, d\Sigma$. Use the relation $d\Sigma=(\vec{n}\vec{r})\, d\bar{\Sigma}=\frac{1}{\lambda}(\vec{n}\vec{R})\, d\bar{\Sigma}$ which means that $d\Sigma$ is defined as a projection of element $d\bar{\Sigma}^m$ onto hyperplane (Fig. 49). Denote $d\Gamma\equiv$

$\equiv d\bar{\Sigma}/\lambda$. The preceding relation then yields

$$d\Gamma = \frac{d\bar{\Sigma}}{\lambda} = \frac{d\Sigma}{|\vec{n}\vec{R}|} \tag{D.8}$$

The left-hand side of (D.8) is independent of the direction of vector $\vec{n}$ and therefore the right-hand side is equally independent of the choice of this vector. Furthermore,

$$d\bar{\Sigma}^m = R^m\, d\Gamma = \frac{R^m\, d\Sigma}{|\vec{n}\vec{R}|} \tag{D.9}$$

This last formula is independent of λ and so remains valid when $\lambda \to 0$, that is for $\vec{R}^2 = 0$. In this case it can be applied to integration over the light cone.

Note also that formula (D.9) can be obtained in an absolutely similar manner by using hyperspheres with timelike radius and, correspondingly, hyperplanes with timelike normals. Vector $\vec{n}$ in the right-hand side of this formula can therefore be absolutely arbitrary (it cannot be only a light vector).

4. Throughout the book we operate with the Minkowski space metric defined by values $g_{00} = 1$, $g_{\alpha\alpha} = -1$, $g_{ik} = 0$ for $i \neq k$. Often the authors use an inverse metrics: $g'_{ik} = -g_{ik}$, so that $ds'^2 = -ds^2 = d\mathbf{r}^2 - c^2\, dt^2$. It will be of use to compare expressions for electromagnetic field parameters written with these two metrics Assume that physical quantities φ and $\mathbf{A}$ are given as functions of coordinates and time. Hereafter we shall prime four-dimensional tensors resulting from the chosen metrics g'_{ik}. Formulas (7.3) give relations between four-dimensional quantities and functions φ and $\mathbf{A}$ in metric g_{ik}. Usually, when metric g'_{ik} is used, one assumes $\Phi'_m = \Phi_m$ and $s'^m = s^m$, whence

$$\Phi'^m = g'^{mn}\Phi'_n = -g^{mn}\Phi_n = -\Phi^m$$

that is $\Phi'^0 = \varphi$ and $\Phi'^\alpha = A^\alpha$. Further,

$$F'_{mn} = \frac{\partial \Phi'_n}{\partial x^m} - \frac{\partial \Phi'_m}{\partial x^n} = \frac{\partial \Phi_n}{\partial x^m} - \frac{\partial \Phi_m}{\partial x^n} = F_{mn}$$

$$F'^{mn} = g'^{ma} g'^{nb} F'_{ab} = g^{ma} g^{nb} F_{ab} = F^{mn}$$

and

$$F'^{m}_{\cdot n} = g'^{ma} F'_{an} = -g^{ma} F_{an} = -F^{m}_{\cdot n}$$

In metric g_{ik}

$$\frac{\partial^2}{\partial x^k\, \partial x_k} = \frac{1}{c^2}\frac{\partial^2}{\partial t^2} - \Delta \equiv -\Box$$

In metric g'_{ik}

$$\frac{\partial^2}{\partial x^k\, \partial x_k} = \Box$$

At the same time, wave equations for potentials have the same form in both metrics, that is

$$\frac{\partial^2}{\partial x^k\, \partial x_k}\Phi^i = -\frac{1}{c}s^i$$

A quantity

$$\theta'^{mn} = F'^{ma}F_a'^{\cdot n} + \frac{1}{4}g'^{mn}(F'_{ab}F'^{ab})$$

$$= -\left\{F^{ma}F_a^{\cdot n} + \frac{1}{4}g^{mn}(F_{ab}F^{ab})\right\} = -T^{mn}$$

where T^{mn} coincides with that in (10.19), is often chosen for the energy-momentum tensor of electromagnetic field. The law of energy-momentum conservation for electromagnetic field takes the form

$$\frac{\partial\theta'^{mn}}{\partial x^n} = -\frac{\partial T^{mn}}{\partial x^n} = 0$$

E. Application of the Fourier transform to wave equations

In this section we consider, as a supplement to § 13, the solution of wave equation by using the theory of functions of complex variables. We shall need formulas of expansion of functions into the Fourier integrals. These formulas are employed in the main text of the book in a number of calculations.

In this section the wave equation will be written in the form more general than in § 13, namely, not for the case of the vacuum but for an arbitrary uniform isotropic medium, with no specific system of units chosen in advance:

$$\left(\Lambda - \frac{1}{v^2}\frac{\partial^2}{\partial t^2}\right)\psi(\mathbf{r}, t) = -g(\mathbf{r}, t) \tag{E.1}$$

Here

$$v = \alpha/\sqrt{\varepsilon\mu} \tag{E.2}$$

With an appropriate choice of the right-hand side, g, function ψ may represent, as in § 13, any Cartesian component of potentials or strengths of electromagnetic field.

Consider a homogeneous equation ($g \equiv 0$). Direct substitution readily confirms that function

$$\psi(\mathbf{r}, t) = A(\mathbf{k}, \omega)\, e^{i(\mathbf{k}\cdot\mathbf{r} - \omega t)} \tag{E.3}$$

in which amplitude A is independent of coordinates and time (frequency ω in this formula may be considered both positive and negative) and is a solution of the homogeneous equation provided

$$\mathbf{k}^2 = \omega^2/v^2 \tag{E.4}$$

Solution (E.3) is called the plane wave with wave vector **k**. It can be assumed for a very wide class of functions $\psi(\mathbf{r}, t)$ that they can be expanded into the Fourier integral:

$$\psi(\mathbf{r}, t) = \int d^3k \int_{-\infty}^{+\infty} d\omega\, A(\mathbf{k}, \omega)\, e^{i(\mathbf{k}\cdot\mathbf{r}-\omega t)} \tag{E.5}$$

If at the same time (E.4) is satisfied (it can be taken into account by including delta function $\delta(\mathbf{k}^2 - \omega^2/v^2)$ into the integrand on the right-hand side of (E.5)), (E.5) will also be a solution of the homogeneous wave equation. This is clear from the formula for delta function (C.6). Further below formula (E.5) will be used to solve other equations as well, when condition (E.4) does not hold.

If function ψ is real, that is $\psi = \psi^*$ (here and in subsequent formulas the asterisk denotes complex conjugation), equation (E.5) shows that necessarily

$$A^*(-\mathbf{k}, -\omega) = A(\mathbf{k}, \omega) \tag{E.6}$$

We shall try to find a solution of the inhomogeneous wave equation for unbounded space in the form of (13.13). Of principal importance here is the property of Green's function G given by equation (13.12). Taking into account that Fourier expansion (E.5) is valid for functions satisfying a very wide spectrum of conditions, we can write Green's function in the form

$$G(\mathbf{r}-\mathbf{r}', t-t') = \int d^3k \int d\omega\, g(\mathbf{k}, \omega)\, e^{i\mathbf{k}\cdot(\mathbf{r}-\mathbf{r}')} e^{-i\omega(t-t')} \tag{E.7}$$

Delta function in the right-hand side of the equation defining the fundamental solution also can be written, via formula (C.14), in the form of the Fourier integral. Substituting these expansions and using (13.12), we obtain

$$g(\mathbf{k}, \omega) = \frac{1}{(2\pi)^4} \frac{1}{k^2 - \omega^2/v^2} \tag{E.8}$$

Function $g(\mathbf{k}, \omega)$ has a singularity when **k** satisfies condition (E.4). Integral over ω in (E.7) has the form

$$I(k) = \int_{-\infty}^{+\infty} \frac{e^{-i\omega(t-t')}}{k^2 - \omega^2/v^2} \tag{E.9}$$

We demand that function G satisfy the causality condition in the form

$$G = 0 \quad \text{for} \quad t < t' \tag{E.10}$$

Recall that t is the time of observation and t' is the time of emission of an electromagnetic pulse by a source. Requirement (E.10) can be used to calculate integral (E.9) via the theory of residues. The integrand in this integral is regular for any complex ω, with the exception of two poles $\omega = \pm vk$. If $t > t'$ and $\omega = \omega_1 + i\omega_2$, where ω_1 and ω_2 are real, then $e^{-i\omega(t-t')} = e^{-i\omega_1|t-t'|}\, e^{\omega_2|t-t'|}$, with $\omega_2 (t - t') < 0$ in the lower halfplane of complex variable ω (i.e. for $\omega_2 < 0$). Hence, if the integration path is chosen as a segment of real axis subtending the semicircle in the lower halfplane, the integral over this semicircle will contain a factor decreasing exponentially when $|\omega_2| \to \infty$ and therefore vanishes in this limit. A similar condition is fulfilled for $t < t'$ for a path with subtending semicircle drawn in the upper halfplane. For condition (E.10) to be satisfied, the integrand must have no singular points inside this upper contour. Therefore we shift the indicated poles by an infinitesimal distance downward of the real axis, assuming $\omega = \pm vk - i\varepsilon$, and calculate the integral in the limit $\varepsilon \to 0$ and for the lower-halfplane semicircle subtending the contour infinitely expanded (Fig. 50). The Cauchy integral formula yields

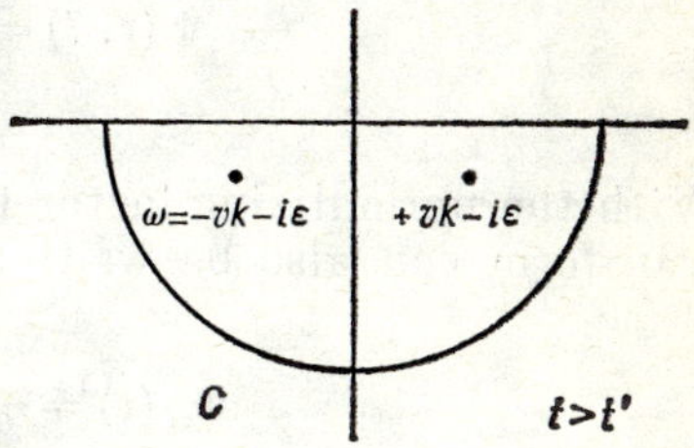

Fig. 50

$$I(k) = 2\pi v \frac{\sin vkT}{k} \tag{E.11}$$

where $T = t - t'$. By using (E.7), (E.8), and (E.11) in the integration over angles, we obtain

$$G = \frac{1}{(2\pi)^4} \int d^3k e^{i\mathbf{k}\cdot\mathbf{R}} I(k) - \frac{v}{2\pi^2 R} \int_0^\infty dk \sin kR \sin vkT$$

Here, as usual, $\mathbf{R} \equiv \mathbf{r} - \mathbf{r}'$. Now we recall that the integrand is even in k, and integrate from $-\infty$ to $+\infty$. By denoting $vk \equiv \xi$, we write

$$G = \frac{1}{8\pi^2 R} \int_{-\infty}^{+\infty} d\xi \left[e^{i\xi(T-R/v)} - e^{i\xi(T+R/v)}\right] = \frac{1}{4\pi R}\, \delta\left(T - \frac{R}{v}\right)$$

Here we have used formula (C.14) and the fact that $\delta(T + R/v) \equiv 0$.

We have arrived at an expression for Green's function completely identical to formula (13.12') derived in § 13 for a particular case $v = c$. The second derivation given above is of interest owing to the application of the methods of the theory of functions of complex variables to express the causality condition (E.10).

Expansion into the Fourier integral (E.5) for solution ψ and source g of the nonhomogeneous wave equation are often used throughout the book. In a number of cases, however, it is sufficient to use only the expansion in time argument. This means that formula (E.5) is written in the form

$$\psi(\mathbf{r}, t) = \int_{-\infty}^{+\infty} \psi_\omega(\mathbf{r})\, e^{-i\omega t}\, d\omega \tag{E.12}$$

With the normalizing factor taken into account, the inverse Fourier transform can also be written:

$$\psi_\omega(\mathbf{r}) = \frac{1}{2\pi} \int_{-\infty}^{+\infty} \psi(\mathbf{r}, t)\, e^{i\omega t}\, dt \tag{E.13}$$

It will be instructive to return to (E.12) and analyze a solution of the inhomogeneous wave equation in unbounded space, obtained, in fact, by separating the variables. By substituting expansion (E.12) and a similar expansion of function g into (E.1) we obtain a differential equation for the Fourier amplitudes $\psi_\omega(\mathbf{r})$ and $g_\omega(\mathbf{r})$:

$$\left(\Delta + \frac{\omega^2}{v^2}\right)\psi_\omega = -g_\omega \tag{E.14}$$

called the *Helmholtz equation*. A solution of this equation must again be sought via an appropriate Green's function $G(\mathbf{r}, \mathbf{r}')$ which in this case must satisfy the condition

$$\psi_\omega(\mathbf{r}) = \int G(\mathbf{r}, \mathbf{r}')\, g_\omega(\mathbf{r}')\, dV' \tag{E.15}$$

with

$$(\Delta + k^2)\, G(\mathbf{r}, \mathbf{r}') = -\delta(\mathbf{r} - \mathbf{r}') \tag{E.16}$$

It can be shown that

$$G = \frac{1}{4\pi R} e^{\pm ikR} \tag{E.17}$$

Consequently,

$$\psi_\omega(\mathbf{r}) = \frac{1}{4\pi} \int \frac{g_\omega(\mathbf{r}')}{R} e^{\pm ikR}\, dV' \tag{E.18}$$

and, according to (E.12),

$$\psi(\mathbf{r}, t) = \frac{1}{4\pi} \int \frac{g_\omega(\mathbf{r}')}{R} e^{-i(\omega t \pm kR)}\, dV'\, d\omega \tag{E.19}$$

If the origin of the time axis is displaced by R/v and we use function $g(\mathbf{r}, t)$ again, formula (E.19) is easily rewritten in the form

$$\psi(\mathbf{r}, t) = \frac{1}{4\pi} \int \frac{g(\mathbf{r}', t \pm R/v)}{R} dV'$$

The values $t + R/v$ are advanced with respect to the observation time t, and so the physical condition of causality necessitates the choice of the lower sign. We have thus derived again retarded solutions of the type already analyzed in § 13.

It should be useful to compare the above derivation with the Fourier expansion in all variables (i.e. t and $\mathbf{r}$), employed earlier in this section. It will be possible, for example, to give a substantiation to formula (E.17). This is offered to the reader as an excercise.

NAME INDEX